12/09

Hope
book a__ ___ many years
of good work at the Fort
and the many years of
our friendship. Happy
memories always!

Anne

A

HISTORY OF

ARMY COMMUNICATIONS AND ELECTRONICS

AT FORT MONMOUTH, NEW JERSEY

1917 - 2007

ISBN 978-0-16-081359-7

Prepared by the Staff of the
Historical Office
Office of the Deputy Chief of Staff for Operations and Plans
U.S. Army CECOM Life Cycle Management Command
Fort Monmouth, New Jersey
2008

Design and layout by Solari Creative Inc.

This history is dedicated to all the men and women, military, civilian, and contractors who have served at Fort Monmouth, NJ; past, present and future.

TABLE OF CONTENTS

*All photos and illustrations are courtesy of the U.S. Army, unless otherwise specified.

The hardships and privations the pioneer soldiers of Camp Vail knew instilled a love in them for the place which was carried over to the period it became Fort Monmouth. Many of them, after their return from overseas in 1918, enlisted as military and civilian employees of the Fort. Many of them are serving today as instructors, engineers, officers, and in other capacities. When asked to explain a loyalty that time had not been able to dim, one of the Camp Vail veterans said shyly, "The place sort of gets into your blood, especially when you have seen it grow from nothing into all this. It keeps growing and growing, and you want to be part of its growing pains. Many of the local communities have become very attached to Fort Monmouth because of the friendship instilled by the wise social counselors of Vail, planted not for just a war period but for as long as...Fort Monmouth...will inhabit Monmouth County.[1]

Rebecca Klang, 1942

FOREWORD

When I first arrived at Fort Monmouth in July 2007 to assume command of the CECOM Life Cycle Management Command, my perspective of CECOM and Army Team C4ISR was that of a customer as I had been the recipient of numerous command and control systems fielded across our Army over the previous 27 years of my career. My respect and admiration for this community were great, but only grew as I realized the scope and breadth of its enormous contributions, and learned about its illustrious history in conjunction with Fort Monmouth's 90th anniversary. To honor this special event, and to capture in one volume accomplishments that not only changed and shaped the Army, but affected world history, this *History of Army Communications and Electronics at Fort Monmouth NJ 1917-2007* was compiled.

Learning about our history helps us to understand who we are and where we're going. It provides us a sense of our organizational identity. Author David McCullough once said, "History is a guide to navigation in perilous times." As we face some challenging years ahead, this history reminds us how far we've come and how much we have accomplished. As you read this history, you will learn that our team has always risen to and overcome every obstacle and challenge which lay in its path, and I have no doubt it will continue to do so as we begin the next 90 years.

To the many veteran members of the CECOM and Army Team C4ISR community, I hope you will enjoy reading about the great events and successes captured in our history, and reminiscing about those you personally helped to shape. To the new members of our community, you are now a part of this wonderful history, and I hope it will inspire you to continue the legacy of achievements for many years to come.

Sincerely,

Dennis L. Via
Major General, USA
Commanding

PREFACE

The name "Monmouth" has been synonymous with the defense of freedom since our country's inception. Named for the brave Soldiers who gave their lives just a few miles away at the Battle of Monmouth Court House (June 28, 1778), this installation has been the site of some of the most significant communications and electronics breakthroughs in military history. Over the last ninety years, over 4,000 patents have been issued to Fort Monmouth inventors.

Scientists, engineers, program managers, and logisticians here have delivered these technological breakthroughs and advancements to our Soldiers, Sailors, Airmen, Marines, and Coast Guardsmen. These innovations have included the development of FM radios and radar, bouncing signals off the moon to prove the feasibility of extraterrestrial radio communication, the use of homing pigeons through the late-1950s, frequency hopping tactical radios, and today's networking capabilities supporting our troops in the Global War on Terrorism.

This history represents an overview of Fort Monmouth's communications and electronics achievements since 1917. Fort Monmouth's Soldiers and civilians have worked tirelessly to develop technologies and field equipment to enable battle command and protect the force. During WWI as Soldiers trained here were charged with establishing phone and telegraph lines on the front lines of Europe, significant strides were being made in the areas of combat photography, pigeon training, meteorology and radio intelligence back at the labs at Camp Alfred Vail. During WWII the necessity for early warning of aerial and submarine attack as well as pinpointing the location of mortar fire became paramount and scientists and engineers here raced desperately to develop and field long range and mortar locating radars - radars that would help win the war. During the Korean War, very high frequency radios, mortar locators and Fort Monmouth trained pigeons all helped to keep troops safer. As America geared up for war in Vietnam, Fort Monmouth experts fielded squad radios and night vision devices, revolutionary items that helped individual riflemen see and communicate.

Our logisticians were among the first civilians to arrive in the war zone during the Gulf War, providing hands-on technical assistance to Soldiers with communications-electronics equipment. Today the team at Fort Monmouth supports critical Command, Control, Communications, Computers, Intelligence, Surveillance, Reconnaissance (C4ISR) and information systems that are making a difference in places like Afghanistan and Iraq.

The team members at Fort Monmouth, who have gone by various organizational names over the years, maintain the current readiness of our Armed Forces and seek new ideas and technologies designed to improve their capabilities. While the tools used to accomplish this mission today are radically different from those used in years past, the nature of this mission has changed very little from the days of wig-wag flags and homing pigeons. This history refers to the various Army organizations responsible for managing this mission today, and similar missions in the past. In the beginning, the Signal Corps here was responsible for everything from training, to materiel development, to procurement. In later years these functions would be dispersed to various organizations throughout the Army and those functions remaining at Fort Monmouth would be primarily responsible for the development, acquisition and sustainment of communications-electronics equipment. The common denominator has always been providing superior support to our fighting men and women. To see a detailed outline of organizational changes here, please see Appendix A.

The contents of this book are not necessarily the official views of, or endorsed by the U.S. Government, Department of Defense, Department of the Army, or Team C4ISR and Fort Monmouth.

Wendy Rejan
Command Historian
U.S. Army CECOM LCMC
June 2008

ACKNOWLEDGMENTS

The Commanding General of the U.S. Army CECOM LCMC, MG Dennis L. Via, and the former Director of the Communications-Electronics, Research, Development and Engineering Center, Gary Martin, provided the impetus for this volume. Thanks to their support, the magnificent accomplishments of the dedicated men and women who have served at this post over the last century will be remembered and cherished for many years to come. MG Via has been a staunch supporter of celebrating and promoting Fort Monmouth's illustrious communications and electronics history. This support culminated in 2007 with the celebration of Fort Monmouth's 90th Anniversary. We must thank the numerous people who contributed to this work, to include command and staff historians as well as archivists, past and present: COL Gregory B. Gonzalez, Operations Officer, PEO IEW&S; Richard DeAtley, Tactical Communications Engineer; Joshua Davidson, Staff Writer; Wanda M. Wohlin, Chief, Human Resources Office; and Emerson Keslar, Chief Knowledge Officer, PEO C3T. Thanks also to those who devoted their time to reading drafts, and contributing ideas and images: Patricia Devine, DCSOPS; Michael Brady, Chief, CEID, and Thomas Cameron, LSS Management Analyst, DCSOPS; John Oltarzewski, Chief Business Operations Team, SEC; John C. Erichsen, Fort Monmouth Fire Chief; Patrick Lyman, Quality Assurance Specialist, LRC; Danielle Oglevee, Procurement Analyst, ACQ; William Meiter, Real Property/GIS Data Custodian, Fort Monmouth DPW; Kashia Simmons, CERDEC Public Affairs Officer; Karen Ryder, Business Development Specialist; Dr. Arthur Ballato, Chief Scientist, CERDEC; Jan Moren, Deputy Director I2WD, CERDEC; Jason Bock, Staff Writer; Denise Rule, Art Director, David Brackmann, Multimedia Specialist and Peter Culos, Graphics Specialist, PEO C3T; Marcus H. Sachs, Executive Director, Government Affairs-National Security Policy; Russell Meseroll and Barbara Guigno, Chenega Technology Services Corporation; and Fred Carl, Director, Infoage. Thanks to the Fort Monmouth employees past and present who donated many of the photos and documents to the Historical Archive which were used to create this work and to the historians and archivists who cared for and preserved them. Thanks to the artists at Solari Creative Inc., Keith Schoeneick who designed the cover and Michael Burke who engineered the artistic design and layout of this history. And finally, thanks to Diane R. Gordon, Ph.D., who compiled the index.

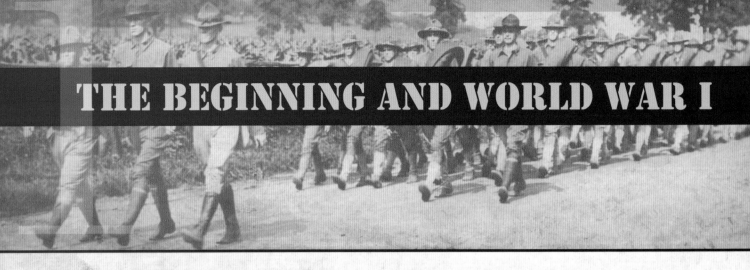

The Army recognized at the outbreak of World War I that the Signal Corps, with a strength of less than 2,000 officers and enlisted men, was incapable of providing the communications support the Army would need, should the United States enter the war. This small service, with personnel obtained chiefly by the detail system, had been designed primarily for border and insular operations.

In October 1916 the Office of the Chief Signal Officer (OC-SigO) asked the executives of American Telephone and Telegraph, Western Electric, Western Union, and the Postal Telegraph Company to recruit from among their trained employees, personnel for a Signal Enlisted Reserve Corps. The response was more than could have been hoped for when 1,400 of the 6,000 male employees of the Bell Telephone Company of Pennsylvania applied for enlistment.

The Signal Corps needed places in which to prepare these citizen Soldiers for service in battle. The history of Fort Monmouth, then, began in 1917 when the Army established four training camps for signal troops. One was located at Little Silver, New Jersey. Fort Leavenworth, Kansas; Leon Springs, Texas; and the Presidio of Monterey, California housed the others. Government-owned land was utilized for all the camps except for Little Silver.

The Little Silver site lay in an area rich in history dating back to the American Revolution. It was near this site, in what became the Township of Freehold, that the Battle of Monmouth Courthouse occurred. There, General George Washington and his Continental Army troops engaged the British forces led by Sir Henry Clinton on 28 June 1778. The British slipped away after dark and reached the safety of the British fleet guns at Sandy Hook. Although victory was inconclusive, the battle did show that the Continental troops had learned to fight on equal terms with the British regulars in open battle thanks to the training of Baron Von Steuben.

The Battle of Monmouth Courthouse became famous as the last major engagement of the Revolution to be fought in the North. It is perhaps best remembered for the alleged exploits of Molly Pitcher, the housewife who, while carrying water to artillerymen, reportedly saw her husband fall wounded and took his place until help could arrive.

Improvements in steamship and railroad transportation later allowed the "Jersey Shore" to become a popular summer vacation retreat for harried New Yorkers during the second half of the nineteenth century. Seven U.S. Presidents favored the seaside resort of Long Branch. Some of the city's wealthier habitués brought horse racing to the area with the construction of Monmouth Park, a one-mile track, in 1870. This park was located in what is now the southern portion of Fort Monmouth, in the vicinity of Patterson Army Health Clinic. The Entrance was located on today's Broad Street, near Park Avenue. An instant success, Monmouth Park flourished for twenty years. In season, two steamboats made daily runs from New York to Sandy Hook. There, patrons could make a connection to the park by rail.[2]

A bigger, fancier Monmouth Park opened on 4 July 1890. It featured a one and one half mile oval track, centered on what later became Greeley Field; a one-mile straight-of-way; a steel grandstand for 10,000 spectators that was reputedly the largest in the world; and a luxury hotel, fronting Parker Creek. The new park encompassed 640 acres – almost all of Fort Monmouth's "Main Post."

Monmouth Park Race Track closed three years later when the New Jersey legislature outlawed gambling. One of the feature races, the "Jersey Derby," moved to Louisville, Kentucky, home of the famous "Kentucky Derby." The deserted grandstand, track, and hotel fell into ruin. The grandstand succumbed to a nor'easter in 1899. The hotel burned to the ground in 1915.[3]

Amidst the turmoil of WWI, Colonel (retired) Carl F. Hartmann, the Signal Officer of the Eastern Department in New York City, tasked Major General (retired) Charles H. Corlett

Monmouth Park Racetrack

to "go out and find an officer's training camp." Corlett recalled his initial discovery of the Monmouth Park land in a 1955 letter addressed to Colonel Sidney S. Davis, Chairman of the Fort Monmouth Traditions Committee. He reported that after examining several other sites, he "finally stumbled onto the old Race Course near Eatontown. I found part of the old steel grandstand with eleven railroad sidings behind it, the old two mile straight away track and two oval race tracks, all badly overgrown with weeds and underbrush." Corlett went on to describe how he arranged a meeting with the owner of the land. "Upon inquiry, I learned that the land belonged to an old man who lived in Eatontown who was very ill (on his death bed in fact), but when he learned my business, he was anxious to see me."[4]

Monmouth Park Hotel

38-H-540

Corlett learned that the owner, Melvin Van Keuren, had offered to give the land to the Army free of charge during the Spanish American War. Van Keuren regretfully informed Corlett that he could no longer afford to do so. He offered instead to sell the land for $75,000.[5]

Corlett returned to his superior officers to report his findings. With authorization of the Adjutant General of the Army, then Lieutenant Colonel Hartmann leased 468 acres of the tract from Van Keuren on 16 May 1917 with an option to buy. The land, which was a potato farm at the time, included 468 acres bounded on the North by the Shrewsbury River, on the West and South by a stone road from Eatontown, and on the East

by the Oceanport-Little Silver Road. Parker Creek, a tributary of the Shrewsbury, traversed the entire property near the northern limits. Notwithstanding the desolation of the site in 1917 – largely overgrown and infested with poison ivy – it afforded the Army significant advantages: six hundred feet of siding on a rail line of Hoboken (a Port of Embarkation) and proximity to the passenger terminal in Little Silver, as well as good stone roads and access by water. The Red Bank Register dated 6 June 1917 reported that the land leased by the government had been "farmed for the past four years by Charles Prothero. He will continue to work the farm south of the railroad tracks but all property north of the tracks has been leased by the government. On this property is a seventy acre field of potatoes. The government will recompense Mr. Prothero for this crop."[6]

The land would be purchased for $115,300 in 1919.[7]

CAMP LITTLE SILVER

The first thirty-two Signal Soldiers arrived at Fort Monmouth in June 1917 in two Model T Ford Trucks. This advance party under 1st Lieutenant Adolph J. Dekker brought tents, tools, and other equipment from Bedloe's Island, New York, to prepare the site on 3 June. By 14 June, they had cleared several acres on which they installed a cantonment, quartermaster facilities, and a camp hospital, all under canvas.[8]

The installation was originally named Camp Little Silver, based merely on its location. General Orders dated 17 June 1917 named LTC Hartmann the first commander. Members of the First and Second Reserve Telegraph Battalions arrived by train the following day. The War Department transferred forty-three noncommissioned officers from Fort Sam Houston, Texas, to meet the need for a cadre of experienced personnel. These men had served on the Mexican border. 451 enlisted men and twenty-five officers were stationed at Camp Little Silver by the end of the month.[9]

Construction of the "old wooden camp" proceeded at this time. Laborers worked overtime to complete a headquarters building, officers' quarters, barracks, transportation sheds, shops, and a warehouse near the railroad siding.

Corporal Carl L. Whitehurst was among the first men to arrive at Camp Little Silver. He later recalled that the site appeared

Camp Little Silver 1917

to be a "jungle of weeds, poison ivy, briars, and underbrush." While remnants of the old Monmouth Park Racetrack seemed to be everywhere, only one building remained habitable. It was there, in that former ticket booth, that he and his comrades sheltered while awaiting the delivery of tents.

Railroads soon brought the tents, as well as lumber with which to build barracks. Unfortunately, most of the lumber was green. According to CPL Whitehurst, "By the time the wood was dried out it was winter, and in December there were cracks you could put your finger through. The winter of 1917-1918 was a tough one, and sometimes the snow would pile up on your blankets, coming through the gaps in the boards."[10]

Colonel Hartmann was succeeded on 13 July 1917 by Major George E. Mitchell. Mitchell organized the Reserve Officers' Training Battalion and two tactical units, the 5th Telegraph and 10th Field Signal Battalions. Instruction of trainees began on 23 July. The curriculum included cryptography, the heliograph, semaphore, wig-wag, motor vehicle operation, physical training, dismounted drill, tent pitching, interior guard duty, map reading, tables of organization for Signal, Infantry, and Cavalry units, camp sanitation, personal hygiene, first aid, and equitation. The troops spent much of their time clearing the area of undergrowth, repairing and extending roads, and digging drainage ditches. Nineteen Soldiers were hospitalized for poison ivy exposure in June; 129 in July.[11]

Map Reading Class, Camp Little Silver, 1917

The Camp sent its first units (the First and Second Reserve Telegraph Battalions) to the Port of Embarkation on 7 August 1917. These units reconstituted in theater as the 406th Telegraph Battalion and the 407th Telegraph Battalion.

CAMP ALFRED VAIL

The camp achieved semi-permanent status and was re-named Camp Alfred Vail on 15 September 1917, just three months after its establishment. Vail, an associate of telegraph inventor Samuel F. B. Morse, was credited with helping him develop commercial telegraphy. Some felt that Vail's great contributions to wire communications merited commemoration of his name at a Signal Corps Camp.[12]

However, LTC Hartmann intimated in a 1955 interview conducted by Dr. Thompson, Chief of the Signal Corps Historical Division, that the Chief Signal Officer actually intended the naming of the Camp to honor his good friend, Theodore N. Vail, Chief Executive Officer of American Telephone and Tele-

graph. In the words of LTC Hartmann, "Recognizing the impropriety of naming the Post for Theodore N. Vail, it requires no stretch of the imagination to figure out why he (General Squier, Chief Signal Officer) came to name it 'Camp Alfred E. Vail'." The impropriety lay in the fact that Theodore Vail was living at the time and was serving as president of AT&T. According to Dr. Thompson, the Signal Corps owed nothing to Alfred Vail, who died a year before the Corps was even established. They did, however, owe a good deal to Theodore Vail, who "helped provide the Signal Corps in World War I with communication company specialists manning the Corps' Telegraph Battalions."[13]

Laying Field Wire, Camp Alfred Vail

Meanwhile, the Signal Corps faced an urgent need for telegraphers and radio operators in France. A six-week intensive training course on foreign codes and languages began at Camp Alfred Vail. The Army sent 223 men to the Camp for training and testing as German-speaking personnel. Additional groups of fifty or more arrived each month thereafter. The need for telegraph operators in France was so great that operators volunteering for overseas duty received bonuses.[14]

The 11th Reserve Telegraph Battalion boarded the train for Hoboken on 18 October 1917. Other units followed in rapid succession – a Radio Operator Detachment and the 408th Telegraph Battalion in November, and the 52nd Telegraph Battalion and the 1st Field Signal Battalion in December. Camp Alfred Vail trained a total of 2,416 enlisted men and 448 officers for war in 1917. The Camp trained 1,083 officers and 9,313 enlisted men in 1918. Between August 1917 and October 1918,

Telegraphy Department, Camp Alfred Vail, 1918

American Expeditionary Forces in France received five telegraph battalions, two field signal battalions, one depot battalion, and an aero construction squadron from Camp Alfred Vail.

The camp was hit particularly hard by the influenza outbreak that struck the nation in September 1918. As outbreaks occurred, units were quarantined, until eventually the entire camp was isolated. By the time the quarantine ended in November, eleven deaths had occurred and the hospital had treated a total of 267 cases.[15]

THE RADIO LABORATORY AND AERIAL TESTING

The particular demands of tank and aerial warfare in World War I necessitated a special Army laboratory devoted exclusively to developmental work. This laboratory would be entirely independent of the commercial laboratories. It would be a place where trained specialists could focus their energies on problems in wireless communication. The existing Electrical Development Division in Washington and the facilities in the Bureau of Standards were deemed insufficient for experimentation. Camp Vail was instead selected as the site.

Construction began in mid-December 1917. It was largely finished by the end of January. In addition to forty-three semi-permanent laboratory buildings in the vicinity of what became Barker Circle, the contractor (Heddon Construction Company) drained and leveled ground for two air fields and built four hangars east of Oceanport Avenue.

Radio Shop, Camp Alfred Vail, 1919

The Army charged the radio laboratory with the development of radio equipment. Research initially centered on vacuum tubes, on circuits of existing equipment, on the testing of apparatus submitted by manufacturers, and on the application of new inventions. A staff of forty-eight officers, forty-five enlisted men and twelve civilians (principally stenographers) accomplished this work.

Within a month, the radio equipment produced required ninety to ninety-five airplane flights a week for testing. This led area residents to mistakenly believe that Camp Vail was primarily an airfield. Two squadrons for the United States Army Air Service were assigned here in 1918. The 504th Aero Squad-

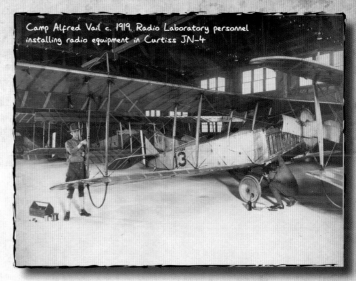

Camp Alfred Vail c. 1919. Radio Laboratory personnel installing radio equipment in Curtiss JN-4

ron arrived here on 4 February 1918, consisting of one officer and 100 enlisted men. The first planes, along with the 122nd Aero Squadron, arrived here in March 1918. The 122nd Aero Squadron consisted of 12 officers and 157 enlisted men. The camp's flying activity reached its peak with personnel of the 122nd Aero Squadron operating a total of twenty aircraft: two DeHaviland 4s, nine Curtiss JN4-Hs, six Curtiss 4-6Hos, and three Curtiss JN-4Ds. This represented the largest number of aircraft ever housed at Camp Vail. The first flights did not take off until May 1918 because the 122nd was quarantined upon arrival due to several cases of measles. On 12 April 1918, the 504th left for Langley Field, VA. Following the signing of the Armistice on 11 November 1918, the Aviation Section was moved from Camp Vail. It had made enormous headway in adapting radio to aircraft for World War I. The 122nd left the Camp on 11 November 1918 for Mineola, Long Island, NY. On 13 December 1918, orders were received to ship all aeronautical equipment from the camp. Only one fatality had occurred during the time of aerial testing at the Camp. Lt. Meril was killed in a crash landing on 18 August 1918.[16]

Colonel George W. Helms, Signal Corps, assumed command of the camp on 28 June 1918.[17]

INITIAL USE OF HOMING PIGEONS

Pigeons have been used for carrying messages since ancient times. Some historians believe they were used by the ancient Egyptians, Assyrians and Babylonians. In the late 1800s the Signal Corps made attempts at using pigeon messengers in the Dakota Territory but the experiments failed, primarily due to Hawks attacking the pigeons. The use of pigeons by the British and French armies impressed General John J. Pershing, Commander of the American Expeditionary Force WWI. He therefore requested such a service be established in the American Army. This was delayed due to the difficulty in acquiring the birds. The service (consisting of three officers, 118 enlisted men, and a few hundred pigeons) finally arrived in France in February 1918. A total of 572 American birds served in the St. Mihiel offensive and 442 in the Meuse-Argonne offensive. Under murderous machine gun and artillery fire during the Meuse-Argonne offensive, the hero pigeon "President Wilson" flew twenty-five miles in as many minutes with a shattered leg and a badly wounded breast. Found dead in June 1929 at the age of eleven, he was stuffed, mounted, and donated to the Smithsonian Institution. WWI hero pigeons on display at the U.S. Army Communications Electronics Museum at Fort Monmouth include Mocker and Spike. Mocker, the last of the World War heroes died at Fort Monmouth in June 1937. On 12 September 1918, heavy enemy artillery fire was blocking the American advance into the Alsace-Lorraine sector of France. With an eye destroyed by a shell fragment and his head a mass of clotted blood, Mocker homed "in splendid time" from the vicinity of Beaumont, with a message giving the exact location of certain enemy heavy artillery batteries. American artillery silenced the enemy guns, saving countless lives. He was awarded the Distinguished Service Cross and the French Croix de Guerre Medal. Spike was hatched in France in 1918. He is credited with carrying a record number of fifty-two vital messages while serving with the 77th Division. He was never injured.

Signal Corps Soldier demonstrates the employment of homing pigeons at Camp Alfred Vail

Perhaps the most famous pigeon of WWI was Cher Ami. This blue check hen was requested by units by name because she had a reputation for reliability. She delivered twelve messages in all from the Verdun Front to the loft at Rampont. She was responsible for saving the Lost Battalion of the 77th Division in October 1918. The Lost Battalion had been pinned down on all sides by the Germans who had surrounded them with barbed wire and machine guns. For five nights they were shelled and machine gunned not just by the Germans but also by Americans who did not know they were there. Four messengers were sent out and they all disappeared. Seven pigeons were sent out and they were all shot down. When Cher Ami was sent out on what would be her final journey, she alighted in a tree and began to preen herself. A Doughboy was forced to expose himself by climbing the tree to shoo her out.

NEW YORK HERALD, THURSDAY, APRIL 17, 1919.

PIGEON THAT SAVED LOST BATTALION ARRIVES HOME; IS TO RECEIVE D. S. C.

Meanwhile, the Commanding Officer of the Lost Battalion, Major Whittlesey received a written proposition from the Germans to surrender. But he had been instructed not to give up any ground without written orders. They had one day of rations left and the men had been reduced to eating leaves and shoots. Fifty percent of the battalion had been killed or wounded. Cher Ami flew twenty five miles in twenty-five minutes and arrived with her message hanging by a tendon and a gun shot wound through her breast. She was awarded the Distinguished Service Cross and the Croix de Guerre. General Pershing personally saw Cher Ami off on her trip back to America and gave strict instructions that she was to be kept in the Captain's quarters and

Cher Ami

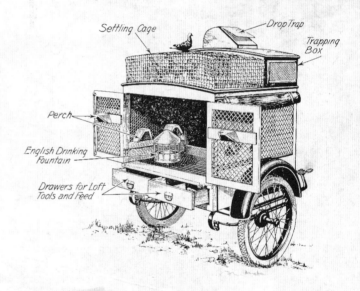

Settling Cage — Drop Trap — Trapping Box — Perch — English Drinking Fountain — Drawers for Loft Tools and Feed

provided with unlimited rations. She died at Fort Monmouth in June 1919 and is mounted in the Smithsonian.[18]

Fort Monmouth's longest lived pigeon was a captured WWI German pigeon by the name of Kaiser. Kaiser was hatched in Cologne in 1917 and lived to 32 years of age, triple the normal lifespan of a pigeon. He outlived many of his mates and offspring. Used by Kaiser Wilhelm's troops, he was captured during the Meuse offensive by U.S. troops. Kaiser's I.D band was embossed with the German Imperial Crown and it was never removed as that would have required amputating the leg. He sired many WWII pigeons and some reports indicate that WWII hero pigeon G.I Joe was a descendant on the maternal side. Kaiser is now on display in the Smithsonian.

The success of homing pigeons in war prompted the Army to perpetuate the service after the Armistice. Squier therefore established the Signal Corps Pigeon Breeding and Training Section at Camp Alfred Vail. The officer in charge of the British pigeon service supplied 150 pairs of breeders. They arrived at Camp Vail, without loss, in October 1919 and resided together with some of the retired hero pigeons of the World War in one fixed and fourteen mobile lofts. Ray R. Delhauer directed the pigeon activities from 1919-1925.[19]

ARMISTICE AND DEMOBILIZATION

Inductions and draft calls stopped with the signing of the Armistice on 11 November 1918. Demobilization began for non-essential units. All flying activities at the camp ceased. The Army shipped all aeronautical property to other locations and directed Radio Laboratory personnel to complete remaining projects. The laboratory decreased in relative importance for a time.

Units assigned to Camp Vail during 1918 included three signal battalions, six telegraph and two depot battalions, two squadrons for air service, and two service companies. A total of 1,083 officers and 9,313 enlisted men served the post that year.

Horse Stables, Camp Little Silver, 1917

The Camp had been dubbed the "best equipped Signal Corps camp ever established anywhere" by the end of 1918. Just nineteen months after its acquisition by the military, 129 semi-permanent structures had been built. The radio laboratories utilized forty-seven of these exclusively. Housing was available for 2,975 Soldiers and 188 officers. Should those men fall ill, there was a hospital equipped to handle forty patients. Two temporary stables could house up to 160 horses. Hard surfaced roads facilitated transportation. One swamp was converted into parade grounds. Another was converted into four company streets, which would be lined by 200 tents.[20]

THE SIGNAL CORPS SCHOOL

In addition to being the year that the Chief Signal Officer authorized the purchase of the land comprising Camp Vail, 1919 was a time of demobilization and transition for the Signal Corps.

Paper Work Class, Camp Alfred Vail

Though initially activated on a temporary basis, the camp survived as an Army installation because the Chief Signal Officer requested in August 1919 that the Adjutant General of the Army move all Signal Corps schools, both officer and enlisted, to Camp Vail. This move standardized signal communications throughout the Army and consolidated Signal Corps installations. The Secretary of War quickly approved the plan. The school was designated "The Signal Corps School, Camp Alfred Vail, New Jersey."[21]

The first school commandant was Colonel George W. Helms. Helms had served as the fourth Commanding Officer of Camp Vail since June 1918. He served concurrently as commandant of the Signal Corps School and as Camp Vail's Commanding Officer until December 1920.

Instruction in the new school began 2 October. The initial curriculum included an officers' division, subdivided into radio engineering, telegraph engineering, telephone engineering, signal organization, and supply. The enlisted radio specialist course consisted of radio electricity, photography, meteorology, gas engine and motor vehicle operation. Electrical students were trained as telephone and telegraph electricians. Operator and clerical courses were also offered.

The school used the hangars as workshops and classrooms since all aerial activity had ceased with the signing of the Armistice. Such use continued past World War II.

POST WAR AND THE 1920s

SIGNAL SCHOOL DEVELOPMENT

The Signal Corps School expanded during this period as demands for communications training grew. Training of Reserve Officer Training Corps (ROTC) personnel developed into a major function of the school in June 1920. Training began for National Guard and Reserve officers the following year.

During 1922, the Officers' Division reorganized its courses into two main sections: a Company Officers' Course for Signal Corps Officers and a Basic Course in signal subjects for officers of other arms and services and newly commissioned Signal officers. Both sections were nine months in duration.[22]

The school, designed primarily for the training of Signal Corps personnel, found itself educating men from several branches of the Army. The name of the school was officially changed in 1921 to reflect this expanded mission. The new designation, "The Signal School" would be retained until 1935 when it would again become "The Signal Corps School."

The school was regrouped into four departments in 1922-23. These were: the Communications Engineering department, the Applied Communications department, the General Instruction course for all officers, and the Department for Enlisted Specialists.[23]

Radio Class, Camp Alfred Vail, 1920

Meteorological instruction was planned and was scheduled to begin in 1919. The repair of equipment damaged in shipment from France delayed the start of classes until 5 January 1920. Photographic instruction began in 1919; however, laboratory facilities did not become available until 1926. Instruction in motion picture production techniques was initiated in 1930. These courses reverted to the Army War College in 1932.

Wire Class, Camp Alfred Vail, 1920

A training literature section was formed in 1921. It supplied the technical and field manuals needed for the instruction of operations and maintenance of Signal Corps equipment. The section remained one of the major departments of the school until 1941 when the Signal Corps Publications Agency assumed its duties.

THE CAMP BECOMES A FORT

The installation was granted permanent status and renamed Fort Monmouth in 1925 in honor of the men and women who fought at the Revolutionary War Battle of Monmouth Courthouse.

Office Memorandum Number 64, Office of the Chief Signal Officer, dated 6 August 1925 stated,

"The station now known as Camp Alfred Vail, New Jersey, is being announced in War Department General orders as a permanent military post and will hereafter be designated as 'Fort Monmouth,' New Jersey. Mail to that post will be addressed to Fort Monmouth, Oceanport, New Jersey."

THE LABORATORY–LEAN YEARS TO CONSOLIDATION

Although overshadowed by the Signal School, the Radio Laboratory remained one of the most important facilities at Fort Monmouth. The Signal Corps quickly concluded after World

War I that adequate research facilities for the design and development of Army communications equipment were necessary, even if at a reduced scale because of budget restrictions.

Code Practice Equipment Class, Camp Alfred Vail, 1922

Research continued, and maximum use was made of the meager budget. The SCR-136, a ground telephone and telegraph set for artillery fire control up to thirty miles, was developed in 1926. Along with the SCR-134, mounted in observation aircraft, the SCR-136 provided air-ground liaison. Other projects included the SCR-131, a light and portable unit designed for infantry division and battalion telegraph with a five-mile range to limit possible enemy interception; the SCR-161 for artillery nets; the SCR-162 for contact between coast artillery boats and shore control points; and the SCR-132, a one hundred-mile telephone transmitter with an eighty foot portable, collapsible mast. Other experimentation was performed on items such as tube testers, crystal controller oscillators, unidirectional receivers, and non-radiating phantom antennas.

The function of the laboratory prior to 1929 had been primarily to design and test radio sets and some field wire equipment. Consolidation of the five separate laboratory facilities of the Signal Corps was planned that year.

The Signal Corps Electrical Laboratory, the Signal Corps Meteorological Laboratory, and the Signal Corps Laboratory at the Bureau of Standards (all in Washington, D.C.) moved to Fort Monmouth in the interest of "economy and efficiency." Conjointly, these laboratories became known as the "Signal Corps Laboratories."[24]

The Subaqueous Sound Ranging Laboratory transferred to Fort Monmouth from Fort H. G. Wright, New York, in 1930. The Signal Corps Aircraft Radio Laboratory at Wright Field in Dayton, Ohio had also been considered for consolidation, but subsequently was deleted. The Aircraft Radio Laboratory and the Photographic Laboratory at Fort Humphreys became the only research organizations not located at Fort Monmouth. These consolidations represented the first time the personnel and facilities needed to handle almost any Signal Corps problem could be found in one location.

The Signal Corps Laboratories employed five commissioned officers, twelve enlisted men, and fifty-three civilians as of 30 June 1930.[25]

SIGNAL CORPS BOARD

The Signal Corps Board was established at Fort Monmouth in June 1924. This was spurred by a suggestion to the Chief Signal Officer from LTC John E. Hemphill, the fifth Commanding Officer of Camp Vail. Hemphill wrote:

". . . the need for a board of Signal Corps officers to be continuously assembled at a center of Signal Corps activities for the consideration of problems of organization, equipment and tactical and technical procedure has long been recognized. Preferably such a board should consist of officers of considerable rank and length of service in the Signal Corps who would be competent to pass on such equations and would also be able to devote their entire time to the duties of such a board. Due to the shortage of personnel it does not appear that it will be practicable to detail such a board in the near future. The best present arrangement would seem to be a board at Camp Vail consisting of the officers at this post who are immediately connected with the administration and supervision of matters relating to general Signal Corps training. Detailed studies, experimental work, or field tests could be delegated from time to time by this board, with the approval of the Commanding Officer, to the proper subordinates at Camp Vail. It is therefore recommended that a permanent Signal Corps Board be constituted at Camp Alfred Vail to act on such matters as may be referred to it by the Chief Signal Officer."[26]

Army Regulation 105-10 (2 June 1924) directed the establishment of such a board. Over the years, typical cases considered by the board included the Tables of Organization, allowances and equipment, efficiency reports, Signal Corps organizations, and Signal Corps transportation needs. The first members of the board included Majors C.N. Sawyer, Frank Moorman, and Harry C. Ingles.[27]

POST ORGANIZATIONS

The 15th Signal Service Company acted as the parent organization for all new recruits, and for camp and school details. The company possessed the longest record of any unit permanently assigned to the Fort. It was activated as Company B, Signal Corps at Camp Wikoff, New York on 27 July 1898 and came to Camp Vail on 4 March 1919. Students at the Signal School were attached to the unit for rations, quarters and administration. Periodically existing as a company, battalion and regiment, the 15th maintained its identity until late in World War II.

The 51st Signal Battalion and the 1st Signal Company comprised two other long-term organizations at Fort Monmouth. Garrison duties or replacement training occupied battalion personnel. Technical subjects such as radio and telegraph operation, electricity, maintenance, line construction, and meteorology consumed the training effort.

FAMOUS FIRSTS

1926: *The SCR-136 and SCR-134 were developed.*

These ground to ground and ground to air radios were the first extended range voice radios put into military production.[32]

The 1st Signal Company, a permanently assigned detached unit of the 1st Division at Fort Devens, Massachusetts, carried out training required of divisional troops and participated to a limited extent in garrison details. Some instructors were furnished to the Signal School.

51st Signal Battalion

The Army, becoming increasingly conscious of the possibilities of mechanized warfare, conducted extensive maneuvers from July-October 1928 in Maryland. The 1st Signal Company conducted experiments with motorized equipment during the exercise. They concluded that radio was a prime means of communications for armored, mobile forces; that wire was useful only in rear areas; and that pigeons were impractical since they could not be trained to home to a moving loft.

The company continued in its prescribed role as a division communications unit, reportedly in a highly satisfactory state of training and morale and with equipment maintained in excellent condition.

The predecessor to the Military Affiliate Radio System (MARS) was born at Fort Monmouth in 1925 when the Signal Corps, working with the American Radio Relay League, organized the Army Amateur Radio Service (AARS). Hundreds of HAM radio operators joined the service that year. They grouped together in Corps Area Nets. Each Corps Area Net had several sectional radio nets, all coordinated by the control station located at Fort Monmouth. AARS had two objectives. The first was to provide a world-wide radio communications capability that could be used if necessary in times of emergency. The second was to provide a ready reserve of skilled radio operators that could be called into service in the event of another war.

AARS was reorganized as a joint Army – Air Force program called the Military Amateur Radio Service in 1948. Subsequently, the word "Affiliate" replaced "Amateur" (reflecting the affiliation

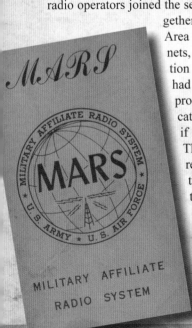

between military and civilian radio operators) and the word "System" replaced "Service" (to better describe the global reach of the MARS networks). The system proved its value in subsequent decades in disaster relief efforts, as well as in relaying messages between service men abroad and their loved ones at home. The MARS station at Fort Monmouth, K2USA, operated around the clock with fifty-one volunteers during Operation Desert Storm. Volunteers included Fort Monmouth's commander, Major General Alfred J. Mallette. K2USA "patched" an average of seventy-five calls a day from Soldiers and airmen in Southwest Asia during this time.

Only fourteen MARS stations existed in the United States in 1995. Though its role in the "psychological support of servicemen" had largely been supplanted by other technologies, K2USA volunteers still handled about thirty calls a week from service members in Haiti. The station continues to provide valuable service when natural disaster disrupts other means of communication. In 2005, the station had approximately seventy-five members.[28]

PERMANENT CONSTRUCTION BEGINS

COL James B. Allison succeeded LTC John E. Hemphill in August 1925 as the sixth Commanding Officer of Fort Monmouth. While he served only one year in the assignment, Allison initiated plans for construction of permanent barracks and a hospital building.

Actual construction did not begin until 1927, during the command of COL George E. Kumpe. Kumpe succeeded COL Allison in August 1926.

Four red brick barracks were completed in August 1927 around what is now known as Barker Circle. These housed approximately 200 men each. The hospital was completed in 1928, with an additional wing completed in 1934. The building, number 209, was known as Allison Hall.[29]

Quarters for field officers, company officers and NCOs were completed and accepted on 15 August 1928. These constituted the second and third increments of permanent construction. Five four-family apartment houses and one BOQ were completed and accepted on 6 August of the following year. The remaining permanent construction would be completed in the 1930s.[30]

COL Arthur S. Cowan succeeded COL Kumpe as the eighth Commanding Officer in September 1929. COL Cowan had served previously as post commander in 1917-18. During his second term he would serve the longest time of any commander, from September 1929 to April 1937.

CITIZENS MILITARY TRAINING CAMP

Authorized by the U.S. National Defense Act of 1920, the Citizens Military Training Camp (CMTC) provided young volunteers with four weeks of military training in summer camps each year from 1921-1941. Approximately 30,000 trainees participated each year. Those who voluntarily completed four summers of CMTC training became eligible for Reserve commissions. The CMTC camps differed from National Guard and Reserve training in that the program allowed male citizens to obtain basic military training without an obligation to call-up for active duty.

MG Charles P. Summerall, commander of the Second Corps Area at Governors Island, NY, discussed the program in 1926, stating, "If a lad attends the complete courses at the CMTC he is of value to the country in time of emergency, for he is then in a position to impart the knowledge he has learned to those who have not taken the courses." Summerall continued, "But the great feature of the camps is that it gives a lad an opportunity to spend a whole month, free of all expense to himself, out in the open, where he is enjoying plenty of athletic exercise and open air sports, and where he is under the supervision of men who have his best interests at heart."

First-year campers generally received basic military training at the camp at Plattsburg, NY. Students who continued with their courses in subsequent summers could train with engineers, field artillery, coast artillery, cavalry, or the Signal Corps, at Fort Monmouth.

Prior to acceptance, the Signal Corps required applicants to demonstrate "a sufficient amount of experience in electrical subjects, either through his school or business training, to insure that he will have the proper skill and knowledge to satisfactorily fulfill the duties of a signal officer in time of war." Once accepted, trainees learned various means of electrical communication, especially telephone and radio. Instructors taught the reserve officer candidates to install small telephone centrals and to construct field wire lines, as well as to operate the different kinds of army radio sets and message centers. Campers even conducted experimental work in the Signal Corps laboratories.

The trainees dabbled in more traditional signal fields as well. The CMTC at Fort Monmouth schooled trainees in the use of pigeons. The birds came in particularly handy during mock battle maneuvers in late August 1928, when one message sent back to Signal Corps Headquarters read, "We are starving, 15 of us without lunch." The *New York Times*, which carried that story, reported that a "chow wagon" was sent to the detachment within fifteen minutes.

Camp consisted of more than schooling and maneuvers, though. Plenty of time for recreation existed, as MG Summerall promised. The day's work ended by mid-afternoon, and tennis, track meets, volleyball, basketball, and boxing were all part of a day's routine. A baseball league was formed in 1927, and the best player on the team won an autographed Babe Ruth bat. Camp chaplain Reverend Clifford L. Miller organized an eleven piece orchestra that year, with plans to play at an end-of-camp dance. Visiting concert parties and movies provided additional on-post entertainment.

The Camp even provided transportation for trainees who wished to spend their free time at the local beaches. Other off-site haunts included the Rumson Country Club, where trainees took in polo matches, and Red Bank, where they watched airplane races.

Students clamored to attend the camp at Fort Monmouth. One, Robert J. Boylan, was so anxious to come here that he paid his own railroad fare from his hometown of East St. Louis, Illinois.

Vice-President Charles D. Dawes (1925-1929) declared, "The CMTC present to the youth of this country today an opportunity which should be seized by every young man. These camps teach the advantages and responsibilities of citizenship. They develop students mentally, morally, and physically. They are an asset to the nation."

The CMTC were disestablished in 1941 as the nation began to mobilize for the possibility of war.[31]

Polo game at Rumson Club '99

Photo snapped by a CMTC trainee enjoying some downtime at the Rumson Country Club, c. 1929

16th F.A. '29

Horseplay at the CMTC c. 1929

POST UNITS-MANEUVERS AND MARCHES

The two tactical units at Fort Monmouth (the 51st Signal Battalion and the 1st Signal Company) were well trained and equipped for field service with the outbreak of war in Europe in 1939. The 51st Signal Battalion had been reorganized in 1933 to prepare for field training on a large scale. Its new missions included providing enlisted instructors and overhead for the Signal Corps School; organizing a provisional radio intelligence detachment; and forming the nucleus of a General Headquarters (GHQ) signal service, to include a meteorological, photographic, and radio intelligence company.

A series of maneuvers kept the tactical units of the Signal Corps in the field much of each summer during the 1930s. In 1934, General Douglas MacArthur, Army Chief of Staff, conducted a GHQ Command Post Exercise centered in the Fort Monmouth – Camp Dix – Raritan Arsenal triangle. The 51st Signal Battalion, with the 1st Signal Company attached, provided signal services for the exercise, staffing message centers, handling radio intelligence, and performing radio, wire, and meteorological functions.

The 51st Signal Battalion installed all communications for the most extensive Army maneuvers held since World War I in 1935. The unit installed the Army corps and umpire nets in the Pine Camp area of New York, using 177 miles of bare copper wire, 126 miles of twisted pair field wire, and 8,260 feet of lead-covered, multiple pair overhead cable.

The 1st Signal Company journeyed to Camp Ripley, Minnesota, to install, operate and maintain signal communications for a phase of Fourth Army maneuvers in the summer of 1937.

Also in 1937, the 51st Signal Battalion was assigned to maneuvers of newly "streamlined" combat divisions in the area near Fort Sam Houston, Texas. As part of its participation, the 51st engaged in a road march from Fort Monmouth to San Antonio, Texas. This represented the longest motor convoy trip of its size in the Army's history. War conditions were simulated as closely as possible. The thirteen officers and 350 enlisted men, along with fifty-five vehicles, departed Fort Monmouth on 21 July 1937 and arrived at their destination on 2 August.

The following year, the 51st journeyed to Biloxi, Mississippi for maneuvers and took part in the Fort Bragg-Air Corps Anti-Aircraft Exercises. The 1st Signal Company participated in the Army War College command post exercise at Washington.

SIGNAL CORPS SCHOOL

The Signal Corps School, the name of which had changed to "the Signal School" in 1921 to reflect its mission at that time, reverted to its original name as part of a reorganization in 1935.

The Signal Corps experienced an acute shortage of trained personnel, particularly instructors, during the Depression years. As a result, advanced courses were offered for selected stu-

51st Signal Battalion, Fort Monmouth, 1931

dents in order to qualify them for the more responsible positions in the Signal Corps. The courses included Tactics and Techniques in Signal Communications; Auxiliary Signal Services in the Theater of Operations; Signal Operating Instructions and Orders; Equipment Studies; Staff Relations; Training Management; War Plans; Expeditionary Forces; Signal Supply; Duties of Corps Area Signal Officers; Historical Studies; and Field Exercises.

Signal Corps Labs Laying Wire, 1935

As a part of its reorganization, the Departments of Communications Engineering and Applied Communications combined into the Officers' Department.

The Enlisted Department adopted new techniques in teaching by converting to individual instruction instead of the classroom method. Courses in the Enlisted Department subdivided in the following year, becoming more highly specialized. They remained basically the same from then until World War II.

As World War II approached, the Signal Corps School functioned with three distinct divisions: The Officers' Department, Enlisted Department, and the Department of Training Literature. Seventy-eight persons comprised the faculty, eleven of whom were officers.

4,618 enlisted men graduated from the school in the decades following WWII. Signal Corps personnel comprised 2,486 of these graduates. The remainder represented sixteen other branches or services, as well as foreign nations.[34]

CORPS LABORATORIES AND RADAR

The newly named "Signal Corps Laboratories," consolidated at Fort Monmouth in 1929, received a new director in 1930. Major William R. Blair, distinguished in scientific and military fields, was appointed and served in this position until illness forced his retirement in 1938.[35]

COL William R. Blair (1874-1962)

Nine crowded wooden buildings constructed in 1918 continued to house the facilities. As a result of constant pressure by Major Blair a $220,000 appropriation was re-

ceived for construction of a permanent, fireproof laboratory building and shops in 1934. This structure was built under contract. It was scheduled for completion 11 November 1934, but was not actually completed and accepted until 1 March 1935. It was named Squier Laboratory in honor of Major General George O. Squier, the Army's Chief Signal Officer from 1917-1923.

Much of the communications equipment used by American forces during World War II was designed and developed at Fort Monmouth during the 1930s. The laboratories completed six field radio sets; readied several artillery pack sets for tests; and fielded the SCR-197, a new Air Corps mobile transmitter. The SCR-300 (the "Walkie-Talkie" radio set) was perhaps the best-known development of the period. In addition, switchboards, field wire, and radio receivers were developed.[36]

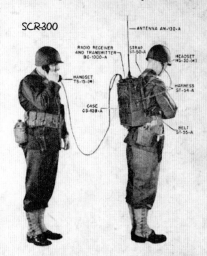

SCR-300

It was MG Roger B. Colton, (successor to Col Blair) who, in 1938, made the historic decision to employ FM in all future military radios. While the labs at Fort Monmouth had been experimenting with FM transceivers since 1936, they had continued the development of AM radios. Though FM signals are of a shorter range than their AM counterparts, FM signals are far less susceptible to interference. Such static free communication is accredited with giving the Allies an advantage during WWII. In a letter from Edwin Armstrong to MG Colton dated 29 June 1944, Armstrong called this decision, "the most difficult decision of the history of radio which anyone was ever called upon to make…I most sincerely hope that when this war is over what your organization accomplished here can be duly laid before the world and properly acknowledged." A Signal Officer in the 1st U.S. Army, Col Grant Williams, remarked, "I feel every Soldier who lived through the war with an armored unit owes a debt he does not even realize to General Colton."[37]

MG Roger B. Colton (1887-1978)

One of the most important pieces of equipment developed during the lead up to war was RADAR (Radio Detection and Ranging). The term RADAR was first coined by the Navy in 1941 and was accepted by the Army in 1942. The term refers to the equipment itself as well as the method by which distant or invisible objects can be detected by reflected radio waves. According to the first Signal Corps Field Manual on the Aircraft Warning Service, "RADAR is a term used to designate radio sets SCR (Signal Corps Radio)-268 and SCR-270 and similar equipment." The SCR-268 and 270 were not in actuality radios at all, but were designated as such for security reasons to protect their identity.[38]

Radar emerged from the defensive need to counter the possibility of massive aerial bombardment as well as submarine attack. Sound detectors suffered from inherent limitations. Experiments with electromagnetic waves during World War I produced interesting experimental results, but no operational equipment was produced.

Numerous tests had been conducted with heat emitted from airplane motors or reflected by airplane surfaces in the 1920s. This work was accomplished by the Army Ordnance Corps from 1926 to 1930. The project was transferred to the Signal Corps in 1930. The research was duplicated by the Army Corps of Engineers for several years due to a misunderstanding. All Army detection development was officially assigned to the Signal Corps by 1936.

> **FAMOUS FIRSTS**
>
> **1937:** *Fort Monmouth laboratories developed and demonstrated a "mystery ray."*
>
> *This was a prototype of the Radio Detection and Ranging (RADAR) sets SCR-268 and SCR-270, which would locate and track airplanes.*

During the 1930s the Labs conducted experiments with infrared, heat detection, radio detection (termed radio optics and now called microwave), and pulse equipment, dubbed "Project 88." Infrared and heat detection equipment were abandoned in 1932 and 1935. A turning point came in 1936 when the detection of aerial and surface targets became a top priority and received extra funding. Work on the "beat" (based on the Doppler effect) and pulse concept intensified. Blair knew that radio waves could reflect off surfaces and that they could be used to locate aircraft in flight. The pulse echo method proved to be the breakthrough as it provided early warning detection of aerial targets at great distances. The range, angular elevation and the azimuth of targets also proved to be reasonably accurate.

Blair proved that radar was capable of detecting and measuring the speed of aircraft through a combined system of heat and radio pulse-echo detection in May 1937. He demonstrated this in front of senior military and congressional officials. A B-10 Bomber was instructed to fly over Fort Monmouth with running lights extinguished as the radar operators tried to pick him up. The operators successfully picked him up in their searchlights each time. The Army Chief of Staff, General Malin Craig, purportedly said that he never would have believed such a thing possible. The Secretary of War, Harry H. Woodring, wrote a letter to then Chief Signal Officer, Major General Allison, stating how pleased he was with the demonstration[39]. The demonstration was such a success that it became one of the nation's best kept secrets, receiving a top secret classification.

According to Dr. Harold Zahl, Director of Research and inventor of the radar's vacuum tube components, these tests before the Secretary of War et al. were not nearly as successful as everyone was led to believe. Zahl relates that when the visitors departed after the "successful" tests, engineer John Hessel noted that it was strange that most of the pickups were made by the number two searchlight. Zahl met the operator of that searchlight, a Corporal who told him, "Remember that low white cloud during the night of the tests, the one hanging over Red

Dr. Harold Zahl

Bank? Well, with the town lights shining on the cloud it was possible for me, with my own eyes, to see the dim outline of the plane before you turned on your light. There I was, tracking the guy with my binoculars, right smack on the cross hairs. Couldn't help but make a direct hit most of the time when you gave the in-action go ahead....that new secret gadget is all right. Why, every time you fellows turned on the control light it was pretty close to the target--almost as good as my eyes." Zahl, crimson-faced, said he made the split second decision to congratulate the Soldier, "I knew well too that this was the period of a never-ending feud between the men on the ground and the men in the air; the men trained with searchlights and sound locators instinctively followed the code of that era, to illuminate that airplane as soon as possible. Take advantage of every break, lest they get you first. They were soldiers, not scientists. And whoever heard of ethics in a foxhole?"[40]

In 1937 there was no school or system of instruction for radar, so students learned by observing and were taught by radar engineers. Early warning SCR-270 classes were conducted at Fort Monmouth in the Squier Laboratory and fire control and search light control SCR-268 classes were conducted at Fort Hancock at Sandy Hook. The first formal class was inducted in 1938. Between 1938 and 1941 about ninety technicians were trained.[41] Students who received training in radar were not permitted to bring books or course materials home because of the security classification. They had to complete all their studying

SCR-270 and 271 installed at Twin Lights, NJ

at school.[42]

The work on radar which had begun at Fort Monmouth was moved to the barrens of Sandy Hook for security reasons. German spies were suspected of observing the work at the Fort's main post.[43] In 1940 the Army contracted with Bell Labs and Western Electric to manufacture the radar units. The Army soon feared German submarine attacks on Sandy Hook and so moved the radar work to the old Marconi Radio Station in Wall Township, NJ. The new lab was named the Signal Corps Radar Laboratory until Intelligence services in Washington pointed out that "radar" was a classified word. It was renamed the

Marconi Hotel, Camp Evans, N.J.

Camp Evans Signal Laboratory in 1942.

The first radar set, the SCR-268-T1, which was demonstrated in 1937, was composed of a radio detector unit, a thermo detector unit and a searchlight. The transmitter, two receivers, a superheterodyne and a supergernerative were located at Fort Monmouth's west gate. Another receiver and search light were located at the other end of post.[44] This set was continually improved upon. Double tracking was applied to provide continuous radio tracking data, more powerful vacuum tubes were installed, and improved antennas were built. The transmitting, azimuth and elevation antennas were redesigned into

SCR-268-T1

a single mount and approved for production as the SCR-268. Western Electric began delivery of the sets in February 1941.[45]

Bureaucratic hurdles in the early stages of radar invention often involved lack of funds and tension between military and civilian personnel when the technical aspects of the scientist's work was not understood. Blair was given $75,000 to continue radar experiments after the successful demonstration. A total of $3 billion had been spent by the end of the war on techniques that grew out of his designs and funding would never again be a problem. Blair could not file for a patent until the day after the war ended in Europe because of the security classification. The law, however, required that a patent be filed within one year of publication of the art. The 81st Congress eventually passed private law 1008 (HR-577) which absolved the prior publication content for Blair.[46] The patent for radar was titled "Object Locating System." The government had a royalty free license to use ra-

dar.[47] When the patent was finally awarded in 1957, the U.S. Patent office declared Blair's patent one of the most important issued this century.[48] A News Release from the Department of Defense said they considered the patent "to be as important and far reaching in its military applications as the first U.S. patent issued on the telephone was to the commercial communications industry."[49]

Many improvements had been made by 1941. Long range detectors, the SCR-270 and 271, were in use.[50] A Fort Monmouth designed radar set warned of the Japanese attack on Pearl Harbor in 1941 but the warning went unheeded according to Blair by "those who were not familiar enough with it to have faith in it."[51] At Pearl Harbor, two U.S. Army Signalmen, Joe Lockard and George Elliott, were on duty with their SCR-270. They were scheduled to close the station at 0700 but since their transport back to the barracks was late, they kept the radar running. At 0702 they picked up echoes that they had never seen before. They checked the equipment, fearing it was malfunctioning. It was not. Their calculations revealed that a formation of airplanes was approaching Oahu at 180 miles per hour. They made a formal report at 0720. The warning went unheeded and the first bombs were dropped at 0755.[52] The radar at Pearl Harbor was the 100 megacycle SCR-270 developed under the command of Major Paul Watson. Six radars had been sent to Hawaii to protect the naval base.

Key to the success of the radars were their vacuum tube components. If radar tubes failed, the enemy could attack without warning. Dr. Harold A. Zahl (1904-1973) received a PhD in physics in 1931 from University of Iowa. Zahl was the Director of Research for the Fort Monmouth laboratories for several years. He conducted pioneering work in acoustics, infrared detection and radar. Zahl held over fifty patents on radio and electronic inventions. He conceived and patented a pneumatic cell detector which was a major component in the Army's first radar set, the SCR-268-T1. He invented the Transmitter-Receiver tube which made single antenna systems possible for Army and Air Corps early warning radars. His engineering designs contributed to the make up of a number of early warning radar sets and his inventions with tubes are credited with helping to shorten the war.[53]

Zahl invented the VT-158, known as the "Zahl Tube." This tube pushed radar into the megacycle operating range. The Zahl tube was the heart of the TPS-3 Lightweight Radar, known as the "Tipsy Three," and the TPQ-3 mortar locating set. Zahl even blew his own glass to make the tubes.[54] The tube consisted of four triode tubes connected in parallel. The tube envelope contained tuned plate and grid lines which made it an oscillator. 250,000 watts of peak power could be extracted during a radar pulse.[55] Until his invention, giant antennas were needed.[56] Zahl was awarded the Legion of Merit in 1946 for his work with radar and vacuum tubes. In a 1963 interview Zahl said, "The next 50 years will be the most interesting and exciting in all human history-if we don't blow ourselves up."[57]

Zahl was in charge of the SCR-268-T1's first service test at Fort Monroe, VA in 1938. The results were so successful the Army formally accepted radar, directed further development and set production requirements. The demonstration also uncovered a new means for navigation. The target bomber had unknowingly drifted far off course over the Atlantic Ocean, obscured by cloud cover. The radar detection trial was converted to a rescue mission as Zahl located the bomber on the radar and charted it back to safety by radio a few moments before the fuel would have run out.

Zahl's GS-4 Transmitter-Receiver tube was used in the radar that gave warning of the attack on Pearl Harbor. Before this tube was invented, a radar set had to put its receiver out of action while the outgoing pulse was being transmitted. A temporary solution was a cumbersome and costly method of using separate antennas for transmitting and receiving. In 1942 the frequency ceiling of radar was 200 megacycles. This limited the efforts to reduce the size of extremely bulky equipment. Zahl was charged with creating the highest frequency device he could without a loss in range and resolution. He was also given responsibility for development of all radar tubes for the Army's ground forces and Air Corps. He assumed development responsibility for entire radar sets, beacons, Identification Friend or Foe (IFF) and other electronic equipment.[58] The first airborne radar was installed in a B-18 flown into Red Bank, NJ from Wright Field. To save weight, the radar engineers had to stay on the ground.[59]

There were many other significant Army soldiers and civilians that contributed to the breakthroughs in American radar. Robert Noyes (b. 1904-) designed and developed the first successful radar transmitter used by the U.S. Army. This was a major contribution to the radio wave reflection system and played an important role in military combat communications.[60] John Slattery (1909-2007) was responsible for the systems design of the first single mount antiaircraft radar equipment in 1937 and for antenna and systems design of numerous Signal Corps radars. He was awarded the Legion of Merit for contributions to military radar.[61]

COL Roger B. Colton's (1887-1978) developments made possible the increased resolution loading to the application of radar to strategic bombing, fire control and guided missiles.[62] Howard Vollum (1913-1986) was awarded the Legion of Merit and Oak Leaf Cluster for work on coast artillery fire control radar and radar for ground forces. His radar development work focused on very short pulses and the shortest wave length available at the time. The part of the radar set in which he specialized was the display or indicator section. This was a specific type of oscilloscope. Vollum's work on coast artillery radar involved locating the mortar that fired the shells by extrapolating part of the trajectory that could be observed back to its source. It was called shell splash radar because one could see the target but also the splash from the shells when they landed around it. The short pulses and high frequencies enabled the operator to tell the shell from the target. According to Vollum, that radar was just shaping up when the war ended.[63]

Another scientist in the Labs, John Marchetti (1909-2003), developed a mortar locator. Zahl said that although it was he that invented the VT-158 tube, it was Marchetti's genius that made it work in the radar set. Marchetti was tasked with extending the range of the early warning radar system guarding the Panama Canal. Protecting the Panama Canal was considered to be of the utmost importance after the attack on Pearl Harbor. Existing radar was almost useless against low flying aircraft. The plan was to mount radars operating at higher frequencies on picket ships patrolling one hundred miles off of each canal entrance. Their test set worked with ranges of one hundred miles on bombers and good coverage at low altitudes.[64]

Initial experimentation with an enemy mortar and artillery locating radar was conducted at Camp Evans in 1944. The objective of this program was to develop and field a radar system capable of detecting and locating hostile weapons with sufficient speed and accuracy to permit rapid and effective counter fire by friendly forces. As U.S. Marines were being slaughtered on Pacific islands, General MacArthur's staff pleaded for a breakthrough technology which could locate mortar firing positions. The breakthrough idea belonged to John Marchetti. He said, "If only we could detect the shells as foliage, say a few hundred feet above the firing point, we could easily extrapolate down to the approximate position where the mortar was before they could change their location. Then we could blast the hell out of them before they could change their location."[65]

Marchetti's idea was to combine a new, powerful, lightweight generator with Zahl's vacuum tube, the VT 158. The tube, which ran at 600 megacycles, sent out radar waves with fifty centimeter wave lengths at its target. A Japanese mortar shell was nearly half this wave length and it sent back a glowing radar reflection. According to Marchetti, the Allies "were taking a tremendous beating by the Japanese mortars....and they

John Marchetti

were firing them at our people indiscriminately."[66] The team at Camp Evans modified the AN/TPS-3 with the new generator and tube. When MacArthur's group found out the problem had been solved, they ordered the equipment immediately. There was no time for contracts or plans; the radars had to be built immediately at Camp Evans. Laboratories were converted into production shops. Twenty engineers and a secretary worked day and night. After ninety-six hours of continuous work, twelve units were fabricated and soldiers were trained to operate them in battle. Then units could be loaded into a landing aircraft and set up in thirty minutes to protect landing beaches from mortar attacks or enemy airplanes. The units were driven to Newark Airport and flown to the Pacific in time for the next island assault. The Zenith Radio Corporation was then charged with the fabrication of 900 units which saw action during the remainder of the war. Twenty-four units were used on D-Day in Normandy.[67] Marchetti also helped British radar operators to better detect incoming German V-1 jet propelled "buzz bombs" increasing the anti kill ratio from zero to eighty-five percent.[68] Marchetti was given the Order of the British Empire, and was decorated by France for his radar contributions. Another Army soldier at Monmouth, Capt Jack Hansen, was later

AN/TPS-3 Radar

charged with traveling to the White Sands Missile range to help with modifying the SCR-584 for the tracking of V-2 missiles.[69]

Radar turned the tide of WWII, affecting the outcome of two key engagements: the Battle of Britain and the Battle of the Atlantic. Radar operators foiled the planned attack on the Allied air base at Malta. The Axis was forced to surrender in North Africa as the Allies had been able to destroy their supply ships by detecting them with radar. This was critical to the outcome of the war. The Japanese in the Pacific theater were at a distinct disadvantage as they were without radar, while the Allied ships were equipped with early warning radar. The Japanese were unable to make a surprise attack or defend their carriers in the Battle of Midway Island. Japanese superiority in night engagements was overcome with gun fire directed from radar sets. The early warning of impending air raids often meant the difference between victory and defeat for the allies.

German radar sets were destroyed and radar cover was provided for ships and landing parties during the Normandy invasion. Radar ruses were used to make the Germans think the land-

ing was going to be at Calais. Allied defeat at Normandy was a significant possibility—overcome by radar navigation and countermeasures. Fort Monmouth designed radar sets landed on the beaches of France on D-Day to help protect soldiers from Luftwaffe fighter attacks. Many historians believe that radar navigation was so important that the landing could not have taken place without it.[70] Earlier in 1944, the SCR-584 radar (developed at Fort Monmouth in 1942) arrived on the beaches of Anzio, Italy just as the Nazis had developed methods to jam the SCR-268. The 584 would transmit a code when it detected a plane and interro-

SCR-584 with IFF equipment

gate the plane's identity. A friendly receiver would respond with the correct code. IFF thus was born.[71]

There were several types of radar by 1945 to include fire control, early warning, ground control of intercept, navigation, gun laying, blind bombing, airborne interception, air to surface vessel, ground control of approach and weather monitoring. Due to the back scattering of water droplets, it was discovered that thunderstorms showed up on radar. Thus operators were able to direct pilots away from storms. After WWII, radars for ship navigation and weather navigation for airplanes became available on the commercial market.[72] In the decade after the war, radar became increasingly used at airports to ensure safe approach and landing. UHF and VHF radars came into use and the cavity magnetron was put into production in 1949.

The average personnel strength of the laboratories between 30 June 1930 and 30 June 1935 was twelve officers, thirty-six enlisted men, and 119 civilians. Civilian personnel strength continued to grow slowly through 1940 with 234 civilians assigned as of 30 June. Officer and enlisted strength dropped slightly to eight officers and fifteen enlisted. The strength increased dramatically, however, within the following year. Civilian manning was 1,227 as of 30 June 1941. Military strength rose to twenty-eight officers, with an additional twenty-nine officers from the Coast Artillery Corps and seven officers from the armored force.[73]

COMPLETION OF PRE-WORLD WAR II CONSTRUCTION

The post's permanent red brick construction, which had undergone its first phase during 1927-29, entered its second phase in 1930 when construction began on three four-family apartments, one Bachelor Officer's Quarters, six double sets of quarters for non-commissioned Officers, and one set of quarters for field officers. These projects were completed in October 1931. Completion of eight double sets of Company Officers' Quarters, seven double sets of NCO Quarters, and one four-family apartment complex followed in June 1932.[74]

The Army's Adjutant General also authorized construction of a post theater. Construction was financed by Army Motion Picture funds. The 574-seat theater opened 15 December 1933 with a showing of "Dr. Bull," starring Will Rogers. The theater (Building 275) was called War Department Theater Number 1. Twenty years later, in December 1953, the building was officially dedicated as Kaplan Hall in memory of Major Benjamin Kaplan, an engineer who served in both military and civilian positions at Fort Monmouth for twenty-five years and was associated closely with the permanent construction program of the 1920s and 1930s. Kaplan Hall eventually became home to the U.S. Army Communications-Electronics Museum.[75]

The final phase of the pre-war permanent construction program was completed between 1934 and 1936 under the Works Projects Administration (WPA). Eleven double sets of NCO Quarters were completed, along with the West Wing and an addition to the North end of the Hospital, in 1934. A blacksmith shop, incinerator, bakery, warehouses, band barracks and utility shops were also completed that year. In 1935, the fire station, guardhouse, and Signal Corps Laboratory (Squier Hall) was

completed as well as three sets of quarters for field officers, and three sets for company grade officers.[76]

The quarters of the Commanding Officer (Building 230) were the last to be completed. Colonel Arthur S. Cowan, then the 8th Commanding Officer, first occupied the quarters. The last of the permanent pre-war construction to be completed was the headquarters building, known as Russel Hall. Construction ended in 1936.[77]

SIGNAL CORPS PUBLICATIONS AGENCY

The growing need for printed training, operational, and maintenance materials gave rise to a Signal School "training literature section" whose mission was to write and publish training manuals, regulations, school texts, and other technical materials. The Joint Congressional Committee on Printing authorized a print plant for the school in 1927. Over the next fifteen years, this requirement evolved into the Fort Monmouth Signal Corps Publications Agency, activated in November 1943. This agency, organized and operated by the Fort Monmouth Training Center, consisted of the School's Department of Training Literature, the Instruction Literature Section of the Fort Monmouth Signal Laboratories, and the Technical Publications Section of the Evans Signal Laboratories. By 15 January 1944, this organization, which occupied sixteen buildings on the Main Post, had five hundred products pending.

Aerial view of Main Post

389th ARMY BAND

The 389th Army Band traces its history back to its 1901 organization at Fort Meade, Maryland as the 13th Cavalry Band. The 389th came to Fort Monmouth in August 1930 as the Signal Corps Band. It was designated the 389th Army Band in 1944, the name it bears to this day. It is the official band of the Army Materiel Command (AMC) and, in that capacity, serves all of AMC's subordinate commands when musical support is required for military and official functions. It also supports Army recruiting and participates regularly in over 200 military and civilian ceremonies and parades each year. During World War II, the 389th Army Band was the motivating factor in a campaign that sold more than $1 million in war bonds for the war effort, earning the band the Meritorious Unit Commendation, 1944-1945[78]. The band left Fort Monmouth in October 1994 for Aberdeen Proving Ground, MD as part of a nationwide realignment of Army bands.[79]

Band honored in farewell ceremony

Continued from page 1

In his remarks, Guenther noted how much support the 389th Army Band provided Fort Monmouth over the 50 years it has been here. He remarked how he had come to know each member of the 389th Army Band" throughout his years of service here. He expressed, also, how much they would be missed at "our Retirement and Award ceremonies, our Changes of Command and our Farewell ceremonies."

Ojeda noted in his remarks the difficulty the band has experienced in leaving their home of 50 years. He also expressed thanks to the many individuals and activities at Fort Monmouth who have provided a great deal of support to the 389th Army Band over the

"In the new Force Projection Army our primary mission is to provide music to enhance unit cohesion and soldier morale," Ojeda said. "Since Fort Monmouth has lost some military units and will lose some more, it is not cost effective for the Army to keep a band here. The band's relocation will save the band from being inactivated in the future."

photo by Greg Brower

Maj. Gen. Otto J. Guenther wishes Spec. Eric J. Matas and other members of the 389th Army Band well during last Thursday's farewell ceremony for the Band.

photo by Greg Brower

Maj. Gen. Otto J. Guenther congratulates Chief Warrant Officer 2 Luis A. Ojeda, commander of the 389th Army Band, after presenting him the Meritorious Service Medal, First Oak Leaf Cluster. Looking on are Command Sgt. Maj. Charles J. Johnson and Lupe Ojeda.

years of his service here.

"In the new Force Projection Army our primary mission is to provide music to enhance unit cohesion and soldier morale," Ojeda said. "Since Fort Monmouth has lost some military units and will lose some more, it is not cost effective for the Army to keep a band here. The band's relocation will save the band from being inactivated in the future."

The members of the 389th Army Band include: Chief Warrant Officer 2 Luis A. Ojeda, commander; First Sgt. Milton M. Belle, Jr.; Sgt. 1st Class Wilson Gonzalez; Sgt. 1st Class Joe E.C. Kirby; Sgt. 1st Class Michael R. Tripp; and Staff Sergeants Raymond S. Becker, III; Stephen G. Crum; William B. Hatter; Gregory Hawkins; Paul C. Hessert; Joseph F. Larry; Eric J. Peterson; James J. Rohrbach; and Elmer M. Smith, Jr.

Sergeants in the 389th Army Band are: Jere R. Bucek; Kurt W. Odendahl, Allison A. Peterson, Jesse L. Hughes and Marc D. Winans.

Specialists who are members of the 389th Army Band are: Theresa M. Palmer, Ty Powell, Stuart A. Whitman, Gwyn C. Bollinger; Hubert W. Holmes, IV, Eric J. Matas; Helena K. McCartney; Sean P. O'Brien; Deloris Payne; Thomas E. Schindler; Angela K. Schoenbeck; Christopher R. Smith; Roger G. Stubblefield; Timothy T. Tipton; and Brian C. Wolfe. ❂

389th Army Band, April 1976

LIMITED EMERGENCY

President Roosevelt proclaimed a state of "limited emergency" on 8 September 1939, following the outbreak of war in Europe. This action significantly impacted Fort Monmouth.

The Army was immediately authorized additional personnel, increasing from 210,000 to 227,000 officers and men. The Signal Corps School curriculum, both officer and enlisted courses, changed to accommodate the increased enrollment. The Commandant, Colonel Dawson Olmstead, was advised that the school would probably be called upon to train 224 officers and 2,455 enlisted men to fill vacancies in newly organized units. Seventy-five officers and 1,300 men would be required annually as replacement. Events would soon prove that these estimates were extremely conservative.

BG Dawson Olmstead

One year following the "limited emergency" proclamation, Congress passed the Selective Training and Service Act providing for one year of compulsory military training. The President simultaneously called the National Guard into Federal service, and the Army increased in size to 1,400,000.

The influx of personnel during World War II (the number assigned to Fort Monmouth peaked at about 35,000 military and 15,000 civilians) produced a severe shortage of housing. To alleviate this problem, the Army, in cooperation with the Federal Public Housing Authority, constructed 265 homes, known as "Vail Homes" in Shrewsbury Township, eighty-two units in Long Branch (the Grant Court Project), and fifty-nine units in Asbury Park (Washington Village), in addition to several dozen residences at Camp Evans.[82]

Colonel Olmstead was promoted to Brigadier General on October 1940, thus becoming the first General Officer to serve as post commander.[83]

SIGNAL CORPS REPLACEMENT CENTER

With the passage of the Selective Service Act, General Olmstead was advised by the Chief Signal Officer to develop a Replacement Training Center at Fort Monmouth where enlisted personnel would receive one year of training. The Signal Corps Replacement Center opened in January 1941. Capacity was fixed at 5,000 men. By December, however, it was necessary to increase the capacity to 7,000 and to reduce the one-year training period to thirteen weeks.

The first Commanding Officer of the Replacement Center was Colonel George L. Van Deusen. He assumed command 14 January 1941.[84]

In April 1941, Colonel Van Deusen was promoted to Brigadier General. He retained his post as Commandant of the Replacement Center until November 1941. By August 1941, he wore two additional hats: that of Signal Corps School Commandant (July 1941 – November 1942) and that of the eleventh Commanding Officer of Fort Monmouth (August 1941 – September 1942).

MG George L. Van Deusen

General Van Deusen initiated the purchase of additional land in view of the increasing expansion of Fort Monmouth. The land, now the Charles Wood area, was considered ideal for replacement training activities for as many as 7,000 men. Adequate space was available for all necessary buildings and a maneuver area.[85]

At Camp Charles Wood, as the area was called in 1942, construction was completed within ninety days on sixty barracks, eight mess halls, nineteen school buildings, ten administration buildings, a recreation hall, post exchange, infirmary, and chapel. The camp was officially dedicated on 14 July 1942.[86]

When the facilities at the Camp Charles Wood Area proved insufficient in meeting wartime needs, negotiations began for leasing numerous properties in Monmouth and Ocean Counties.

In 1941, Olmstead leased the National Guard Encampment at Sea Girt for $1 a year, plus $125 a day for power, gas and water. The 1st Signal Training Battalion moved from the main post to the new camp at Sea Girt by April 1942. The land was designated Camp Edison in honor of Charles Edison, governor of New Jersey and son of the famed inventor.

Camp Charles Wood

The Replacement Training Center was in operation at three locations by mid-1942: Fort Monmouth, Camp Charles Wood, and Camp Edison. The Army acquired two noncontiguous field training areas near the communities of Allaire and Hamilton.[87] Field bivouac and maneuvers utilized these wooded tracts extensively.

By the spring of 1943, the recruits underwent a program that began with three weeks of basic training at Camp Edison, continued with four days of field operations at Allaire or Hamilton, and culminated in an overnight march to Camp Wood for final specialist training. The recruits marched to Allaire, six miles distant, where a bivouac was established and security detachments were posted. During their first day at Allaire, the trainees were conducted through the field expedient sanitation area, hasty field fortification area, and camouflage area. They watched a demonstration squad lay a mine field, set booby traps, and destroy a mock-up wooden tank with Molotov cocktails. The next day they were given practice in extended order drill and taught the elements of individual security, scouting, and patrolling. The third day was given over to similar practical instruction and a demonstration of field cooking by the Mess Specialists' Division. On the third night a "raiding party" of cadremen descended upon the bivouac, trying to surprise the trainee outposts and capture the model command post. On the fourth day, the trainees rose early, drew "C," "D," and "K" rations to suffice them for the whole day and proceeded in groups to the combat conditioning course, a circular four-mile route comprising fifteen different training phases. Each group started at a different section of the course and devoted the entire day to completing the circuit. A typical group might encounter successively a gassed area; a cadreman playing the part of a Japanese sniper whom the group had to dislodge; an anti-tank emplacement; a snipers' nest in an abandoned house; a machine gun nest; a 60- millimeter mortar; a "no-man's land" laced with barbed wire and pocked with "shell-holes"; a tank-trap area; a "Nazi village" complete with mines and lurking snipers; a booby-trap house; another sniper emplaced in a tree; an "enemy parachutist"; a party of "enemy officers" studying combat orders, which the trainees must try to overhear; a stream which must be crossed by means of an improvised cable bridge; a "barbed-wire and overhead fire zone"; a series of enemy outposts which must be taken, and, finally, a first-aid station. Of particular interest on

this course were the Nazi village and booby-trap house. The former, constructed by soldier-craftsmen working with salvaged materials, offered the appearance of a small European hamlet recently subjected to a barrage, and planted with mines and booby-traps by retreating Germans. The trainees were directed to gain control of the village, and as they made this effort they encountered barbed wire, mined areas, and cross-fire from the hidden snipers. The booby-trap house was a 4-room structure, completely furnished and provided with an extensive assortment of traps for the unwary. A sign on the door warned, "When the bell rings, St. Peter says hello." Each of the traps, comprising furniture, souvenirs, food, doors, and windows, was wired to a bell instead of an explosive, but the lesson learned was nevertheless vivid. The trainees spent thirteen hours - 0700 to 2000 - on this course and then returned to the bivouac area to break camp. Shortly after midnight they marched to the Hamilton area, where they set up a new bivouac. The following morning they proceeded to Camp Wood to start their specialist training.[88]

Camp Charles Wood 1 Year Anniversary

All officers at Fort Monmouth, both student and staff, were required to cover an infiltration course at the Allaire Combat Training Area beginning October 1943. The course involved eighty-five yards of crawling under machine gun fire, through terrain heavily intersected with barbed wire. Small land mines and fire crackers were exploded to simulate battle conditions.

A unique innovation at the Allaire Field Training Area was the Mental Conditioning Chamber which was called the "Lunk Trainer," patterned after a similar device at Fort Benning, Georgia. A covered dugout, unlighted and unventilated, served as a simulated command post for signal communications operations. The trainees performed their specialist functions with high-power fans directing wind, sand, and water into their faces, and smoke pots providing the smell of burnt gunpowder. Climactically, a horse cadaver, in an advanced state of decomposition, was used to approximate the odor of decaying human flesh. This feature was later removed upon the advice of medical authorities.

The Replacement Training Center was finally deactivated in November 1943. The center produced more than 60,000 Signal Corps specialists during the thirty months of its existence. The enlisted cadre peak was 1,157, with 250 officers and civilians also assigned.[89]

LOCAL SITES LEASED

Other local sites leased to incorporate wartime expansion included Asbury Park's Convention Hall, a beachfront landmark and New Jersey treasure, located on Ocean Avenue between Fifth and Sunset streets in Asbury Park. The Hall was designed by Grand Central Station architects Whitney Warren and Charles Wetmore in 1923. COL Willard Matheny, Assistant Executive Officer at Fort Monmouth, revealed the Army's plan to lease the Convention Hall in August 1942. Matheny explained that the expansion of the Signal Corps at Fort Monmouth had been so rapid that there was no time to build sufficient facilities on post. The Hall would serve as a training facility for officers.

Convention Hall, Asbury Park, NJ

Photo courtesy of the National Park Service

The Army divided the Convention Hall into sixteen classrooms, including two large rooms that each provided seating space for 150 students at map tables. While the classrooms were being partitioned off, instruction was conducted in the Asbury Park YMCA gymnasium, also leased by the Signal Corps. The YMCA, completed in 1929, primarily housed the Eastern Signal Corps School Enlisted Department.

The Army also leased another popular Asbury Park site, the Marine Grill at Ocean Avenue and Deal Lake. This restaurant served as a mess facility for 325 officers housed in Asbury Park. Many of those officers training at the Convention Hall and dining at the Marine Grill lived at the plush Kingsley Arms Hotel or the Santander Apartments, both in Asbury Park.

The eight story Kingsley Arms apartment hotel faced Deal Lake. The nine-story Santander was located between Deal Lake and the ocean. The Army displaced eighty-eight families when it took over the site, then owned by Prudential Real Estate Company, in the fall of 1942.

The Signal Corps later leased the 150 room, oceanfront Grossman Hotel on Ocean Avenue in Bradley Beach. Leased in early 1943, the hotel was used for office space for Signal Corps procurement functions. Utilization of this facility consolidated functions previously scattered between Fort Monmouth and its

satellite site, Camp Evans. Approximately 400 Army and civilian personnel worked at the Hotel.

The Sea Girt Inn was also leased in the spring of 1943. The facility, once closed by the local authorities for staging "indecent performances," was used for far more serious purposes by the Signal Corps Engineering Laboratories. USAF COL (Ret) Albert C. Trakowski later recalled, "It was an old nightclub that the Signal Corps had rented for the purposes of doing remote experimentation…Rawinsonde. I did most all the work on developing how to use that equipment…there at the Sea Girt Inn." The rawinsonde was an adaptation of radar that proved especially helpful to air operations by allowing the measurement of wind velocity and direction at high altitudes without the necessity for optical tracking previously required.

Fort Monmouth leased dozens of other local sites during the war, some famous and some nondescript, some still standing and some demolished long ago. Nonetheless these wartime activities left an indelible footprint on the surrounding communities.[90]

THE OFFICER CANDIDATES

Fort Monmouth's other wartime training focused on officer candidates. The Officer Candidate Department was activated within the Signal Corps School on 2 June 1941. The first class commenced 3 July 1941. That class of candidates was comprised of 490 students chosen from warrant officers and enlisted soldiers based on leadership, communications knowledge and prior service. A total of 335 graduated newly commissioned second lieutenants after three months' training.[91] Subsequent classes averaged about 250 men, but gradually grew to 1,000 men per class.

The curriculum of that first class included physical training, dismounted drill, military law, sanitation and first aid, military courtesy and customs, interior guard duty, defense against chemical attack, pistol marksmanship, supply, administration, mess management, map reading, signal communications, motor transportation, inspections, and training methods.

12 Weeks Specialist Course, 1940 - Group A, Wire Specialty
1st row: Achenbach, Aikele, Akerstrom, Bagnall, Barrows, Brady, Burt, Cain, Christofk. 2d row: Fite, Goodrich, Helton, Higginson, Hill, Hirte, Jasczmult, Kelly, Koroly. 3d row: Kurth, Leitman, Latta, Lavrakas, McCaffrey, Nicholson, Petit, Placko, Prickett. 4th row: Rielly, Van Harlingen, Vaughan, Waterman, Wingo, Wood.

The Signal Corps School was re-designated the Eastern Signal Corps School (ESCS) on 20 June 1942.[92] As such, its department for Officer Candidate Training was renamed the Officer Candidate School (OCS). All training functions at Fort Monmouth consolidated into the Eastern Signal Corps Training Center (ESCTC) in October 1942. The Officer Candidate School extended from three months to four. This provided one month of field work in addition to the academic instruction. Thirty-six officers of the Women's Army Corps enrolled in the School's message center course in December 1943, becoming the first women to be accepted for this training at Fort Monmouth.

By 1943 the size of the OCS had significantly increased and the interval between classes had been cut to just two weeks. Courses in electricity and mathematics, as well as field exercises, had been added to the curriculum. The school initiated a sixteen-hour exercise simulating signal company support of an infantry division and offered training in message center and messenger procedures, wire construction, and radio and wire communication. Command posts were established for the forward and rear echelons of a division headquarters and three combat teams. The officer candidates moved from one to another, alternating duties among the four phases of communications.

Enrollment at the OCS peaked in January 1943 then steadily declined. The 30th class was admitted in September 1943 with just 151 enrollees. Nearly fifty percent were ROTC students. Only overseas veterans and members of the Electronics Training Group were also being admitted. The interval between classes had been extended to seventeen weeks with one month added to the curriculum to include an eighteen-day field exercise at Camp Misery, a sub-post of Fort Dix.

The following year the War Department authorized expansion of the OCS to provide more opportunities for troop unit and overseas candidates. By July 1944, the quota per class had increased to 500. The OCS graduated its last class (Class 55) on 17 October 1946. The OCS at Fort Monmouth admitted 27,387 officer candidates between 1 July 1941 and 17 October 1946. Of these, 21,033 received commissions.

WOMEN'S ARMY AUXILIARY CORPS

Manpower shortages during WWII as well as the increasingly complex work of communications and electronics necessitated women's entry into the Army. So desperate were the circumstances that Chief of Staff General George Marshall told the War Department in November 1941, "I want a women's corps right away, and I don't want any excuses!" The 77th Congress established the Women's Army Auxiliary Corps (WAAC) with Public Law 554 on 14 May 1942, which allowed women to serve "with," not "in," the Army. Key members of the Signal Corps advocated the use of women early in the conflict. This contrasted with much of the legislature, many military personnel, and a high percentage of the general public, all of whom considered a woman's place to be in the home. The Signal Corps, by 1942, had identified 2,000 jobs suitable for WAACs. Hundreds of auxiliaries soon descended on Fort Monmouth and other Signal Corps posts.

The Commanding Officer of the 15th Signal Training Regiment at Fort Monmouth, Colonel Frank H. Curtis, claimed of these women, "their pedagogical training has given them a well diversified background...They grasp their work rapid-

Famous Firsts

1940: *Dr. Zahl invented the GA-4 Transmitter Receiver Tube.*

This tube made it possible for the first time for early radars to transmit and receive from the same antenna.[110]

1941: *The Zahl tube (VT-158) was developed, making possible the use of a high 600-megacycle frequency, thus allowing for smaller antennas.*

Dr. Harold Zahl had worked on this project for several years, reportedly blowing his own glass and shaping hundreds of tubes.[111]

1941: *The SCR-510 FM backpack radio was developed.*

This development was an early pioneer in frequency modulation circuits, providing front line troops with reliable, static-free communications.[112]

1943: *Multichannel FM Radio Relay AN/TRC-1, developed at the Signal Corps Laboratories was first fielded.*

1946: *The Automatic Mortar Locating Radar was developed in 1946.*

The AN/MPQ-3 and the AN/MPQ-10 became standard equipment and both were used as major electronic weapons during the Korean War.[113]

1948: *The first Weather Radar was developed at Fort Monmouth and observed, for the first time, a rainstorm at a distance of 185 miles.*

The radar was able to track the storm as it passed over the Fort.[114]

1948: *The first synthetically produced large quartz crystals were grown by researchers at Fort Monmouth.*

The crystals could be used in the manufacture of electronic components, and made the U.S. largely independent of foreign imports for this critical mineral.[115]

1949: *A technique for assembling electronic parts on a printed circuit board, developed by Fort Monmouth engineers, pioneered the development and fabrication of miniature circuits for both military and civilian use.*[116]

Although they did not invent the transistor, Fort Monmouth scientists were among the first to recognize its importance (particularly in military applications). They did pioneer significant improvements in its composition and production.[117]

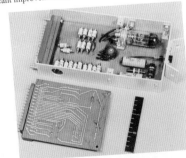

ly, and have a keen sense of loyalty. We're glad to have them working with us." This type of praise provided a marked contrast to the ongoing, public slander campaign against WAAC morality, which became so widespread that First Lady Eleanor Roosevelt, Secretary of War Henry L. Stimson, and General Marshall all publicly denounced it.

Fort Monmouth received its first contingent of WAACs (167th and 168th WAAC) on 30 April 1943 and welcomed them at the Little Silver train station with a brass band and a formal reception at the Service Club. The detachment comprised four officers and 344 auxiliaries. The headline of the *Signal Corps Message* newspaper that week read, "Vanguard of WAAC arrives to plan invasion of Fort." The article began by exhorting men to straighten their ties and comb their hair, "press those uniforms more carefully than ever because a new impetus has been added to Army life at Fort Monmouth." By July 1943 the detachments had been redesignated the WAAC Detachment, Eastern Signal Corps Training Center. The WAACs were assigned as follows: ninety-four with the 15th Signal Battalion, sixty-seven with the 803rd Signal Battalion, sixty-nine with the Post Quarter Master, nineteen in the Station Hospital, fifteen in Special Services, nine in the School Headquarters, eight in the Motor Pool, six in the Post Headquarters, five as Chaplain's Assistants, six in the Photographic Laboratory, two in the Post Exchange, one in the Training Center Headquarters and twenty-four as overhead.

Members of the WAC attached to the Signal Corps, 1943

Overcoming significant obstacles, the WAAC gave way to the Women's Army Corps (WAC) in 1943. This made women a part of the Army as opposed to an auxiliary thereof. Signal Corps Colonel Harry O. Compton outspokenly supported this decision, declaring, "They [women] are particularly adept at work of a highly repetitive nature, requiring light, manual dexterity…Where men grow tired and bored, women's efficiency remains unimpaired." WAC Director Oveta Culp Hobby also recognized women's usefulness to the Signal Corps, stating, "From the inauguration of the WAC, the potential usefulness of members in carrying out Signal Corps duties was recognized."

The Signal Corps would be the first agency of the Army Service Forces to request Women's Army Corps personnel and utilized a higher percentage of female replacement communi-

cators than any other technical service, except the Chemical Warfare Service. Signal Corps WACs in Europe represented 23.5 percent of all the WAC personnel in the Communications Zone, exclusive of the United Kingdom. This meant approximately one Signal Corps WAC for every fifty-five Signal Corps men. The other services could claim only one WAC to every 234 men.

Telephone Switchboard Operators

During World War II, women went through basic training at one of four bases, Fort De Moines, IA, Daytona Beach, FL, Fort Devens, MA, or Fort Oglethorpe, GA. During basic training, the top women in each of the camps were selected by the Signal Corps to continue for advanced training at the Signal School at Fort Monmouth, N.J. These women would come in groups of forty-five per class and were trained along side male Signal Corps Officers, to do the same work. WACs worked in all the major activities on the main post as well as at the Camp Wood and Camp Edison sites. One former Fort Monmouth WAC, H. C. Wiese described her training as including instruction in basic mathematics, code cracking, telephone switchboards, radio equipment, and even truck driving skills. Weise said when the training was complete; "the top three spots in the class were taken by enlisted women." After their training at Fort Monmouth, the Signal Corps women were then deployed as detachments to both the European and Pacific theaters of action. Five Fort Monmouth WACs were deployed on a secret trip to the sixth War Conference in Quebec in August 1943. The WACs were specially selected to handle communications for the Joint Chiefs of Staff.

A Signal Corps board convened at Fort Monmouth in 1948 eventually contributed to the Army's decision to retain women in peacetime. The Board deemed women "more adaptable and dexterous than men in the performance of certain specialties." It was not until 1978, however, that the Army abolished the WAC and fully integrated women into the Regular Army. The Signal Corps, with its "home" at Fort Monmouth, had championed women for the past sixty years. Its use of civilian female telephone operators during WWI represented one of the ways the Army first cautiously used women in what it considered gender acceptable roles, outside of nurses and camp followers. Breaking down the barriers, these pioneering women answered the call that would integrate by gender what historians have called the "most prototypically masculine of all social institutions," the United States Army.[93]

WARTIME LABORATORIES

The Signal Corps established three field laboratories during 1940 and 1941. Field Laboratory Number One, later designated the Camp Coles Signal Laboratory, was located at Newman Springs and Half Mile Roads west of Red Bank, New Jersey. There, 46.22 acres of land allowed for observing and measuring pilot balloon ascensions. Right-of-way for the land was obtained in April 1941, with subsequent purchase by the government in June 1942 for $18,400. The Chief Signal Officer earmarked more than $700,000 for building construction at the site.[94]

Camp Coles

Field Laboratory Number Two, later designated the Eatontown Signal Laboratory, required an experimental area on which to construct antenna shelters. The laboratory received 26.5 acres of a 200-acre tract west of Eatontown, which had been leased as part of the expansion of training activities (part of the Charles Wood Area).[95]

Field Laboratory Number Three originated in the Radio Position Finding section of the Signal Corps Laboratories and resided temporarily at Fort Hancock, New Jersey. It later became the Evans Signal Laboratory located south of Fort Monmouth on land which the Army began purchasing in November 1941. The purchase included land and buildings originally owned and developed by the Marconi Wireless Telegraph Company of America. A three-story brick building, dedicated in 1914 as the Marconi Hotel and meant to house the firm's unmarried employees, served as the main administration building of the Evans Signal Laboratory. Two one-story brick cottages also constructed by the Marconi Company and located directly across the street from the hotel served as quarters for military officers.

A number of brick buildings were constructed at the Evans Signal Laboratory from 1941-1942. Four long, rectangular, one-story buildings connected by enclosed wooden walkways were the first to be completed. These comprised a large laboratory complex. Two brick boiler houses with oil-fired boilers were also constructed.

A group of three research and development laboratories with an office; two smaller laboratories, each with a separate boiler house; another laboratory and boiler house; and a shop facility were also constructed in 1941-1942. All one-story brick structures housed a research center for the Signal Corps radar program.

A large number of wood buildings were also constructed at the Evans Laboratory site during World War II. This included two groups of radio antenna shelters designed to house radar units. They resembled tall, one-story structures with exterior wood post buttresses. Several remained intact for years although most of these structures were later altered to accommodate other functions.

The Army also organized Squier Laboratory on post into the Signal Corps General Development Laboratories (SCGDL).

SCR-510

The laboratories at Fort Monmouth developed the SCR-510 in 1941. This was the first FM backpack radio. This early pioneer in frequency modulation circuits provided front line troops with reliable, static free communications. Multichannel FM radio relay sets (such as the AN/TRC-1) were also fielded in the European Theater of Operations as early as 1943. FM radio relay and RADAR, both products of the Labs at Fort Monmouth, are typically rated among the four of five "weapon systems" that made a difference in World War II.[96]

In 1939 a series of radio demonstrations had taken place which included General Electric and Edwin H. Armstrong. The Signal Corps witnessed these tests and, as a result, MG Colton directed Maj. William S. Marks, an Electrical Engineer in the labs, to design the first mobile FM radio set. According to Marks' own written history, the military requirements of mobile radio necessitated his quick action to translate, "into technical characteristics and to rush thru to completion development models to be thoroughly service tested before large scale production could begin." According to a 1953 Signal Corps Engineering Laboratory (SCEL) document, Marks "developed the first military radio relay sets used successfully in World War II."

The initial manufacturing contract for radio relay (referred to as Antrac in Europe and VHF in the Pacific), was given to the Link Corporation. The informal radio relay program, first used in the North African campaign, had been adapted from Galvin (later Motorola) police radios, procured by the Signal Corps. The formal program was designed by engineers at the Camp Coles lab. The Signal Corps worked together with Bell Telephone and RCA to develop later models of radio relay.[97]

In December 1942, the War Department directed the Signal Corps General Development Laboratories and the Camp Evans Signal Lab to combine into the Signal Corps Ground Service (SCGS) with headquarters at Bradley Beach, New Jersey (Hotel Grossman).

The laboratories had a personnel strength of 14,518 military and civilian personnel in December 1942. The War Department, however, directed the Signal Corps Ground Service to cut the total military and civilian personnel to 8,879 by August 1943.[98]

ENHANCED USE OF HOMING PIGEONS

In 1925, the Pigeon section had a breeding base with seventy-five pairs of breeders, two flying lofts with one hundred birds for training and maneuvers, and one stationary loft with thirty long-distance flyers. Available facilities permitted the breeding of a maximum of 300 birds per season. That number was banded and held available to fill requisitions from the eighteen lofts scattered throughout the United States and its possessions. Signal School maneuvers and ROTC courses used the birds for instruction. The Officers' Division featured twelve hours of pigeon instruction.

Fort Monmouth's pigeon handlers successfully bred and trained birds capable of flying under the cover of darkness in 1928. By the outset of World War II, they had also perfected techniques for training two-way pigeons. The first test was conducted in May 1941. Twenty birds completed the approximately twenty-eight mile round trip from Fort Monmouth to Freehold in half an hour.

The Pigeon Center at Fort Monmouth had at that time an emergency breeding capacity of 1,000 birds a month. This represented about one fourth of the Army's maximum anticipated requirements. American pigeon fanciers supplied by "voluntary donation" 40,000 of the 54,000 birds that the Signal Corps furnished to the Armed Services during World War II.

The Pigeon Breeding and Training Center briefly relocated to Camp Crowder, Missouri, in October 1943. The Center returned to Fort Monmouth, along with the long-lived "Kaiser," "G.I. Joe," "Yank," "Julius Caesar," "Pro Patria," and "Scoop" on 20 June 1946 with more than two dozen other heroes and heroines of World War II.

Pigeons proved particularly useful in WWII, especially in places like the Pacific jungles where it was difficult to string wire. Perhaps the most famous pigeon of WWII was G.I. Joe. Hatched in March 1943 in Algiers, North Africa, G.I. Joe was responsible for saving the 169th Infantry Brigade from becoming victims of friendly fire by the British 56th Brigade. The British had occupied the city of Colvi Vecchia near Cassino in Italy ahead of schedule in October 1943. Radio attempts to cancel the bombing had failed. G.I. Joe accomplished the most outstanding flight of any pigeon during WWII, flying twenty miles in twenty minutes to deliver his urgent message. The Lord Mayor of London presented him the Dickin Medal, awarded by the People's Dispensary for Sick Animals (a veterinary charity) and known as the Animal's Victoria Cross or Medal of Honor. Joe was flown from NJ to London via Paris to receive the medal. Corporal Norman Caplan had been responsible for training Joe in Italy. When the pigeon service was deactivated, G.I. Joe was sent to live at the Detroit Zoo. He died in 1961 at the age of 18 and now resides in the U.S. Army Communications- Electronics Museum at Fort Monmouth.

Col Clifford Poutre releases pigeon

Also on display in the museum is Jungle Joe. Hatched in 1944, Jungle Joe parachuted behind enemy Japanese lines in Burma and was the only means of communication for those units with their Headquarters. He flew 225 miles over very high mountains when he was only four months old. Jungle Joe died in 1954.

Pigeon training at Fort Monmouth, 1923

Otto Meyer was the resident pigeon expert at Fort Monmouth after WWII. He received his first pair of pigeons at the age of seven, though he was a musician by training. He was called up in 1941 and took command of the Army Pigeon Service Agency in 1943. In this position he was responsible for pigeon training all over the world including 3,000 handlers. In 1947 he became Technical Adviser of the Fort Monmouth Pigeon Breeding and Training Center to further develop and perfect the art. After the pigeon deactivation he provided instruction to the zoos on how to care for their hero pigeons.

Fort Monmouth pigeons also served during the Korean War, where they proved particularly useful to covert operatives in enemy-controlled territory. Hundreds of pigeons were attached to the 8th Army and were sent with agents 75-200 miles behind enemy lines. No messages were ever lost.

During the Cold War Era, a German pigeon, Leaping Lena, made a fantastic flight through the Iron Curtain from Pilsen, Czechoslovakia in 1954. Attached to her leg was an anti-communist message addressed to Radio Free Europe[99] Headquarters in Munich. The message read:

> *"We plead with you not to slow down in the fight against communism because communism must be destroyed. We beg for a speedy liberation from the power of the Kremlin and the establishment of a United States of Europe. We listen to your broadcasts. They present a completely true picture of life behind the Iron Curtain. We would like you to tell us how we can combat "bolshevism" and the tyrannical dictatorship existing here. We are taking every opportunity to work against the regime and do everything in our power to sabotage it."*
>
> *Unbowed Pilsen*

Leaping Lena was brought to the United States where she was greeted by four hero pigeons from Fort Monmouth. Pigeons released in her honor carried a copy of the message to President Dwight Eisenhower and Henry Ford II, President of the Crusade for Freedom.

Despite significant fame and success, Field Manual 100-11, entitled "Signal Communications Doctrine" (22 July 1948), stated, "The widespread use of radio in conjunction with the airplane to contact and supply isolated parties has rendered the use of pigeon communication nearly obsolete." Chief Signal Officer, MG James D. O'Connell ordered the disbanding of the pigeon service at the end of 1956 and the service was finally discontinued by 1957. In essence pigeons had been superseded by the vacuum tube. When news of the end of the pigeon service reached the public, protests filtered in from all over the country and made their way to the Pentagon. Many people cited the unreliability of radios in combat and the pigeons' exemplary combat records. Despite such protest, the deactivation went forward. After donating the fifteen living hero pigeons to zoos in various parts of the country, Fort Monmouth sold the remaining birds (about a thousand of them) for $5 per pair. The Public Information Office at Fort Monmouth received 1,500 requests for information on the pigeon sale and people came from all over the country, almost every single state including Canada and Mexico and stood in line for over six hours. Those buying the pigeons were told that if their pigeons homed back to Fort Monmouth it was their responsibility for the return shipment.

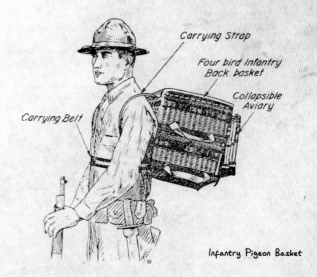
Carrying Strap
Four bird Infantry Back basket
Collapsible Aviary
Carrying Belt
Infantry Pigeon Basket

Pigeons had proved so useful over the years, however, that they were debated for reactivation during the Vietnam conflict. In 1968, Fort Monmouth received a letter from a Signal Corps major suggesting that pigeons be reactivated in the Saigon vicinity to serve as a back up means of delivering high priority requisitions, and as an emergency means of communication. The Major argued this would lessen the burden on critically needed aircraft and vehicles, bridge the communications gap, alleviate traffic loads on radios and phone circuits, reduce the exposure of aircraft, and there was no method of interception. The letter was forwarded to the Commanding General of the U.S. Continental Army Command where the issued was debated but declined.[100]

FIGHTING FALCONS

While the Army's use of homing pigeons has been well publicized, little is known about the use of the "fighting falcons." Falconry, by some accounts, dates back to the first century B.C in the Middle East. Falconry was a sport of the nobility up through the Middle Ages and was considered an important part of a gentleman's education in Europe.

Thunderbolt

A feature in the *New York Times* on 8 August 1941 proclaimed that Fort Monmouth had begun training a falcon "draftee" who had been caught in the Hudson River Palisades. Signal Corps Officers hoped to train the bird, named Thunderbolt, and other falcons to "blitz enemy carrier pigeons and fight parachute Soldiers by tearing into their umbrellas with some sort of secret weapon."

The *New York Times* reported that Fort Monmouth planned to experiment with 200 to 300 falcons. Lt. Thomas MacClure, head of the falcon program, explained that the falcons would be donated by falconers from all over the country and would also be caught locally in Manhattan. MacClure, described in some accounts as a noted ornithologist, consistently declined to answer questions about the bird's secret weapons.

However, one report later stated that the falcons would have steel blades attached to their bodies. An article from September 1941 described the falcon's role as "rendering enemy carrier pigeons hors de combat." As a result of the newspaper publicity of the new program, the Staten Island Zoo offered to donate five birds of prey to the new Falcon Squadron in August 1941. Three members of the Fort Monmouth Falcon program were sent to Staten Island to inspect the birds being offered. One Prairie Hawk and two Sparrow Hawks were selected.

The Fort Monmouth Falcons, dubbed "the First Pursuit Squadron," were put on display in Times Square along with homing pigeons in August 1941 in support of the American Flying Services Foundation, which helped cadets to become pilots.

The publicity surrounding the falcons garnered significant protest from the falconry and conservationist communities who were particularly shocked and opposed to McClure's proposal to use falcons to suicide attack enemy planes.

Despite high hopes, the Fort Monmouth falcons never gained the same notoriety as the pigeons.[101] A *New York Herald Tribune* article in January 1942 described the cause of the programs' downfall: a U.S. Army pilot and falcon enthusiast who returned from overseas assignment got wind of the program and convinced the powers that be in Washington that it was wholly impractical.[102]

END OF THE WAR

Wartime training quickly subsided. Reductions began in May 1943 with orders to inactivate the Replacement Training Center. This was later partially revoked. The capacity of the Officer Candidate School was set at 150 in August 1943. Classes entered at seventeen-week intervals. Enrollments fluctuated thereafter.

Former Italian prisoners of war called "Signees" arrived at Fort Monmouth in June 1944 to perform housekeeping duties. A Lieutenant Colonel and 500 enlisted men became hospital, mess, and repair shop attendants, relieving American Soldiers from these duties. The Signees, originally taken prisoner in Africa, were not subject to the Geneva Conventions as regular prisoners were, but fell under the American Military Articles of War. However, if they abused their privileges, they were returned to prison camps as ordinary prisoners of war and were treated and punished as such. The community was instructed by Post Commander Col James B. Haskell that the Signees were, "co-belligerents in the war against Germany and do not hold the status of prisoners of war. These former Italian army soldiers volunteered for work in the American war effort... The units were formed as a result of military necessity to relieve a shortage of manpower. The 'Signees'—as they are called since they have signed a pledge to work for the Allied cause—have been accepted by the United States, Great Britain and Russia as co-belligerents. The rate of pay is almost the same as that of regular prisoners of war who work, Col. Haskell said, and the only extra reward they receive for their contribution to the war effort is additional freedom. This takes the form of sight-seeing or educational tours and recreational visits to nearby communities under American military supervision." Fort Monmouth personnel were permitted to invite the Signees to their homes provided that they submitted a written request to the Commanding officer of the 309th Italian Quarter Master Battalion.[103]

Brigadier General Stephen H. Sherrill became Commanding Officer of the Eastern Signal Corps Training Center on 3 January 1945. He served only until the end of that year, when he was succeeded by Brigadier General Jerry V. Matejka.

Most of the functions of the Enlisted Department of the Signal School transferred to Camp Crowder, Missouri with the decline in requirements for trained replacements within the Signal Corps.

Watson Laboratories

The Eatontown Signal Laboratory transferred from the authority of the Chief Signal Officer to that of the Commanding General, Army Air Forces, on 1 February 1945. It was renamed Watson Laboratories and moved to Rome, New York in 1951.

A Redeployment Branch was established as a separate function of the Unit Training Center with the end of the war in Europe on 8 May 1945. A redeployment program was carefully established to retrain personnel before deployment to the Pacific. However, with the war against Japan ending shortly thereafter, the redeployment initiatives changed to meet the challenge of speeding Army discharges in the New York, New Jersey and Delaware areas. A Separation Center was established at Fort Monmouth in September 1945. The Center separated more than one thousand men from the Army every day until 31 January 1946.

Camps Edison and Wood deactivated and were almost abandoned after the War. The Eastern Signal Corps Training Center, too, deactivated, in April 1946. Brigadier General Jerry V. Matejka, the Center commander since the end of 1945, became the fourteenth commander of Fort Monmouth. He succeeded Colonel Leon E. Ryder, who had served in the assignment since November 1944.[104]

PAGE TWO   THE SIGNAL CORPS MESSAGE, FRIDAY, JUNE 2, 1944.

Italian Service Battalion Stationed

Ex-POWs To Do 'Housekeeping' Chores at Post

They Are On Our Side Now

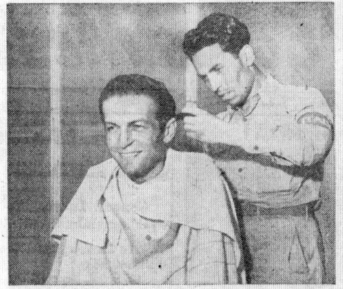

If you've been wondering about the GIs with the green brassards and unusual cap patches—and couldn't get to first base with them in conversation, it's because they are members of a newly-arrived Italian Service Battalion.

The outfit, which arrived at the Post Saturday, is made up of former Italian prisoners of war taken in Africa who, now that Italy is a declared enemy of Germany, have taken the opportunity given them by the United States to volunteer for specially organized service units.

Organized along American lines to perform various service functions for our forces, the men will do general "housekeeping" duties here for the Officers' Mess and the 803d and 15th Signal Training Regiments. They cannot be used for actual combat duty.

The Italians officially are known as "signees" because of the oath of loyalty to the United States they must sign when they volunteer. Like American military personnel, they are subject to the Articles of War.

They are staffed by Italian officers and noncoms, but are under the ultimate command of American officers. Officers and enlisted men wear GI uniforms with a green brassard bearing the word Italy and a red and green circular patch with the word Italy on the cap.

Under the Geneva Convention governing prisoners of war, they are paid 80 cents a day, but receive only one-third of that amount in cash. They receive the remainder in Px coupons.

The battalion is housed in the Area D. Like American GIs, they arrived loaded down with musical instruments and all sorts of gadgets. Their area seems to be

LIKE OUR OWN GIs, the Italians have plenty of extra-curricular pastimes. Upper left, two are shown doing their own barbering. At the right, a man who was a violin-maker in the old country seems little affected by the war. Their own band, below, plays swing in addition to "Funiculi, Funicula." And you don't need three guesses to figure out what's cookin'. It's spaghetti, all right.

Hand Grenades 'Damned Handy' Things In Battle

PROJECT DIANA

Research in radar technology continued at the Evans Signal Laboratory despite the end of World War II. The Belmar (later, Wall Township) site witnessed a milestone in scientific history on 10 January 1946. Signal Corps scientists, under the direction of LTC John J. DeWitt (1906-1999), used a specially designed radar antenna (called the Diana Tower) to suc-

Diana Radar Antenna

cessfully reflect electronic signals off the moon. The project was named in honor of Diana, Greek goddess of the moon.[105] The Diana antenna focused a beam of high frequency energy at the moon, traveling at the speed of light (186,000 miles per second). Success was achieved shortly after moonrise when an audible ping came over the loudspeaker of the scientist's receiver, signaling the return of the radio wave just two and a half seconds later. Continuous recordings were made at regular 2.5 second intervals.[106]

LTC John J. DeWitt

LTC Dewitt had conceived the idea as an amateur astronomer in 1940. He said it occurred to him that it might be possible to reflect ultra short waves from the moon which would open up study of the upper atmosphere and possibly mean a new method of world communication. His first unsuccessful try was with a 138 megahertz transmitter and receiver. Dr. Walter McAfee (1914 - 1995) made the mathematical calculations to make this idea a reality. McAfee graduated with a PhD in physics from Cornell University in 1949. He joined the Army Signal Corps Radar Laboratory at Camp Evans in 1942 in the theoretical studies unit as a physicist. He wrote studies on radar coverage pattern including diffraction around the curved surface of the earth.[107]

The Diana antenna was a forty foot square bed spring model mounted on a one hundred foot tower. At the time, the handful of scientists working on the project did not feel it was of great long range importance towards putting a man on the moon. The first thought, according to McAfee, was to see if it was possible to bounce an electronic signal. If that could be done, he concluded that it would be useful in propagating sound waves. People were not thinking of going to the moon in those days, according to McAfee.

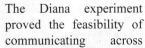

Dr. Walter McAfee in lab, 1946

Because of the antenna's limitations, experiments could only be conducted at moonrise and moonset when the moon was almost horizontal with the earth. McAfee said the experiment led to some understanding about the nature of the surface of the moon and to our knowledge about the space between the earth and the moon.

The Diana experiment proved the feasibility of communicating across vast distances of space. Using the moon as a reflector suggested another way to transmit line of sight radio messages over the horizon to distant terrestrial points. The radar techniques used in Project Diana also led to the more accurate measurement of distances in the solar system. This experiment ushered in the development of satellite communications and missile guidance systems. Newspaper reports at the time put the feat into the same category as the development of the atomic bomb.[108]

DIANA RADAR

COMMUNICATIONS IN SPACE STARTED HERE ON JAN. 10, 1946, WHEN THE U.S. ARMY SIGNAL CORPS MADE RADAR CONTACT WITH THE MOON.

Celebrity Notes:

Eleanor Roosevelt

(1884-1962), former first Lady and civil rights advocate, made a tour of inspection of Fort Monmouth in March 1943. She was escorted by MG Van Deusen.[124]

Sgt. George Baker

(1915-1975) was stationed at Fort Monmouth from 1941 to1942. It was here that he drew his first Sad Sack comic strip. His yet unnamed comic featured the private that would be Sad Sack doing KP (kitchen police) duty. Baker sent the cartoon to an Armed Forces cartoon competition, where it won first prize. The same cartoon attracted widespread attention when it was published in *Life* Magazine. The Executive editor of *Yank* magazine, Major Hartzell Spence, saw Baker's KP cartoon and eventually hired him on at the staff of *Yank* in Manhattan in June 1942. *Yank* was the most widely read publication in the history of the U.S. military and Sad Sack was its first permanent feature. The comic followed Sad Sack's undistinguished journey through the bewildering bureaucracy of Army life and epitomized the experiences of the common Soldier.[118]

Garson Kanin

(1912-1999), writer and director of *Tom, Dick and Harry* fame, was drafted and assigned to Fort Monmouth. However, despite being a well paid Hollywood Director with several successful films under his belt, his rank of private prohibited him from directing Signal Corps films. There was no regulation at the time that prohibited enlistees from being Directors; however, the prospect of privates giving orders to officers rendered the idea impractical. This became a huge problem for the Army as the desperate demand for training films increased and drafted Hollywood directors sat idle, victims of arcane military tradition. The draftees were eligible for a commission after six months provided they had a college education or equivalent. This predicament so concerned the leadership at Fort Monmouth that it made headlines in the *New York Times* as the heads of the Film laboratories elevated their concerns to the War Department.[122]

Jean Shepherd

(1921-1999), TV and radio personality, was stationed at Fort Monmouth with the U.S. Army Signal Corps during WWII. Although most well known for his narration of the 1983 film *A Christmas Story*, it was his time with the Army that gave him much of the fodder he used in his radio shows.[125]

The New York Giants

baseball team played an exhibition game at Fort Monmouth's Camp Edison on 10 June 1943. The score was 7-1 in the Giant's favor. The Giants dined at the Officer's Club prior to the game. The Commander of Camp Edison changed the show time to 5:15p.m. that day so all the men could have an opportunity to attend the game. BG Van Deusen and BG E. L. Clewell attended the game.[123]

Bill Hewlett

(1913-2001) co-founder, with David Packard, of the Hewlett-Packard Company (HP) worked at Fort Monmouth for the Signal Corps Laboratories in 1941.[120]

John W. Rogers

(1916-1999), Associate Producer of the *Mission Impossible* television series, joined the Signal Corps and was stationed here in 1942. He came to the Fort despite being offered a lucrative project by United Artists Studio.

The Boston Red Sox

played an exhibition game at Fort Monmouth's Camp Charles Wood on 21 June 1943. The All Star team, made up of members from each team in the Signal Corps Diamond League played the entire roster of the Red Sox team. The score was 8-0 in the Sox' favor.[119]

Bob Hope

(1903- 2003), legendary entertainer, broadcast almost every one of his weekly *Pepsodent* Shows from an Armed Forces installation in the United States. One of Hope's first five broadcasts in 1945 was aired from Fort Monmouth. Hope's dedication to our nation's troops would eventually lead to his being named an honorary veteran by Congress in 1997.[121]

Tony Randall

(1920-2004), radio, television, screen and stage star, best known for his reoccurring role as Felix Unger in *The Odd Couple*, served at Fort Monmouth during World War II from 1942-1946. He served as a private in the Signal Corps, attended the Officer Candidate School, and was discharged as a Lieutenant.

Korea and the 1950s

Although overall military strength decreased rapidly following the end of World War II, the need for trained signal personnel continued throughout the post-war period. Fort Monmouth remained intact as the "Home of the Signal Corps." Personnel strength, however, had dropped to a total of 11,419 by January 1948. This included 700 officers, 3,221 enlisted men, 3,867 students, and 3,631 civilian personnel.[126]

Things soon changed as world tensions increased with the Cold War and the Berlin Airlift. Enlarging the capacity of every activity on Post again became necessary to sustain the Army's worldwide commitments. Camp Charles Wood, which had been placed in temporary caretaker status in 1945, was rehabilitated to facilitate an increase in personnel for the Signal School. Post strength climbed to 15,296 by mid-November 1948, representing an increase of nearly 4,000 in less than a year.

SIGNAL CORPS CENTER

The Signal Corps Center was established at Fort Monmouth in August 1949 as a Class II activity under the jurisdiction of the Chief Signal Officer. The Center consisted of the Signal Corps Engineering Laboratories, the Signal Corps Board, the Signal School, the Signal Corps Publications Agency, the Signal Corps Intelligence Unit, the Pigeon Breeding and Training Center, the Army portion of the Armed Services Electro Standards Agency, and all Signal Corps troop units stationed at Fort Monmouth. Fort Monmouth was re-designated "the Signal Corps Center and Fort Monmouth" concurrently with this 23 August 1949 action.[127]

The President quickly received the authorization necessary to call the National Guard and organized reserves to twenty-one months of active duty with the onset of hostilities in Korea in June 1950. He also signed a bill extending the Selective Service Act until 9 July 1951.

The Officer Candidate School was reestablished at Fort Monmouth. Its first class began 24 September 1951. The school continued until 27 April 1953, graduating twenty-four classes for a total of 1,232 second lieutenants. The number of military personnel at Fort Monmouth nearly doubled in the period from 1947 to 1953, increasing from 9,705 to 17,358.[128]

The fighting in Korea highlighted the need for new techniques in modern warfare. The use of mortars by the enemy and the resultant need to quickly locate and destroy the mortar sites resulted in development of the Mortar-Radar Locator AN/MPQ-3 and AN/MPQ-10 in August 1950. Between 1951 and 1954, 485 sets were delivered to the Army. The equipment was used to form a single operating station and produced a three dimensional location of the target.[129] Radar worker William D. Townes remembered that spec writers were working seven days

a week and two nights a week when the war broke out in Korea. He worked on the AN/MPQ-10 which he said was needed so desperately that the security classification was lowered so they could field it faster.[130]

AN/MPQ-10

The development of new equipment, however, required the Signal Corps to provide increased numbers of trained electronics personnel to work in the fire control and guided missiles firing battery systems. The Army therefore established Signal Corps Training Units (the 9614th and 9615th) at Aberdeen, Maryland and Redstone Arsenal, Alabama. These units provided instruction on electronics equipment used in the Anti-Aircraft Artillery and Guided Missile firing systems.[131]

Student loads increased in all classes of the Signal School at Fort Monmouth. Night classes were established for some of the enlisted courses, particularly Radar, as a result.

The Signal Corps Laboratories and sub-installations employed approximately 4,500 scientists and support personnel between 1951 and 1953. Responsibilities included production engineering of equipment designed since World War II. A total of 250 of the 274 pieces of major signal equipment moving to the

Students study operation of Radio Set AN/GRC-26A at USASCS, Fort Monmouth NJ

field were improved over their predecessors by 1952. Detection equipment was among those improved upon. Significant advances were made on smaller and lighter forward-area equipment, wire communications, meteorological and photographic equipment, nucleonics, radar, and thermionics.

A new research and development engineering laboratory was constructed at Camp Charles Wood in order to centralize work formerly conducted at Evans, Coles and Squier Laboratories, and the Watson Area. This was Building 2700, later dubbed the Hexagon, and eventually christened "the Albert J. Myer Center." The first increment of the building was completed in September 1954. Dedication ceremonies occurred 30 September 1954.

CONTINUED CONSTRUCTION

Fort Monmouth welcomed its sixteenth commander, Major General Kirke B. Lawton, on 20 December 1951. At this time, plans had been drawn and contracts would soon be let on $25,000,000 in new construction.

Six new permanent, 500-man barracks were completed for the Signal School by 1953. This included buildings 1200 through 1205, located north and south of Hemphill Parade Ground (Abbey-Whitsell Avenues). The 1200 area was located on previously undeveloped land in the western end of the post. A new Administration Building for the Signal School (Building 1207) housed the school library, reading and reference rooms, classrooms, theater, cafeteria, a post exchange, book store, barber shop, cleaning concession and a laundry (Buildings 1208-1210). Also constructed in this area was Building 1206, an auditorium with an outdoor amphitheater.[132]

Demolition of World War II buildings began in 1954 to dispose of wooden structures that had fallen into disrepair. The work removed ten structures from the area of Squier Hall and nine from the area of Russel Hall. Three buildings in the 500 area along Allen Avenue came down to make room for a new three-story barracks building. This barracks building (Building 360), completed in 1956, was built to house sixty bachelor non-commissioned officers.

Construction completed two major warehouse buildings (Buildings 975-976) in 1954 and replaced World War II troop housing in the 900 area. Approximately fifty World War II buildings in the 1000 area, located in the southern part of the post, were demolished to make way for a new hospital (Building 1075). The new hospital was completed in 1961.[133]

Two permanent barracks buildings replaced the remaining World War II structures located in the 200 area around Allison Hall in 1965. These three-story brick structures served as Bachelor Officer Quarters (BOQs). Two other BOQs, both similar in design, were built in this area between 1968 and 1971.

Many of the World War II buildings in the 800 area were demolished in 1970 to make way for the present Post Exchange,

1200 Area Construction

cafeteria, post office, and bank complex.

Several major development programs were completed in the Charles Wood area in the years following World War II and the Korean War. The housing program was initiated in 1949 with the construction of eleven officers' family housing units. These two-family houses were constructed west of the Officers' Club, along Megill Drive. Ten additional units were constructed in 1951 on a circular drive with access to Megill Drive. An additional eleven housing units were constructed in 1955 west of Hope Road on Hemphill Road.

Fifty-two Wherry Housing units were constructed in the Pine Brook area of Camp Charles Wood to provide additional quarters in 1953. This housing project, named Eatontown Gardens, was built in three funding increments with a total cost of $6,000,000. It was completed in December 1954.

West Gate 1951

A program of housing construction financed by the Capehart Housing Act began in 1955. World War II cantonment camps around Colin Kelly Field and Frawley Field in Camp Charles Wood were demolished to make room for the new housing. Actual construction, however, did not begin until 1958. Thirty-six housing units were completed between 1958 and 1959. Each structure contained either four or eight two-story apartment units. The final group of Capehart housing units to be built on the World War II cantonment area was completed in 1960. Ten years later, in 1970, seventeen additional units were constructed along Tinton Avenue in the Charles Wood area.

Very few buildings were constructed at the Evans Signal Laboratory post- World War II. The Signal Corps did construct several warehouses, storage buildings, and small test structures.[134]

UNIT MOVEMENTS

1954 witnessed an exodus of almost 1,300 military and civilian personnel as two organizations transferred to Fort Huachuca, Arizona which, in February of that year, was redesignated a Class II installation under jurisdiction of the Chief Signal Officer and placed in an active status.

The Signal Corps Electronic Warfare Center, activated at Fort Monmouth in 1950, was the larger of the two organizations to make the westward trek. The other was the 9460th Technical Service Unit, Signal Corps Army Aviation Center, activated at Fort Monmouth in 1952 to evolve and test aviation support to Signal Corps activities near Fort Monmouth and to meet the growing needs of Army aircraft in modern military communications, electronics, and photography. Originally based at Red Bank Airport, the Signal Corps Army Aviation Center later moved its operations to Monmouth County Airport. It subse-

quently moved to Fort Huachuca.

Two smaller units transferred to Fort Monmouth from Maryland as these organizations departed. The 9463rd Technical Service Unit, Radio Propagation Unit transferred from Baltimore and was re-designated the Signal Corps Radio Propagation Agency on 8 January 1954. Seventeen instructor personnel from the Signal Supply School (which had been discontinued at Fort Holabird, Maryland on 31 January 1954) also arrived. Most of this group was assigned to the Officers' Department of the Signal School to form a Supply and Maintenance Division.

FORT'S DARKEST HOUR

Fort Monmouth entered perhaps the darkest period of its history at the end of the Korean War, when it was for a time the object of congressional opprobrium and public notoriety. Julius Rosenberg, executed with his wife for spying in June 1953, had worked for the Signal Corps labs during the Second World War as a radar inspector. He was dismissed early in 1945 when it was learned that he had formerly been a member of the Communist Party, but not before he reportedly gave the Soviet Union the secret of the proximity fuse.[135] Having received word of possible subversive activities from Fort Monmouth's Commanding General, MG Kirke B. Lawton, Senator Joseph McCarthy (the Chairman of the Senate Committee on Government Operations) launched an inquiry on 31 August 1953 designed to prove that Rosenberg had created a spy ring that still existed in the Signal Corps labs. In the pursuit of this inquiry, McCarthy came into possession of an Air Force Intelligence document which claimed that an East German defector had seen and heard about classified Evans lab documents that had made their way to Russia. This issue became the focus of McCarthy's investigation.

The investigation into lax security at Fort Monmouth pre-dated McCarthy's inquiry however; as Lawton, who was known for his preoccupation with security, had testified that he knew the Evans lab was vulnerable when he was the Deputy Chief of Staff of the Signal Corps at the Pentagon. When he arrived at Fort Monmouth, Lawton put stricter security measures into practice and was known for periodically ransacking employee's desks looking for inappropriately stored materials. Security measures had become so strict by the time of the McCarthy hearings that several of the supervisors and employees testified that the new system was interfering with getting work done. Edward J. Fister, Chief of the Meteorological Branch at Evans testified, "…it is so difficult now to get and to read a classified document that you spend a good percentage of your time that you would like to spend in working, so that you can read it. So that a lot of information that should be passed around among people working in the field is not being passed around, just because it is so difficult to get it."[136]

According to historian David M. Oshinsky, author of *Fort Monmouth and McCarthy: The Victims Remember*, Lawton had also personally criticized engineers who had gone to "communist" universities like Harvard, the City College of New York (CCNY), Columbia and the Massachusetts Institute of Technology.[137] In fact, college association would become a hot topic at the McCarthy hearings as it was revealed that many of the

General Lawton Welcomes Washington Officials to Fort Monmouth in October 1953

Fort's engineers had attended CCNY and some of them had even been in the same classes as Rosenberg and had socialized and attended meetings with him. Fort employees were investigated and suspended under the authority of President Eisenhower's Executive Order 10450 (27 April 1953) which called for "unswerving loyalty to the United States" and allowed for reinvestigations of federal employees who were suspected of having subversive connections. By the time of the McCarthy hearings, dozens of employees had already been suspended. Many of those suspended had yet to see the charges against them at the time they were asked to testify. When the committee asked them to guess why they had been suspended they offered reasons ranging from having a subscription to *Consumer Reports* to having reprimanded subordinates who may have taken revenge by branding them as subversive. Fred B. Daniels, a physicist and Chief of the Electromagnetic Wave Section at Evans told the committee, "I am one of the few people around there who seem to have guts enough when a person is unsatisfactory to try to get rid of him, transfer him out of the section or do something. Most supervisors are getting so intimidated they are afraid to do that. And there are other people that we have gotten rid of recently that might make accusations against me."[138]

McCarthy initially conducted the Fort Monmouth hearings at the Federal Courthouse, Foley Square, in Manhattan. Some of the hearings were also conducted at Fort Monmouth. The Army's counsel from 1953-1954, John G. Adams, recalled that the Fort Monmouth hearings were held in a "windowless storage room in the bowels of the courthouse, unventilated and oppressively hot."[139] The hearings were initially held behind closed doors, but were opened to the pub-

lic on 24 November 1953. The Fort witnesses subpoenaed to testify were asked whether they knew the Rosenbergs. They were all asked whether they had ever been a member of the communist party (The Mundt-Nixon Act of 1950 prohibited communist party affiliation for Department of Defense employees). The sealed transcripts of these hearings were released on 6 May 2003 and revealed that close to one hundred Fort Monmouth personnel had testified, including MG Lawton and many witnesses were brought back two and three times. Many of the witnesses were questioned by Roy Cohn, McCarthy's Chief Counsel. They were quizzed on their relationship with those accused of subversive activities and whether or not they had discussed politics with the accused. According to William Jones, an engineer and one of the employees suspended, personnel were never permitted to confront their accusers.[140]

McCarthy ultimately failed to prove the existence of a Communist conspiracy at Fort Monmouth and the probe eventually led to his downfall. His actions nonetheless brought notoriety to the Signal Corps Labs and grief to the employees who were dismissed from their jobs on mere suspicion. Charges against suspended employees included attending a benefit rally for Russian children, having a relative who was associated with the communist party or any organization thought to be subversive, being a member of the American Veterans Committee or American Labor Party, allowing employees to read a copy of the *Daily Worker,* and belonging to a C.I.O. local thought to be subversive. Ultimately, forty-two employees, mostly engineers, would be suspended, not for espionage, but for posing security risks. Forty were reinstated and two resigned. All of the reinstated received back pay. The last six to get their jobs back rejoined the workforce in 1958. The attorney representing many of the suspended workers, Ira J. Katchen of Long Branch, believed the suspensions were primarily anti-Semitic. He noted that of the forty-two employees dismissed, thirty-nine were Jewish. Lawton was questioned about this for a 1954 *New York Times* article and stated, "We never distinguish between Jew, Gentile and Colored." Many of the engineers that testified during the McCarthy hearings told the committee that they believed there was an anti-Semitic problem at the Fort as evidenced by the statistically high number of Jewish suspensions.[141] One African American employee, William D. Townes, declared that it was most def-

Impact

C

Asbury Park Press
Sun., Oct. 2, 1983

Three weeks in October: When McCarthyism ran wild at Fort Monmouth

By ERLINDA VILLAMOR
Press Staff Writer

THEN AS NOW it was the nerve center for high-tech electronic research. At its Signal Corps and other laboratories, hundreds of scientists and engineers toiled to develop and perfect radar, and research in satellite communication was taking off.

But Fort Monmouth in the early '50s also seethed with reports of a spy ring. It was suspected of harboring subversives and the likes of executed atomic bomb spy, Julius Rosenberg. In an era marked by the wildest red-baiting that ever gripped this nation, Fort Monmouth was also seen as a hotbed for communists.

Thirty years ago, the Army launched its infamous Fort Monmouth purge in which more than 60 civilian employees — scientists and engineers — were suspended on suspicion of being security risks, but mostly without charges.

Finally in October 1953, Sen. Joseph R. McCarthy, whose name had become synonymous with witch-hunt, charged into Fort Monmouth, bent on purifying the Signal Corps laboratories of their "communist" taint.

FOR SOME THREE WEEKS, the junior senator from Wisconsin as journalist Edward R. Murrow would brand him — held sway over a modern-day inquisition that produced sensational headlines of treason and espionage.

But little much else.
If anything, the Fort Monmouth attack proved to be McCarthy's "fatal confrontation," capping as it did a series of failed anti-communist crackdowns.

Most of the employees who got caught in the net were young engineers and physicists from Monmouth County. Many were fresh from serving their country in the war. Some were just starting their careers and raising families. Most of the charges were found to have no basis and several of the accused were reinstated.

The storm McCarthy stirred up 30 years ago has died down. But the fear he struck into the hearts of the dismissed employees and their relatives lives on.

Several of the employees still live in Monmouth County and were contacted by the Asbury Park Press. They expressed mixed feelings about being interviewed on what they consider the darkest period of their lives.

ONE OF THE engineers had been reinstated after losing his clearance without charges. A month later he was charged for having a brother who was believed to be a registered member of the American Communist Party, suspended and, despite his pleas of innocence, dismissed. He now lives in Long Branch and is in private industry. He turned down an interview.

"I don't want to raise old ghosts; besides I

Ira J. Katchen thumbs through one of the many books he has on McCarthyism, John G. Adams' "Without Precedent."

the Signal Corps was expanding it laboratories and they needed engineers. This time they didn't care about color, race or religion."

After serving two years in the Air Force, he

can be caught in paranoia of a witch-hunt because of what he calls "a narrow" view of life.
"I believe this thing could happen again," he said. "Because of the mediocre quality of education, we are producing people who are extremely susceptible to this type of pressure. Our education is too narrow. Courses are too concerned with working with buttons and typewriters and not on why we have typewriters or computers. There's no effort to teach how to think, to evaluate information intelligently.

Now 66, he says his only regret at leaving Fort Monmouth is his pension.
"Other than that I had no desire to go back to that kind of organization again," he said. His partner refused altogether to talk about the incident. "I don't want to rehash all this," he said on the telephone.
He said he hardly regrets his decision to quit. "Our business has prospered. It has been a very satisfying work for me. Had I stayed at Fort Monmouth I would be restricted to one narrow specialty which I would still be doing now. Here we get a chance to do everything."

IRA J. KATCHEN, Long Branch, one of the three area lawyers who represented a number of the suspended employees, is still convinced the McCarthy attack at Fort Monmouth had an anti-Semitic flavor. Now 62, Katchen still maintains a general law practice in Long Branch.
"It wasn't a coincidence that most of the scientists and engineers (involved) were Jewish," he said in an interview in his office. "Of the 60 accused, 58 or 59 were Jewish."
He said he doesn't believe a similar crackdown could reoccur. "Not in the present atmosphere. But then again, you have to be on the alert that it doesn't happen again."
Katchen knew he was risking his career and standing when he took the 60 cases. "But I chose to take on the burden, because these were innocent men," he said.
Katchen now has difficulty recalling the details of the investigation and even the names of some of his clients. "I should have written a book about it, but I never got around to doing it," he said, producing a stack of books on McCarthy.

HE DOES RECALL, however, that one day in October 1953, "I received a series of phone calls from friends in B'nai Brith asking to see me on matter of great urgency."
Katchen at the time was specializing in real estate cases.
"Later, there were some 40 people from Fort Monmouth who came to my office to ask me to represent them," he remembered.
He called his friend and colleague, Harry Green, of Little Silver, to help him defend the group. Another lawyer, Charles Frankel, Ocean Township, represented two.

initely a racist and anti-Semitic ordeal. He recalled that when the last worker was cleared of wrong doing and hired back, it was announced over the PA system. However, he said, many of the workers only came back because they were required to work one year in order to receive their back pay, which he said usually would equal what they had had to spend on legal fees. Townes commented on Katchen and said that he charged high retainers ($5,000 and more) because "he knew his business was gone the minute he took those cases."[142]

Electronics Technician Harold Tate recalled "My boss, one of the best engineers that I ever worked for, was Harold Ducore. He got kicked out of a job for a long time and was stigmatized. He got cleared eventually. I was getting called two and three times a day by Army intelligence and the FBI. 'Come up front and answer some questions,' they'd say, and so forth. Just you and the FBI in a room. It was more harassment than anything else… The people up at MIT got sucked into that thing, too, because they let people go. The president of MIT fired a number of good people. It was not until quite a long time after that that Edward R. Murrow really started the downfall of McCarthy. That was a very, very dark time in the history around here." [143]

In his subcommittee deposition, David Greenglass (Julius Rosenberg's brother-in-law) claimed that Rosenberg had created a spy ring at Fort Monmouth. Although McCarthy was not to able to prove it at the time, following the arrest of the Rosenbergs in 1950, two former Fort Monmouth scientists, Joel Barr and Alfred Sarant, defected to the Soviet Union. Under new KGB identities, they worked on Soviet Defense projects and later on pioneering work in the Russian computer industry. History has yet to reveal if this could have been the connection between the East German defector who claimed to have seen top secret radar documents from the Evans lab. At the time, the Army claimed that these documents could have been associated with war-time lend-lease agreements from the time when the two nations were on friendly terms. They also noted that Soviet Soldiers had been stationed at the Fort during the war with official access to classified documents.[144]

In his two days of testimony, MG Lawton was questioned by Francis Carr, Staff Director, McCarthy and Cohn. When Lawton refused to answer how many people were being investigated at Fort Monmouth , McCarthy retorted, "I think it is ridiculous to the point of being ludicrous to think that an army officer cannot tell the American people how many communists or disloyal people he has gotten rid of." Although he went on to say "I think that your suspicion of people, and the removal of them from handling top secret work, is an excellent thing. And I think it would be a good thing if the people knew that. But I am not going to order you to answer it." Lawton replied that he agreed with everything McCarthy had said. He remarked that he had been working on investigating the "security risks" for years and was disgusted with his own people for security violations which often involved the inappropriate storing and tracking of classified documents.

The questioning of Lawton primarily concerned the alleged documents seen by the defector. Lawton repeatedly claimed he did not recall receiving the Air Force Intelligence report. McCarthy charged that someone in the Pentagon ordered that the report not be investigated.[145] Some of the testimony follows:

Mr. Cohn: *I want to ask you a few questions there. Is there, to your knowledge, an espionage ring operating at Monmouth now?*

Gen. Lawton: *No, sir.*

Mr. Cohn: *Has there ever been?*

Gen. Lawton: *Not to my knowledge, no, sir.*

Mr. Cohn: *Do you know of any papers which you had information concerning to the effect they had gone from the Evans Signal Laboratory into the hands of the Russians?*

Gen. Lawton: *No, sir.*

In the televised press conference after Lawton's second day of testimony, McCarthy praised Lawton's "house cleaning" abilities and assured him that he had a Secretary of the Army who would back him up on all the suspensions.

McCarthy was censured by the Senate on 2 December 1954 for bringing the Senate into dishonor and disrepute. MG Lawton retired in July 1954.

On the condition of anonymity, several of the dismissed employees spoke about this terrible time in their lives with Oshinsky. One of the employees, Mr. A., worried about bills—he and his wife had just bought a new house. He received hate mail and lost friends. But he said, there were also remarkable acts of kindness, "I can remember one evening when a technician from my lab, a Black fellow, came to our house with an affidavit he had written about my good character. He told me to use it at my hearing. I'll never forget that." Decades later Mr. A. said, "I can't forget what they did to me. I mull it over today. I brood over it and I worry about the consequences for my children. I have a son in law school in Boston who needed a security clearance for a job with the U.S. Attorney's office. I thought 'Oh No! They'll find out about us and take it out on him.' It turned out all right but we lost a lot of sleep over it." He remarked that what affected him the most was that no one ever apologized and admitted they had made a mistake.

In an oral history interview, Walter McAfee recalled intervening early to help a person he knew who had been accused. He advised the individual, who he knew had been an altar boy, to write to the security people and tell them that he had been raised Catholic and subscribed to the Catholic Church's position on communism. The person took the advice and was spared. Many employees who were not fired, but were in one way or another, connected to the accused or suspected, were placed in what was referred to as the "Leper Colony" on Watson Avenue. They lost their security clearances and were assigned menial work.

The removal of so many engineers and scientists significantly affected the research and development program at the Fort. Two years after the hearings, the *Bulletin of Atomic Scientists* reported, "Damage done to the work of the Fort Monmouth laboratories has been substantial." Fifteen of those removed were section chiefs, fourteen were working in radar, and ten were working on the continental defense against air attack.

During his testimony, Alan Sterling Gross, Assistant Chief of the Electro Magnetic Wave Propagation Section at Evans said, "…I consider the fact that I am not cleared as very detrimental to the government. I mean, I am not an egotist, there are a lot better engineers than I am, but I have a lot of knowledge, and they have to have it for certain problems, and if I am not cleared, they do not have it, period. And that goes for Dr. Daniels and quite a few other people. You get to be a specialist, and you can't replace a specialist in six months or a year. It takes a long time to do it."[146] According to Oshinsky, it was not only the removal of brilliant engineers and those who quit in disgust that affected the Fort's work. He said, "The Army found it impossible to convince people at other installations to transfer to Fort Monmouth. No one wanted to work there. As a result, the Evans Signal Laboratory lost more than twenty-five percent of its upper level scientists and engineers."[147]

SIGNAL CORPS ENGINEERING LABORATORIES

A new era of accelerated progress began at the Signal Corps Engineering Laboratories following the Korean conflict. Personnel concentrated their efforts on solving electronics problems. This paved the way for future developments.

Important work in radar, countermeasures, physical sciences, and electron devices proceeded at the Evans Signal Laboratory. A total of 300 of 349 major signal items in production existed of modern vintage, improved for the most part in speed, integration and flexibility, by 1954.

Developments of the laboratories included a lightweight field television camera with a back-pack transmitter; a personal atomic radiation dosimeter that clipped in the pocket like a fountain pen; an ultrasonic quartz saw; a high accuracy mortar locator; and super-small experimental field radios.

The laboratories formed new equipment training teams to train units in the installation, maintenance and operation of the new equipment. Some 200 Soldier specialists conducted training in the United States, Alaska, and Japan.

The Signal Corps Engineering Laboratories also played an important role in the International Geophysical Year (IGY) in 1957-58, cooperating in research efforts by ninety-six member countries. The laboratories' involvement concerned upper air research and measurement of winds and temperatures by means of rockets. Support was also provided in the earth satellite program. Scientists developed instrumentation for meteorological measurements. They also developed instruments

Famous Firsts

1950s: *Named Project Cirrus, Signal Corps Scientists (working with Dr. Langmuir of Syracuse University) developed the ability to manipulate weather by seeding clouds.*

Some reports indicate that the technology was used effectively for clearing away cloud cover in Vietnam.

1953: *Engineers here developed one of the first electronic spy gadgets—the Dick Tracy transistor watch radio.*

Weighing in at just over two ounces, the radio could pick up radio stations from here to New York City. A hearing amplifier was connected to the radio by a wire concealed in the wearer's sleeve. The mercury battery of the first model lasted about ten hours and a knob on the face of the watch allowed the wearer to select a frequency.

1957: *A method of measuring polar ice using radar was developed by the Fort Monmouth Laboratories.*

This technique significantly aided in the study of the Polar Regions.[160]

1958: *Solar cells were developed for satellite power in space.*

These cells powered the Vanguard I satellite for more than five years.[161]

1958: *The first communication satellite, Project SCORE, was developed in 1958.*

Launched on December 18, Project SCORE (Signal Communications via Orbiting Relay Experiment) broadcast President Eisenhower's Christmas greeting, proving that voice and code signals could be relayed over vast distances using satellite communication technology developed at Fort Monmouth.[162]

1951: *Fort Monmouth developed a new miniatuized radar beacon to aid in the observation of weather.*

The beacon was carried into the atmosphere by rocket. The beacon relayed signals to and from radar sets on the ground.[156]

1953: *Fort Monmouth scientists discovered transistor action in silicon and invented a floating zone refining process to produce inexpensive, chemically pure silicon crystals.*

By 1964, silicon was the Army's principal transistor material.[158]

1959: *The first Weather Satellite, the Vanguard II, was launched on 19 February 1959, equipped to map the earth's cloud patterns by a varying infrared scanning device.*

The electronics for the satellite were developed by Fort Monmouth.[163]

1950-1953: *A major success of this era was the introduction of Automatic Artillery and Mortar Locating Radars AN/TPQ-3 and AN/MPQ-10, both products of the labs at Fort Monmouth.*

Other developments of the period included a lightweight field television camera with a backpack transmitter; a pocket dosimeter for detecting radiation; an ultrasonic quartz saw; and super-small experimental field radios.[157]

1957: *After much litigation, COL William Blair finally received his radar patent, U.S. Patent Number 2,803,819.*

He is remembered as the "father of American radar."[159]

for "Cloud Cover," a satellite launched on 17 February 1959 to survey the earth's global cloud paths.

Fort Monmouth scientists developed a method for measuring polar ice by using radar in 1957. This technique greatly aided the study of the Polar Regions.[148]

Personnel strength at Fort Monmouth totaled 15,859 as of December 1957. This reflected overall growth since the Korean War and included 1,156 officers and warrant officers, 7,503 enlisted personnel and 7,200 civilian employees.

The Army re-designated the U.S. Army Signal Corps Engineering Laboratories as the U.S. Army Signal Corps Research and Development Laboratory (USASCRDL) in April 1958.[149]

Armed Forces Week May, 1950

The laboratory placed increased emphasis on internal research and created an Institute for Exploratory Research in the Office of Research Operations in fiscal year 1958. Exploratory Research Divisions were also created in each of the three operating departments. The consolidation of internal research efforts was completed when the Institute for Exploratory Research achieved department status and the three Exploratory Research Divisions transferred from departments to the Institute. This was the final step in fostering a research organization free from the pressures that characterized development activities. A Computational Analysis Division was also established within the institute to provide a mathematical and computational service.[150]

An Astro-Electronics Division was established in the Communications Department to give proper recognition and priority to astro-electronics projects. The division embraced Astro-Instrumentation, Astro-Observation and Analysis, and Astro-Communications Branches.[151]

PIONEERING AFRICAN AMERICANS

The Signal Corps at Fort Monmouth offered unique opportunities for minorities in the 1940s and 1950s. African American electrical engineer and retired SES Thomas E. Daniels, who worked at Fort Monmouth for thirty-five years stated, "Fort Monmouth was known as the Black Brain Center of the U.S." Daniels affirmed that this post "provided a place where black scientists and engineers could finds jobs and advance their careers," while other research facilities closed their doors to African Americans. Engineer William Jones described the three types of civilian vacancies that were available when he started at the Fort during WWII: radio mechanics which required no college education, engineering aides which required some college training and engineers which required a degree.

These employment opportunities did not inoculate the Fort's African American employees from the culture of discrimination and segregation that marked this period in the country's history. The Army itself institutionalized discrimination until President Harry S. Truman signed Executive Order 9981 on 26 July 1948, ending segregation in the United States Armed Forces. Jim Crow ruled in the private sector, and even in New Jersey Ku Klux Klan chapters marched in the streets. Despite its best efforts, the Signal Corps could not ensure uniformly equitable treatment for African American Army personnel. Jones commented, "You have to be good, but you also have to have a chance and you had to have some nice people around you once in a while. I think there were some decent people. I would be less than honest if I said they were all racist, but I can say for every good person, there were three racists then…"

The experiences of the Fort's early African American employees illustrated the dichotomy between the Signal Corps' progressive hiring policies and the day-to-day lives of the African Americans benefiting from that progressiveness. Harold Tate began working as a civilian electronics technician at Fort Monmouth in 1942. He recalled in a 1993 oral history interview that the government paid him $2,600 a year, "quite a bit of money in those times." When Tate completed his undergraduate degree, his unit chief tapped him to train as a radar officer at Harvard and the Massachusetts Institute of Technology (MIT). Tate remembered, "there were only two Blacks in the whole class of somewhere in the neighborhood of about 200 officers. The other was fair skinned and passing for white."

Tate encountered discrimination and segregation despite the opportunity afforded to him by his Fort Monmouth supervisor. He remembered difficulties in finding off post restaurants, transportation, and housing. As he recalled, "Blacks had to go scrounging around for a place to sleep." His experience at Fort Monmouth was so good, comparatively speaking, that when offered the option to return to the Fort after his university training, he said, "I'd be tickled to death to be sent back to the lab."

Ever mindful of the obstacles impeding African Americans' success, Tate and his fellow African American engineers and technicians started programs to educate their coworkers and even local high school students. According to Tate, these "tutorial programs for black Fort Monmouth employees did a respectable job of trying to get people better educated."

According to Tate, Dr. Walter McAfee helped teach these tutorials. McAfee, a renowned physicist, held numerous supervisory positions during his four decades at Fort Monmouth. McAfee recalled in a 1994 oral history interview that several government agencies initially rebuffed his attempts to gain employment, based on his race— something he couldn't hide during the application process, as most applications required a photograph. Fort Monmouth's application, however, did not. McAfee received instructions to report to Fort Monmouth almost immediately after submitting his paperwork. He resigned from a steady teaching job in order to do so, despite fears that he would be fired when Fort officials discovered his race.

Those fears dissipated when McAfee arrived and found a number of African Americans already at work. Upon arriving in

New Jersey, he, like Tate, noted that off post segregation and discrimination made it difficult to obtain housing and meals. Interpersonal discrimination occasionally reared its head on base as well, sometimes making it difficult for African Americans to receive promotions. As McAfee described it, "…if they had one position and had a black man and a white man competing for it, the white man got it. Mainly, there's less friction that way. The black man isn't going to fight that hard. Of course, today you wouldn't say that. I guess we were just getting into the jobs." McAfee recalled his initial interview at the Fort, "…the man who was in charge of personnel for the

White House ceremony: Dr. Walter S. Mc Afee (right) is shown with Dwight D. Eisenhower in 1956 shortly after the president presented the Fort Monmouth physicist with one of the first of the Secretary of the Army's research fellowships, which provided for post-doctoral study at Harvard University and at laboratories in Europe and Australia.

general engineering branch asked me what my education was. I told him that I had a Master's Degree in theoretical physics and had practically finished the course requirements for a Doctorate. He said, 'Theoretical physics, oh, good. Tell me some of the courses you've had.' I gave the courses and he says, 'By, God, Dr. Carroll wants somebody like that and we've been getting people who want to pass themselves off as mathematicians or mathematical physicists and they don't have anything. Now we've got one.'"

McAfee's section chief, like Tate's, championed him, and McAfee received his initial promotion. Many more followed, to the point that he said he "never had trouble on an Army base after '48. So, when Truman integrated the Army, my rating was high enough they didn't bother me. I went to places and they gave me good quarters. They gave me the best- I usually got the VIP quarters." That experience was in stark contrast to his first travel experiences in 1946, which required McAfee to carry a booklet denoting accommodations open to African Americans. He also recalled during the war period that the only instance of socializing with co-workers was between Blacks and Jews. MacAfee's critical mathematical computations that allowed for the successful Diana radar moon bounce were not mentioned in the initial press coverage of the experiment at the time. He said this "really annoyed" him because the people that were mentioned all went on to land very good paying jobs after the war.

African American women also found non-traditional opportunities at the Fort and were hired as draftspersons, engineers, computer analysts, laboratory technicians and illustrators. They were trained in the civilian training school and assigned to labs to prepare engineering reports and technical manuals. Constance Wright, a junior professional assistant recalled being required to build her own radio and taking night courses on wave guide theory at nearby universities. Mary Tate was

an engineering aide who recalled working at her first job in an office where no one spoke to her. Despite this, she was promoted to Mathematician and was quite certain she was the only black female mathematician at the time. She worked on Top Secret nuclear testing projects. Her motto was, "I don't know how to do it but I'll do it anyhow."[152]

Jones, who noted the Signal Corps' reputation for accepting qualified African Americans, shared in a 1993 oral history interview that forty-eight private employers and the U.S. Air Corps rejected him before Fort Monmouth hired him. His opportunity to work here, like so many others', was marred by the fact that, as he put it, "you couldn't eat anywhere else in the county" other than the African American boarding house. He endured repeated humiliation when he could not dine with co-workers at local eateries. Luckily, the man for whom Jones worked early on in his career, radar pioneer John Marchetti, "didn't care what kind of people worked on (his projects) as long as they were qualified."

As Jones rose through the ranks, he, too, refused to select employees based on the color of their skin, despite the past slights he himself had endured. According to Jones, "…I didn't want an 'all Colored' department. I could see no advantage in spite of the arguments put forth by the Colored who wanted me to do so 'in order that we could show them (the whites) what we can do'….there was and is no advantage from self-segregation."

Daniels, who pointed out the post's reputation as a "Black Brain Center," addressed the imperfections of the opportunity afforded him by the Signal Corps, saying, "Working at the Fort, you still had a microcosm of people who really didn't think Blacks were smart enough, but you still had opportunity."

After the war, many African American women experienced the same type of disenfranchisement being experienced by white women as the fighting men returned from overseas and they were turned out of their jobs. One African American worker, William D. Townes commented, "After the war they were prejudiced against sex, gettin' rid of those women. When they were finished laying off there wasn't a woman left in the machine shop…"

The trail blazed by these World War II era pioneers paved the way for more recent African American leaders here, like Mary Pinkett, the first black woman to attain GS-14, or Emmett Paige, the first black general in the history of the Signal Corps. Paige served at Fort Monmouth as Commander of the Communications Systems Agency, as the first two-star Commander of the Communications Research and Development Command, and as commander of the Electronics Research and Development Command. Other notable African American leaders here have included Albert Johnson, the first black Colonel in the Signal Corps, who served several tours here; John Patterson, the first black Chief of Staff of CECOM; and CECOM Commander MG Robert Nabors (September 1998- July 2001), Fort Monmouth's first African American Commanding General.[153]

MG Emmett Paige

DEAL SITE TRACKS SPUTNIK

When the Russians launched Sputnik I in October 1957, the Deal Area was the first government installation in the United States to detect and record the Russian signals. The Army acquired the site in 1953 for use by the Signal Corps, here. The property is a 208 acre parcel of land in Ocean Township, Monmouth County, NJ. It is two miles inland from Deal and bound by Deal Road on the south, Whalepond Road on the east, Dow Avenue on the north, and private property to the west. The U.S. Army Corps of Engineers, District of New York leased the property; however the using agency was first the Signal Corps and, later, the U.S. Army Electronics Command (ECOM), both headquartered at Fort Monmouth. Approximately eight miles separated the Main Post of Fort Monmouth from this test site.

The Deal Test Area often made the news in the late 1950s and 1960s because of its excellent facilities and because of its performance in monitoring satellites. It was, for a period, one of the prime tracking stations of the North Atlantic Missile Range. A total of 273 orbits of Sputnik I were observed and recorded, covering approximately 500 hours of continuous monitoring. Dr. Harold Zahl, Fort Monmouth's Director of Research, later reported that "we had no legal project set up for tracking Russian satellites…but within our own laboratory, we had an immediate potential, and duty called desperately." He recalled that a "small select group of Signal Corps R&D personnel at Fort Monmouth," eventually dubbed the "Royal Order of Sputnik Chasers," vowed to work without overtime pay twenty-four hours a day, seven days a week, "until we knew all there was to be learned from the mysterious electronic invader carrying the hammer and sickle." According to NASA, the Cold War era launch of that electronic invader "ushered in

Deal Test Site, 1966

new political, military, technological, and scientific developments. While the Sputnik launch was a single event, it marked the start of the space age and the U.S.-U.S.S.R space race." Fort Monmouth personnel were there when the gun sounded the beginning of that race.

An elaborate (and this time official) monitoring facility was set up in time to monitor Sputnik II, launched in November 1957. Once again, Deal was the first American station to record the signals. According to Dr. Zahl, the Command's Technical Information Staff lacked official permission to release the news of Sputnik II's entry over North America. Being only 0250, the Pentagon "was not yet ready for the space maze." So, George Moise of the Technical Information Division "quite innocently" called one of the news wire services and asked, "Did anyone hear the new Russian satellite over the U.S. before our reception at 0250 hours?" The very excited gentleman on the other end of the line reportedly exclaimed, "What! No- but,

but- you've got a story!" Moise avoided possible reprimand when a co-worker, Len Rokaw, managed to awaken enough high-level officials to authorize an official news release.

Other space achievements followed the Sputnik launches rapidly. Deal personnel monitored and logged all satellites, both American and Russian, as well as all missiles launched from Cape Kennedy. Deal's space availabilities dropped off gradually as NASA and the Air Force set up their own monitoring and tracking facilities, but as of 1960 Dr. Zahl declared that the "Deal library, of all satellite recordings, is the most complete in the world." In compliance with Army and DoD directives to abandon excess leased real estate, the Electronics Command terminated their lease with the Deal site owners effective 30 June 1973. Most Deal facilities and personnel moved to the government-owned Evans Area. The land was deemed safe for non-military use and uncontaminated by explosives or other toxic material. Ocean Township and the state of New Jersey joined to purchase the land as a nature preserve in 1973. The site was entered into the Ocean Township Parks system in 1980 and today it is known as Joe Palaia Park.[154]

SATELLITES

The Signal Research and Development Laboratory accomplished a major satellite payload contribution with the launch of Vanguard I on 17 March 1958. This project demonstrated the feasibility of solar converters for satellites. The laboratory developed solar-powered devices, consisting of six cell clusters, to power one of the two radio transmitters in the 3-1/4 lb, 6.4 inch sphere.

The Deal Test Station of the laboratory picked up the Vanguard I's signals three minutes after its launch from Cape Canaveral in Florida. Vanguard I traveled 409,257,000 miles in 11,786 orbits in the first three years of its existence. Its radio voice never failed, and the satellite proved itself invaluable in scientific computations. Vanguard I had a predicted life of 200 to 1,000 years and its solar cells, and perhaps its radio, were expected to operate as long as it circled the globe.

Vanguard Satellite

The second major satellite payload contribution was the complete electronics package for Vanguard II, launched on 17 February 1959. This satellite, with infrared scanning devices to provide crude mapping of the earth's cloud cover and a tape recorder to store the information, operated perfectly during the entire twenty-day life of the battery power source.

The first communications satellite, Project SCORE (Signal Communications via Orbiting Relay Experiment), was successfully launched on 18 December 1958. It broadcast a Christmas message from President Dwight D. Eisenhower to people around the world. The experiment effectively demonstrated the practical real-time feasibility of worldwide communications in delayed and real-time mode by means of relatively simple active satellite relays. SCORE was a project of the Advanced Research Project Agency (ARPA) conducted by the Signal Corps. The Air Force provided the Atlas launching vehicle.

In addition to its work with satellites, the laboratory developed and tested equipment to fit into the new concept of rapid and flexible communications.

Scientists at Fort Monmouth participated in Project WOSAC, or the World Wide Synchronization of Atomic Clocks, from 1959 to 1960. The project, carried out with the aid of the U.S. Navy, U.S. Air Force, Harvard University, and the British Post Office, established a global standard for time measurement.[155]

Celebrity Notes:

Edward Charles "Whitey" Ford

(1928-) a Baseball legend, had the best winning percentage (.690) of any twentieth century pitcher. He won the prestigious Cy Young award in 1961, and was elected to the Baseball Hall of Fame in 1974. He has the most career wins of any New York Yankee. But before any of this, Whitey Ford served in the United States Army as a Private in the Signal Corps at Fort Monmouth. On April 17, 1951, Whitey Ford tossed out the first pitch against the Red Sox at Yankee Stadium dressed in an Army uniform. He would then spend the 1951 and 1952 seasons in the service during the Korean War. Known as the "money pitcher" he played in twenty games as pitcher and twenty games as an outfielder while here. According to Ford, "Army life was rough. Would you believe it, they actually wanted me to pitch three times a week."[164]

Adam West

(1928-) better known as television's Batman, served two years in the Army. During those two years he worked on television, first at San Luis Obispo, CA, and eventually at Fort Monmouth.

Joe DiMaggio

the Yankees' famed Clipper, (1936-1951) made remarks at the 1959 Colin Kelly Field dedication in the Charles Wood Area. He also demonstrated his famed swing for a crowd of enthusiastic Junior League ball players.

AMC/ECOM & VIETNAM

REORGANIZATION

A reorganization of the Army in 1962 resulted in some significant changes for Fort Monmouth.

In response to a study directed by the Secretary of Defense, the Army reviewed its managerial practices in order to achieve more efficient and economical operation and eliminate unnecessary overlap and duplication of effort.

One segment of the Army study analyzed the Technical Services, one of which was the Signal Corps. As a result, the Signal Corps and the other Technical Services ceased to exist. Their functions transferred to new commands. Signal Corps functions, for instance, would no longer fall under the purview of the Chief Signal Officer. Management of Signal Corps personnel was assigned to the Office of Personnel Operations (OPO); signal training was transferred to the Continental Army Command (CONARC); signal doctrine and combat development to the Combat Development Command (CDC); and signal materiel development and procurement to the Army Materiel Command (AMC).[165]

AMC stood up 1 August 1962 as the first centralized logistics command to exist in peacetime. A subordinate element of AMC, the U.S. Army Electronics Command (USAECOM, or ECOM), was established at Fort Monmouth that same day.[166]

ECOM exercised integrated commodity management of assigned materiel within the concept of cradle-to-grave management.

The command was responsible for research, design, development, product and maintenance engineering, industrial mobilization planning, new equipment training, wholesale inventory management, supply control, and technical assistance to users in the commodity areas of communications, electronic warfare, combat surveillance, automatic data processing, radar, and meteorological materiel.

Major General Stuart Hoff was appointed the first Commanding General of ECOM, effective 1 August 1962. He simultaneously became the 22nd Commanding Officer of Fort Monmouth.

The initial effort at reorganizing Army electronics materiel management carried with it a major organizational deficiency. Field agencies previously reported to the Office of the Chief Signal Officer, the only staff interposed between them and the Department of the Army (DA). The reorganization aggravated this situation by creating two levels between the units and DA: ECOM and Headquarters, AMC. A study was initiated almost immediately to design a better organization for ECOM.

A restructuring of the command was implemented in July 1964. It was a logical continuation of the U.S. Army reorganization of 1962 that made ECOM a cohesive operating command of AMC. The objectives of the restructuring were to consolidate missions and eliminate command and staff layering; to collocate principal mission and operating functions of research and development, procurement and production, and materiel readiness; and to establish ECOM as the primary authoritative point within the Department of Defense for integrated life-cycle management of assigned commodities.[167]

The ECOM reorganization essentially established a directorate-type organization that combined the former headquarters staff with the operating elements of corresponding functional areas.

Major organizational changes within ECOM's research and development operations were accomplished in 1964 and 1965. Initially, a supervisory research and development staff was eliminated and staff supervision within the U.S. Army Electronics Laboratories was streamlined. The laboratories were designated the U.S. Army Electronics Laboratories in July 1964 and authorized a personnel strength of ninety-four officers, 143 enlisted personnel, and 2,725 civilian employees.[168]

A laboratory for Combat Surveillance and Target Acquisition was organized as an element of the Electronics Laboratories in January 1965. The following month an Avionics Laboratory was organized, also as an element within the Electronics Laboratories.[169]

As a result of an ECOM study, other major areas of research and development were organized into laboratory-type organizations within a few months. This included communications, electronic warfare, and atmospheric sciences.

Fuel cell operations for AN/PPS-4 on 14 July 1966

The Electronic Laboratories were then discontinued on 1 June 1965. Six separate laboratories emerged: the Electronic Components Laboratory, Communications/Automatic Data Processing (ADP) Laboratory, Atmospheric Sciences Laboratory, Electronic Warfare Laboratory, Avionics Laboratory, and Combat Surveillance and Target Acquisition Laboratory. A Directorate of Research and Development (R&D) and an Institute for Exploratory Research were also organized.[170]

The new organization was designed to provide greater efficiency and responsiveness in meeting the ECOM R&D mission. The new Directorate of R&D was authorized eleven officers, eighty-seven enlisted personnel, and 1,102 civilian employees.[171]

VIETNAM

The new command responded quickly to the exigencies of war in Southeast Asia during the Vietnam conflict, supplying and supporting the most advanced radios, switches, teletypewriters, and telephones any Army had ever seen.

With the Uniform Communications/ Strategic Army Communications Systems (UNICOM/STARCOM) program, ECOM bought the equipment and services needed to build an infrastructure in Southeast Asia and the Pacific for efficient, reliable telephone and data communications. That effort culminated in the 1965 award to Page Communications Engineers of what was then the largest contract ever negotiated by ECOM or any of its Signal Corps predecessors. This was a contract to install, maintain, and operate the Integrated Wideband Communications System. This system and associated switching centers provided the backbone for what was the first conscious attempt to create an Army area telecommunications networking tactical arena. Mobile satellite terminals supplemented the network's troposcatter and cable links across the Pacific. The two channel link from Tan Son Nhut to Hawaii, established in August 1964, was the world's first operational satellite communications system.

General Frank Moorman, ECOM commander, ordered the new, transistorized FM radios of the AN/VRC-12/PRC-25 families shipped to Vietnam in July 1965 in response to General Westmoreland's complaints about the AN/PRC-10. The new, transistorized FM radios of the AN/VRC-12/PRC-25 families soon became the mainstay of tactical communications in Southeast Asia. ECOM awarded competing production contracts to sustain the flow. ECOM's next commander, General William Latta, personally browbeat contractors to ensure timely delivery of a dependable product. The Command delivered 20,000 VRC-12 and 33,000 PRC-25 radios to Southeast Asia in three and a half years. The PRC-25 was, according to General Creighton Abrams, who commanded military operations in the Vietnam War from 1968-72, "the single most important tactical item in Vietnam."[172]

The first AN/GRC-163 arrived in Vietnam in January 1968. ECOM developed this trailer-mounted four-channel multiplexed radio to support communications in airmobile operations. It replaced the AN/MRC-69, which was too heavy to fly even in the downsized "34-and-a-half" version.

ECOM delivered a new squad radio to replace the AN/PRC-6 "walkie talkie" in 1967. Troops in Vietnam had found the AN/PRC-6 too awkward for use in combat. The new radio consisted of a helmet-mounted receiver, the AN/PRR-9; and a shirt-pocket transmitter, the AN/PRT-4. The contractor, Delco, produced sets for 47,000 infantrymen through 1971.

During Vietnam, transistors and integrated circuits replaced tubes. Communications equipment became smaller, lighter, more dependable, and more versatile. It reached lower into the ranks and accommodated a much larger volume than ever before, providing more information to more people more of the time.

ECOM supplied combat troops with a number of other high-technology commodities during the war. These included night vision devices, mortar locators, aerial reconnaissance equipment, surveillance systems, sensors, and air traffic control systems.

Cpt Edward Boyd, First Cav Air Mobile Div, establishes communications during operation Long Reach, RVN 1965

Second generation night vision devices (image intensification technology) replaced the first generation "sniper scope" (near infrared technology) of World War II. The Small Starlight Scope AN/PVS-2, the Crew Served Weapons Sight AN/TVS-2, and the medium range Night Observation Device AN/TVS-4 all saw service in Vietnam. The Night Vision Laboratory, which was attached to ECOM in 1965, began development of these products in 1961. Production of the AN/PVS-2 began in 1964.

The war provided the first test of the improved counter-mortar radar AN/MPQ-4 in a tactical environment. The AN/MPQ-4, which had existed in the Army inventory since 1960, was deployed to Vietnam in 1965 and proved particularly useful in the defense of fixed installations. During the war, ECOM scientists devised operational schemes that permitted effective scanning over 360 degrees.

Starlight scope that intensifies moonlight, starlight, or skyglow, eliminating need for infrared light source, 21 May 1968

Mortar Locating Radar AN/MPQ4-A, Vietnam, 1967

ECOM developed the AN/PPS-5 man-portable surveillance radar to replace the AN/PPS-4 and AN/TPS-33. The ninety-five pound set had a 360 degree scan capability. It could detect personnel within five kilometers and vehicles within ten. ECOM awarded the production contract in April 1966, following evaluation of Engineering Development models in Southeast Asia. There were more than 350 sets in the theater by the end of 1970. Though often deadlined for lack of repair parts, the set was popular with the troops because it reduced the need for hazardous surveillance patrols. According to one commander, "One AN/PPS-5 in operating condition is worth 500 men."

ECOM scientists in Project SEACORE (South East Asia Communications Research) developed a number of electronic sensing devices originally intended for use in the McNamara Line. These included sensors that could be emplaced by artillery, listening devices, and seismic detectors. Some sensors were cleverly disguised as dog feces. While the McNamara Line concept was impractical, the sensors proved useful in the perimeter defense of Army compounds.

ECOM supplied, managed, and supported nearly half the line items in the Army's materiel inventory during the 1960s. The items ranged in size (from a transistor to a sixty-foot parabolic antenna), complexity (from two-strand twisted wire to airborne surveillance systems), and technologies. The range exceeded that of any other AMC commodity command. Supporting this materiel in the theater involved unique problems and solutions. For example, ECOM was hard pressed to find producers who could deliver quality batteries in sufficient quantity. The command additionally had to worry about how the batteries were stored in the torrid climate of Southeast Asia.

ECOM addressed problems of supply and support through a variety of means. Commodity Management Offices (Avionics/Navigation Aids, Electronic Systems, Combat Surveillance/Night Vision/Target Acquisition, Communications/Automatic Data Processing, Intelligence Materiel, Electronic Warfare/Meteorology, and Test Equipment/Power Sources) provided intensive management of critical items. Established when the Command was organized in 1962 and staffed by some of ECOM's best people, the Commodity Management Offices survived in one form or another until 1971.

General Latta established the twenty-seven man Operational Readiness Office in 1965. Its sole mission was to monitor the progress and detect the problems of every ECOM project or activity relating to Southeast Asia. ECOM established and staffed the Aviation Electronics Agency and the Avionics Configuration Control Facility in 1966-1967 to address the unique problems associated with the installation of ECOM equipment in Army helicopters.

Soldier Operating Squad Radio

1960: *The Tiros-1 satellite, developed under the technical supervision of the Fort Monmouth Laboratories . . .*

sent the first televised weather photographs of the earth's cloud cover and weather patterns to the giant sixty foot "Space Sentry" antenna at Camp Evans.

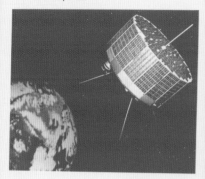

1960: *A Mobile Digital Computer (MOBIDIC) developed at Fort Monmouth was the world's first mobile, van-mounted computer for use at Field Army and theater levels.*

This computer was the first experiment in automating combat support functions and would be the prototype of the computers the Army used in Vietnam to automate these functions in artillery, surveillance, logistics, and battlefield administration.

1960: *The Courier Satellite, developed and built under the supervision of the Fort Monmouth Laboratories . . .*

was an experimental communications satellite that proved high-volume communications, up to 100,000 words per minute, could be relayed through space.

1961: *Signal Corps scientists from the U.S. Army Signal Research and Development Laboratory at Fort Monmouth, in conjunction with New York University's College of Engineering, developed a method to measure rain in 1961.*

The device was designed to be placed outdoors where it would size and collect information on all the raindrops falling in a 1 ¼ inch square. Knowing the size and number of raindrops was critical to perfecting radar measurements of water particles in the atmosphere.[175]

1962: *A "weather radar on wheels" was developed that could track a storm as far as 400 miles and monitor the movement of clouds resulting from nuclear bursts.*

It provided more than twice the detection capability of earlier models. The radar could survey a 600,000 square mile area and allow operators to track the progress of a storm and predict to within minutes when it would strike any point in that area. The density of precipitation, structure and shape of the storm could all be shown on TV tubes inside an Army van.[176]

1962: *An experimental ten-pound hand-held radar unit was developed which used the latest micro-miniaturization technology.*

It was able to spot moving targets more than a mile away. This prototype model, developed at Fort Monmouth, was the first step in the production of light, hand-held tactical electronic equipment. It also was the prototype for the radar units used by all police departments across the United States to detect speeding motorists.[177]

1964: *A Morse Code Readout device, developed at Fort Monmouth, plugged into any Army radio and transformed the dots and dashes of Morse Code into letters formed by a light-emitting diode (LED).*

This device allowed a Soldier with no knowledge of Morse Code to be able to receive coded messages.

1965: *A single pencil-size laser beam simultaneously relayed all seven of the television channels broadcast from New York's Empire State Building between two points at Fort Monmouth.*

The volume of information, believed to have been a record, further demonstrated the high capability of laser as a means of relieving over-crowded radio channels. Efficient communications networks were a key to winning the Cold War era race to perfect the best electronics.

1965: *The ECOM Labs developed the PRR-9 helmet-mounted receiver and PRT-4 hand held transmitter for squad-level use in Vietnam, exploiting technology to provide miniature, all solid-state radio designs with lower power consumption.*

Follow-on products included the AN/PRC-68 in the 1980s, the AN/PRC-126/127 in the 1990s, and civilian counterparts (e.g., hand-held police radios). This "individual soldier radio system" could be considered a precursor of the Army's "Soldier System."[178]

1968: *Fort Monmouth developed and deployed passive night vision devices to Vietnam that, by using image intensifier tubes, made targets almost as visible at night as in daylight.*

ECOM also fielded a small omni directional radio ground beacon, the AN/TRN-30, for Army aircraft. The beacon was for use at remote airstrips and landing facilities.

ECOM instituted a Direct Exchange/Repair and Return program for nineteen critical items, mostly avionics equipment, in August 1965. Under this program, spares were exchanged for damaged equipment in the theater. Defective components were then returned to the U.S., usually to the Sacramento Army Depot, for repair and eventual return to the field. As repair requirements changed during the war, so, too, did the number and kind of items on the repair and return list. Defective modules were arriving at Sacramento Army Depot at the rate of 5,000 a month by 1969. Noting that many modules were damaged or misplaced in shipment, General Latta had the labs design and issue padded, pre-addressed envelopes called "jiffy bags."

The RED BALL Express, instituted by the Army in December 1965, provided emergency supply of critical repair parts and air delivery to Vietnam. ECOM handled 27,000 RED BALL requisitions in 1967, filling 99.2 percent within thirty days (the AMC average during the same period was 97.8 percent).

The National Inventory Control Point at ECOM established a permanent office in South Vietnam in January 1968. Civilian supply technicians replaced military expediters to locate equipment in the depots.

ECOM instituted a Technical Assistance Program in Vietnam in 1965 to solve the most troublesome maintenance and support problems on site and also to provide feedback information for correcting design and support deficiencies. One civil servant and thirty-three manufacturer representatives worked the Technical Assistance Program. Latta then organized a formal ECOM Area Office in Vietnam in February 1966. Three years later, the office had a staff of 141 civilian engineers and technicians. Most of the staff was assigned to support Military Assistance Command Vietnam (MACV) Headquarters, the 1st Signal Brigade, the 1st Logistical Command, and the 34th General Support Group.

ECOM deployed the R&D Technical Liaison Team to Vietnam in January 1967 at the request of the 1st Signal Brigade. The team typically consisted of six or seven people: a team leader and representatives of the R&D Technical Support Activity and the various ECOM laboratories (Avionics, Electronic Components, Combat Surveillance/Target Acquisition, Night Vision, and Communications/Automatic Data Processing). Team members typically served three-month tours in theater (leaders, six months) to observe the operation of ECOM equipment, identify deficiencies in design or performance, provide quick-fix solutions, and acquire first-hand knowledge of field conditions. More than eighty ECOM scientists and engineers served on the team between 1967 and 1972. Several served more than once. The team also supported AMC's Vietnam Laboratories Assistance Program.

Soldier operating the AN/VRC-12

Military and civilian personnel of the ECOM New Equipment Training teams conducted more than eighty missions in direct support of the war in Southeast Asia from 1965 through 1968, including fifty-one missions in theater. More than half of all the missions supported avionics equipment.

In all, from 1967 to 1972, seventy-four ECOM civilian scientists and engineers, including one female, served as members of the Technical Liaison Team. Five of the seventy-four volunteered their services twice, including one person who went once as a representative of the Electronics Technology and Devices Laboratory (ETDL), then again, a year later, as a representative of the Communications/ADP Laboratory. One member of the Avionics Laboratory served three tours between April 1968 and May 1970; one member of the ETDL served four tours between January 1967 and September 1970. The R&D team functioned as a point of contact in Vietnam for units that received developmental equipment, systematically visited tactical units to identify problems that could be resolved either on the spot or in the laboratories, and supplied United States Army Vietnam (USARV) with up-to-date information on the status of work in the laboratories. Its achievements included identification of the need for transient voltage suppressers in AN/VRC-12 radio sets, development of a power supply for the AN/GRA-39, and development of the "Scotchcast" quick cable splicing technique.

The Vietnam War was tapering off and priorities had shifted by the time ECOM observed its 10th anniversary in 1972. Research and Development received increased emphasis for the design and development of the next generation of the military's electronic needs.

ECOM's personnel strength reached over 1,350 military and 10,250 civilians as it entered its second decade. The majority of the personnel (approximately 7,200 civilian and 900 military) worked at Fort Monmouth, with the remainder dispersed amongst ECOM Philadelphia and Fort Belvoir, Virginia; among other smaller contingents.[173]

THE SIGNAL SCHOOL TRANSFER

The war in Vietnam, like the Korean War fifteen years before, brought to Fort Monmouth a dramatic increase in its number of students. As late as 19 April 1965, the Signal School planned to enroll 4,290 students in enlisted courses during the coming fiscal year. CONARC increased the required enrollment to 8,806 in increments by the end of November 1965. The required and anticipated officer enrollment at that time was 1,185.

Congress authorized construction of three new, permanent classroom buildings to make room for the influx of students: Building 292 for the Officer School's Department of Command Communications (Tactical Division), Building 814 for the Photographic Laboratory and Building 918 for the Radar Laboratory. In breaking ground for these buildings on 19 August 1966, Congressman James J. Howard declared, "This ceremony is symbolic as a reassurance to the people of Monmouth County that the Signal School is here to stay."

On 25 November 1966, the Commandant of the Signal School, Brigadier General Thomas D. Rienzi, presented a special diploma to the 200,000th graduate of the School's Enlisted Department. PFC Lloyd B. Hansen of Minot, ND, had completed the twenty-eight week microwave radio repair course.

At that time, the School anticipated fiscal year 1967 enrollments of 18,194 enlisted personnel and 2,124 officers. Many of the courses operated with three shifts a day to accommodate these students. As the war in Vietnam wound down, so, too, did Signal School enrollments. The School admitted 14,139 students, enlisted and officer, in 1970.

These numbers notwithstanding, the majority of the Signal Corps' enlisted personnel trained during the war at Fort Gordon, not at Fort Monmouth. Fort Gordon also hosted the Officer Candidate School. The Army placed its branch schools under the jurisdiction of the newly-created Training and Doctrine Command (TRADOC) in July 1973. The following year, TRADOC began consolidating Signal Corps training at the Southeastern Signal School, Fort Gordon. TRADOC redesignated this school the "The U.S. Army Signal School" on 1 July 1974. The Signal School at Fort Monmouth continued to operate as "The U.S. Army Communications-Electronics School" while equipment and personnel transferred.[174] The movement of the school involved the transfer of only eighty-nine civilians. More than 700 others received reassignment to other agencies on post or retired. Fort Monmouth's last class in signal communication graduated on 17 June 1976.

Celebrity Notes:

Ft. Monmouth Holds Valachi Under Guard

JOSEPH VALACHI

Post Protects Informer From Underworld

FT. MONMOUTH—This Army post was disclosed today as the secret hideout of Joseph Valachi, the former mobster reportedly facing an underworld death sentence for informing on a nationwide crime syndicate called "Cosa Nostra."

Valachi's presence here under heavy guard was confirmed by Army information officer Peter Hoffman.

Valachi for more than a year has been telling all he knows about the nation's $40 billion annual crime industry to Justice Department agents.

They have sworn to keep him alive despite a reported $100,000 underworld offer to whoever kills him for violating crime's code of silence.

The Justice Department declined to confirm that Valachi is at Ft. Monmouth.

Tried Secrecy

Justice officials, who consider the 60-year-old Valachi the most important informer they have ever had in the war against organized crime, have tried to keep his whereabouts a closely guarded secret.

"If we let him out on the street he'd be dead in half an hour," a spokesman has said.

Valachi has described the blood oath taken by members of Cosa Nostra (Our Thing), saying — "If I talk, I'm dead."

He has a date to tell Senate investigators his story that "Cosa Nostra" controls the nation's underworld through a ruling council of crime "families," headed by narcotics boss Vito Genovese, Valachi's former cellmate in the Atlanta federal penitentiary.

Genovese formerly lived in Atlantic Highlands.

Heavily Guarded

Barbed wire and a maximum security guard watch over Valachi here, the New York Daily News said in a copyrighted story. Ft. Monmouth is heavily guarded because of secret areas, including electronic laboratories.

The post is the home of the U.S. Army Signal Engineering Laboratory and has a satellite tracking station.

The date of Valachi's appearance before the Senate investigations subcommittee, headed by Sen. John L. McClellan, D-Ark., is being kept secret and elaborate precautions are planned to protect him in the crowded hearing room.

"Nothing will happen to Valachi at that hearing," a subcommittee source has said.

Told Secrets

Valachi began telling the secrets of the underworld while in the Atlanta penitentiary, where both he and Genovese were serving narcotics sentences. Genovese is still there.

Valachi thought Cosa Nostra had marked him for death in the mistaken belief he had already talked. When another convict approached him one day, Valachi thought it was his assassin and killed him with a length of pipe.

It turned out to be a case of mistaken identity and Valachi, to save himself from a possible

See VALACHI Page 2

VALACHI

From Page 1

death sentence, asked to talk to federal agents. He later got a life sentence for the slaying.

Valachi has told of Cosa Nostra crime families operating in New York, Detroit, Buffalo, Philadelphia, Chicago, Miami, Kansas City, New Orleans, Pittsburgh, St. Louis, Cleveland, Los Angeles, San Francisco, Providence, Boston and various cities in New Jersey.

Sought Support

He said the now famous Apalachin, N. Y., "crime convention" several years ago was called by Cosa Nostra and that Genovese sought support for killing racketeer Frank Costello, was was later wounded in an attempt on his life, and gangster Albert Anastasia, who was later slain. Valachi has told details of other gangland slayings.

Valachi was hustled out of the Atlanta penitentiary after he began talking and since then has been in hiding while federal agents continued detailed questioning. Agents have said his story has been corroborated by other evidence.

The news said the FBI turned Valachi over to the Army for protection after it was learned that crime leaders were offering $100,000 for his death.

Valachi's story has given federal agents an unprecedented glimpse into organized crime and led investigators to other weak spots in the barrier of secrecy surrounding it.

"We're going to keep him alive," one federal source said recently, "We don't know what we're going to do with him, but we're going to keep him in a safe place."

Asbury Pk Press
14 Aug 1963
WED

Joseph Michael Valachi

(1903-1971) was a member of Lucky Luciano's mob family and turned informer in 1962. The government incarcerated him in the stockade at Fort Monmouth from January-September 1963. In 1962, Valachi was serving time in the Atlanta Federal Penitentiary on charges of heroin trafficking. He became convinced that mafia boss Vito Genovese (1897-1969) had mandated his death. Valachi went so far as to kill another inmate whom he believed to be an assassin, beating the man to death with a two-foot length of pipe. According to author Peter Maas (1929-2001), who chronicled the Valachi saga in his best-selling book *The Valachi Papers*, paranoia motivated Valachi to turn informant. He was relocated from the Atlanta jail to the more secure Fort Monmouth. Here, handpicked guards from the Federal Correctional Institute at Lewisburg, Pennsylvania protected him. Fort Monmouth originally placed the gangster in a tiny enlisted man's cell surrounded by bars. According to Maas, this shoddy treatment enraged Valachi. He was transferred the next day to a more private area in the stockade normally reserved for officers. Here, Valachi confessed to forty-two years of a life of crime in a secret society that he called the Cosa Nostra. Valachi was originally scheduled to spend just three weeks at Fort Monmouth. He stayed over six months, as U.S. Attorney General Robert Kennedy (1925-1968) was painstakingly debriefed on the situation. Valachi finally testified to Senator John McClellan's (1896-1977) Permanent Investigations Subcommittee after being smuggled off post disguised as a Military Policeman. He ultimately expressed his desire that Congress pass a law against participation in organized crime. Valachi's wish was granted. Four years before his death, Congress passed the Racketeer Influenced Corrupt Organization (RICO) Act, Title 18, United States Code, Sections 1961-1968. The Maas book was made into a movie in 1972. The film starred Charles Bronson (1921-2003) as Joseph Valachi. The cast even included Syl Lamont who played the Commander of Fort Monmouth who was then MG Stuart S. Hoff.[181]

THE 1970s

By the time the Vietnam War was winding down, several of the projects started either at the beginning of, or during the period of the conflict were becoming ready for the troops. Based on the "second generation" starlight scopes and other image intensifier devices, the new AN/PVS-5 night vision goggles were undergoing operational testing, with production and fielding set for 1975.

In 1973, based on information learned in Southeast Asia, development of lithium batteries was well under way to provide a more reliable and longer lasting power source for tactical field radios.

By 1974, ECOM personnel established the feasibility and completed the designs for a new family of mini-laser range-finders for individual and crew served weapons. The devices, one of which could be mounted on the sights of an M-16 rifle, weighed less than a pound and could accurately range in on a target up to a distance of one kilometer.

The following year, ECOM and the Avionics Laboratory successfully fielded the Proximity Warning Device (PWD) placing it in over 2,000 Army helicopters. The device was used to reduce the number of mid-air collisions at the Army Aviation School at Fort Rucker, Alabama, where increased use of training airspace was becoming hazardous to the student pilots. The unit was so successful at Fort Rucker that there resulted an urgent requirement to equip all Army helicopters at Fort Hood, Fort Bragg and Fort Campbell, as well as those at Fort Rucker. Two years after the start of the program, there were no mid-air collisions of helicopters equipped with the PWD. Also during this period, ECOM produced the prototype of the first Lithium "D" cell battery. This battery was able to last fifty percent longer than any previous Lithium battery and three-hundred percent longer than conventional carbon batteries.

By the end of FY76 and the beginning of FY77, the AN/TPQ-36 and AN/TPQ-37, mortar and artillery locating radars had passed all testing and were ready for production and fielding. The AN/TPQ-36 exceeded all performance tests at the Yuma Proving Grounds and provided the Army with a greatly enhanced capability to locate mortar and artillery fire. The success of both of those programs was attributed to the accomplishments of the Combat Surveillance and Target Acquisition Laboratory's advances in computer simulation techniques, radar cross-section analysis, and automatic height correction concepts. Also during this period, ECOM's Electronics Technology and Devices Laboratory produced the first of this Command's successes in the development of Very High Speed Integrated Circuit (VHSIC) technology. The laboratory developed the technology to produce a high speed, high density, low power, computer chip operating at 50 Mhz at 10 volts. An offshoot of this technology introduced into the commercial market the "CDP 1800 family" CMOS micro-processor, an early eight-bit central processor.[182]

AN/TPQ-36

FIRST ROUNDS of hostile mortar launchers are detected, tracked, and their trajectories are back-plotted to pinpoint the firing site by the U.S. Army's new Mortar Locating Radar, AN/TPQ-36. In the small, highly-transportable operations shelter, the location is displayed to the operator as a point of light on a cylindrical map board. Easily moved by aircraft, helicopter or ground vehicle, the two lightweight modules that make up the entire system can quickly be deployed close to the forward edge of the battle area. A small crew can set the unit up for operation in minutes. Developed by Hughes Aircraft Company's ground systems group, Fullerton, Calif., the radar has successfully competed the first phase of live-firing tests at the Army's Yuma Proving Grounds in Arizona.

Public Rel.... Department
.... Group
.... Company
........rnia 92634
.....xt. 4631

AN/TPQ-36
MORTAR LOCATING RADAR SYSTEM
(BLOCK II CONFIGURATION)

AN/TPQ-37 Firefinder
Artillery Locating Radar

Famous Firsts

1972: *Developed in cooperation with doctors from Patterson Army Hospital...*

the Defibrillator Pacemaker regulated the heartbeat but, in addition, could detect the start of fibrillation (wild tremors of the heart's muscle) and briefly stop the heart to allow a normal beat to resume.

1973: *Fort Monmouth developed the Carbon Dioxide Communications Laser.*

This was an air-cooled dioxide laser communications system with a range of five miles.

1974: *Fort Monmouth scientists and engineers were testing lithium batteries...*

with potentially four times the life of carbon-zinc and twice the life of magnesium batteries.

1974: *The AN/APR-39 Radar Warning System was an analog system that identified and provided directional information of radar threats, thereby permitting avoidance flying and increasing the probability of survival.*

About 5,000 had been employed as of 1990. There is currently an improved, programmable digital version aboard most Army aircraft.[179]

1974: *Fort Monmouth developed a Laser Mini Rangefinder.*

This small rangefinder, weighing less than one pound, could be mounted on small arms. It was accurate up to distances of one kilometer.

1975: *Mortar and Artillery Locating Radars...*

AN/TPQ-36 and AN/TPQ-37 were developed.

1975: *The solid state AN/TTC-38 Automatic Telephone Central Office was smaller, lighter, faster, and more easily maintained than manual switch systems.*

It gave the user touch-dialing capability to anywhere in the worldwide military telephone system.

1975-1976: *The AN/ALQ-144 was developed.*

This on-board infrared jammer can defeat incoming missiles. It has been employed effectively world-wide and is also used to protect civilian VIP aircraft.[180]

ECOM TO CERCOM/CORADCOM TO CECOM

In 1973, the Secretary of the Army charged the Army Materiel Acquisition Review Committee (AMARC) with finding ways to improve "current Army organization and procedures for materiel acquisition," and doing so within one hundred days.

The committee's report, released 1 April 1974, concluded that the Army's standard commodity command structure, with its emphasis on "readiness," limited flexibility and impeded the acquisition process. It recommended that research and development (R&D) functions be separated from readiness functions within the Army Materiel Command (AMC) and that the disparate and scattered R&D activities of AMC be consolidated in six development centers.

This meant a simple two-for-one split for most of the major subordinate commands within AMC. The picture was more complicated, however, for ECOM. AMARC concluded that the breadth of ECOM's responsibilities "tended to defocus the organization's responsiveness to modern mission-oriented needs." A splintering, not a split, was proposed. This transferred the Avionics, the Combat Surveillance, and the Electronic Warfare R&D missions to development centers not headquartered at Fort Monmouth.

The recommendation proved unpopular. Within days, community leaders joined Fort Monmouth personnel in a vigorous "Save the Fort" campaign. Campaigners sent more than 50,000 letters to the Secretary of the Army. These letters attracted White House attention and twice obliged the Army to reassess its reorganization plans. The letter writing campaign

had some effect. The Army's initial plan, announced 1 April 1976, would have cost the Fort 780 jobs. The final plan, announced 13 July 1977, left the Electronic Warfare mission at Fort Monmouth and resulted in the elimination or transfer of only 418 personnel.

As of that date, much of AMARC was already implemented. The Aviation Systems Command, soon to become the Aviation Research and Development Command, had assumed operational control of the Avionics Laboratory and PM Navigation Control Systems (NAVCON).

The Electronics Research and Development Command (ERADCOM), established provisionally 30 March 1977, assumed operational control of its assigned elements on 15 July, as did the Communications Research and Development Command (CORADCOM).

ERADCOM was headed by MG Charles B. Daniel and was headquartered at Adelphi, MD. With a combined military and civilian strength of 4,200, it was made up of the following organizations: PM Firefinder, PM Remotely Monitored battlefield Sensor System (REMBASS), PM Standoff Target Acquisition System (SOTAS), PM Control Analysis Centers (CAC), the Harry Diamond Laboratories (Adelphi, MD), the Electronics Technology and Devices Laboratory (Fort Monmouth, NJ), the Electronic Warfare Laboratory (Fort Monmouth, NJ), the Signals Warfare Laboratory (Vint Hill Farms, VA), the Night Vision and Electro Optics Laboratories (Fort Belvoir, VA), the Combat Surveillance and Target Acquisition Laboratory (Fort Monmouth, NJ) and the Atmospheric Sciences Laboratory (White Sands Missile Range, NM).[183]

CERCOM Activation Ceremony, 3 January 1978

CORADCOM was established provisionally under Brigadier General William J. Hillsman, who was the Project Manager for Army Tactical Data Systems (ARTADS). He led the task force that planned the organization of the new command.

It was the organizations of CORADCOM which produced breakthroughs in the fiber optic transmission technology, over long distances and under extreme battlefield conditions. The techniques developed for interfacing the fiber optic cables with existing multiplexer switches was the beginning of the Army's move to fiber optic transmission for most of its battlefield tactical communications, and provided civilian companies with the technology to produce fiber optic data transmission lines for business communications.

The Army Materiel Command was re-designated the U.S. Army Materiel Development and Readiness Command (DARCOM), with no change in mission.

Activation of the new commands—CERCOM (Communications-Electronics Readiness Command), CORADCOM and ERADCOM—was initially planned for 1 October 1977. The date moved to 1 January 1978, partly to permit review of revisions imposed on the CERCOM organization concept by Major General John K. Stoner, the ECOM Commander; and partially to accommodate additional planning necessitated by a DA-imposed reduction of 500 spaces in the Headquarters Installation Support Activity (HISA), along with a reduction in average grade and a reduction in the number of high-grade positions permitted in the two new commands.

Activation ceremonies for the new commands occurred 3 January 1978 in the Field House. DARCOM Commander General John R. Guthrie officiated, handing the CERCOM flag to Major General John K. Stoner and the CORADCOM flag to Major General Hillman Dickinson.

The two commands made significant contributions in the ensuing three years and four months of their operation. They also encountered problems.

The separation of acquisition from readiness gave the research and development community the visibility AMARC thought the communities needed. However, it was costly. Separation meant duplication. Each command required an administrative staff, which, at that time of constrained resources, meant the diversion of personnel from mission activities. There also was a duplication of effort in the mission activities, and overlapping areas of responsibility that used manpower simply to ensure a coordination of effort. Such duplication affected performance most severely in integrated logistics support, in initial fielding, and in long-term field support. It was also apparent in production engineering and product assurance.

AMARC was an experiment, not a solution. The AMARC committee itself insisted on periodic review, updating and revitalization of the measures it proposed to improve materiel acquisition. It was only natural to "revisit AMARC" when, in February 1979, DARCOM Commander General Guthrie voiced concern about the impact of continued manpower reductions on the mission performance of DARCOM commands.

Review of the Army electronics community began in August 1980. A marked improvement was noted in the electronics R&D capability, but the review committee found the readiness capability weakened. They attributed this to diverging workload and fixed resources. The review team, addressing this imbalance, decided there was a need for greater economy and greater flexibility in the use of existing manpower resources. This could be achieved by pooling the resources of the two commands headquartered at Fort Monmouth, a move which would eliminate duplication. Control could be assigned to one commander with the authority to move personnel as required to meet the most pressing needs. A decision was announced in December 1980 that CERCOM and CORADCOM would merge and become the Communications-Electronics Command (CECOM) effective in May 1981.

Gen. Hillman Dickinson to head CORADCOM

The Army has announced that Maj. Gen. Hillman Dickinson will assume command of the new Communications Research and Development Command (CORADCOM) at Fort Monmouth.

General Dickinson is expected to arrive here Sept. 19 to replace Brig. Gen. William J. Hilsman, prvisional commander of CORADCOM, but recently named to take command of the Army Signal Center and School, Fort Gordon, Ga., in December.

General Dickinson, 51, is a 1949 graduate of the United States Military Academy. He is a native of Independence, Mo. He holds a master's degree in physics from Columbia University, New York, as well as a doctorate in physics from Columbia University, New York, as well as a doctorate in physics from Stevens Institute of Technology, Hoboken.

Since 1974, Gen. Dickinson has been with the office of the Army's deputy chief of staff for research, development and acquisition in Washington. Before this Pentagon assignment, he was stationed at Fort Knox, Ky., and became deputy commanding general of the Army Training Center there in 1973.

A graduate of the Army War College, Carlisle, Pa., Gen. Dickinson was a squadron commander of the 11th Armored Cavalry Regiment in Vietnam in 1967-68 and a plans officer in the United States Military Assistance Command there in 1968-69.

He then was assigned to Washington for two years and became deputy director of the Defense Communications Planning Group, Joint Task Force 728. He returned to Vietnam from May 1971 to May 1972 as senior adviser with the 1st Regional Assistance Command of the Military Assistance Command.

In addition to the Army War College, Gen. Dickinson is also a graduate of the Army Armor School's basic and

MAJ. GEN. HILLMAN DICKINSON

advanced courses, and Command and General Staff College.

Gen. Dickinson wears the Legion of Merit with three oak leaf clusters; the Bronze Star Medal with "V" Device with oak leaf cluster, the Meritorious Service Medal, and Air Medals with "V" Device.

General Dickinson and his wife, the former Miss Nancy Cameron, have a daughter, Cynthia, 25, and a son, Stuart, 23.

CECOM was to be structured to ensure that materiel acquisition was not totally submerged in the new command as it had been in the pre-AMARC commodity command. The Development Center of the new command would have a General Officer in charge, also serving as Deputy Commander for Research and Development, to ensure that R&D at CECOM retained the visibility obtained under AMARC.

The 1980s

CECOM

Major General Donald M. Babers, CERCOM and Fort Monmouth commander since June 1980, became the first Commanding General of CECOM. He continued in his role of post commander. Colonel (P) Robert D. Morgan became Deputy Commander for Research and Development and Commander of the Research and Development Center. Colonel Robert G. Lynn (P) became Deputy Commander for Materiel Readiness. Both men received promotions to Brigadier General on 31 July 1981.[184]

MG Robert Morgan's Staff, May 1987

Essentially, CECOM was charged with the research, development, engineering and acquisition of assigned communications and electronic systems and management of all materiel readiness functions associated with these systems and related equipment.

Research facilities of the command included the Center for Tactical Computer Systems (CENTACS), which conducted research and development in computer science and systems, including hardware and software for diverse applications; the Center for Communications Systems (CENCOMS), which researched programs to produce advanced communications technology, equipment and systems; and the Center for Systems Engineering and Integration (CENSEI), the Army's system engineer for Tactical Command, Control and Communications.

CENSEI aimed to produce a well-engineered, affordable and evolutionary system design.

A Program Manager (PM) directed the Test, Measurement and Diagnostic Equipment (TMDE) modernization effort. Two product managers reported to this PM; one, for Test, Measurement and Diagnostic Systems; and one, for Army Test, Measurement and Diagnostic Equipment Modernization.

In addition, eight project managers (PMs) existed within CECOM. These included Army Tactical Communications System (ATACS)/Mobile Subscriber Equipment (MSE); Position Location Reporting System/Tactical Information Distribution System (PLRS/TIDS); Satellite Communications (SATCOM/SATCOMA); Field Artillery Tactical Data Systems (FATDS); Single Channel Ground and Airborne Radio Systems (SINCGARS); Operations Tactical Data Systems (OPTADS); Multi-Service Communications Systems (MSCS); and Firefinder/Remotely Monitored Battlefield Sensor System (REMBASS), which was transferred to CECOM from the Electronics Research and Development Command (ERADCOM) on 30 March 1984.

Position Locating Reporting System

The Army established a series of Program Executive Offices (PEOs) in 1987 in order to consolidate and better manage the vast array of Program Managers responsible for major acquisition programs in the inventory. The Army created PEOs for Communications Systems, Command and Control Systems, and Intelligence/Electronic Warfare and Sensors to manage all of the electronics programs. A PEO/CECOM association existed due to the nature of their missions. The PEOs received significant technical, logistical and program management support from CECOM, but reported directly to the Assistant Secretary of the Army for Acquisition, Logistics and Technology (ASA ALT). For an in depth discussion on the creation of the PEOs, see Chapter Twelve.

The command added a Software Development and Support Center in October 1984. Located in Building 1210, a former Signal School classroom building, the center conducted software development and life cycle software support activities associated with the Army communications equipment.[185]

Field offices in various parts of the United States and Europe supported CECOM's research and development efforts and procurement and readiness functions. TASA, CECOM's Television-Audio Support Activity at Sacramento, California, was the Army life-cycle manager for non-tactical, commercial broadcasting and television equipment for the Army forces. This subsequently transferred to the U.S. Army Information Systems Command.

A number of separate agencies within CECOM were responsible for supporting all the systems in CECOM's inventory during the 1980s. CECOM's National Inventory Control Point (NICP) played a key role in keeping fielded communications and electronics equipment in a high state of readiness. This task included worldwide materiel management of communications-electronics systems and support items. Complimenting the NICP was the command's National Maintenance Point (NMP), which provided maintenance and engineering expertise on maintainability of communications-electronics materiel from conception to obsolescence.

Certain CECOM activities were managed at locations aside from Fort Monmouth. The Communications Security Logistics Agency (CSLA), based at Fort Huachuca, Arizona provided commodity management of communications security equipment, aids, and accountable spare parts.

The Electronics Materiel Readiness Activity (EMRA) at Vint Hill Farms Station, Warrenton, Virginia, furnished commodity management and depot-level management for signal intelligence/electronic warfare equipment and systems. EMRA supported the Army Intelligence and Security Command (INSCOM) and other Signal intelligence and electronic warfare units and activities worldwide.

The CECOM Logistics and Readiness Center (LRC) stood up on 10 November 1987 to act as an overseer to all communications-electronics logistics functions within CECOM. Its mission was to support the U.S. Army by providing integrated, timely, cost-effective, and high quality worldwide logistics support to include fielding, new equipment training, operations, maintenance, and sustainment. In addition, the LRC was responsible for all Foreign Military Sales (FMS), communications security programs and the management of Level II and Level III programs, having completed their initial development and fielding.

POST IMPROVEMENTS

The physical area of Fort Monmouth during the 1980s encompassed the main post area, the Charles Wood Area, and the Evans Area nine miles to the south. All of the other sub-installations had been closed or were released to the General Services Administration for disposal. The last large area identified for disposal was the Coles Area on Newman Springs Road west of Red Bank. It was declared excess in March 1974 and officially closed 1 January 1975.

Bowling Center, 1985

The post continued to grow with the construction of new facilities through the years. An interdenominational Chapel was dedicated in July 1962; a Bowling Center opened in December 1965; dedication of the Post Exchange complex took place in February 1970; the Commissary opened in April 1971; Green Acres, the CECOM Office Building, officially opened in November 1973; the Credit Union Building and the Post Exchange Service Station and Convenience Store in the Charles Wood Area opened in March 1975; and the post library opened in June 1974. The library was dedicated as the Van Deusen Library in 1977 in honor of the 1941-42 post commander and Signal School Commandant.

CECOM Office Building, 1984

Multi-million dollar projects in the 1980s upgraded and modernized the Myer Hall complex and barracks in the 1200 area; the Communications Center (Vail Hall); Russel Hall, and Squier Hall. A modernization program began at the Hexagon (Building 2700) in July 1982. Major objectives of a three-phase Hexagon modernization program included the installation of air conditioning; installation of energy-saving wall and window insulation; accommodations for the handicapped; installation of additional elevators; replacement of existing communications equipment; and alteration of building elements to conform to health safety and fire codes.

Commissary, July 2005

A new NCO/Enlisted Club (Lane Hall) opened 10 November 1983. The first phase of construction for the club, built in the area between the post service station and Husky Brook Pond, provided a facility with fast food service and a bar. Subsequent construction added a kitchen and dining room.

TENANT ORGANIZATIONS

JOINT TACTICAL COMMAND, CONTROL AND COMMUNICATIONS AGENCY

The Joint Tactical Command, Control and Communications Agency stood up at Fort Monmouth on 10 September 1984 with Major General Norman E. Archibald as Director. The DoD chartered this agency to ensure interoperability among tactical command, control and communications systems used by U.S. Armed Forces and to develop and maintain a joint architecture, systems standards and interface definitions for tactical/mobile command, control and communications systems.

The agency, headquartered in Russel Hall, united four former defense elements under the leadership of a single director: the Joint Tactical Communications Office and the Joint Interface Test Force, both at Fort Monmouth; the Joint Test Element, Fort Huachuca, Arizona; and the Joint Interoperability of Tactical Command and Control Systems Program, Washington. This office was later reorganized into the JIEO (Joint Information Engineering Organization) and was separated into various other organizations over the next fifteen years.[186]

U.S. ARMY CHAPLAIN CENTER AND SCHOOL (USACHCS) AND CHAPLAIN BOARD

The Army's Chaplain Center and School, the Army's only training center for the clergy, moved to Fort Monmouth in 1979 from Fort Wadsworth, N.Y. It conducted resident training for over 1,000 students per year, including 700 enlisted chaplain activity specialists and 300 chaplains in both the officer basic and advanced courses.

Chaplain Center and School, 1993

The school, which transferred to Fort Jackson, South Carolina during the 1990s, was headquartered in Watters Hall (Building 1207, formerly Myer Hall and later Mallette Hall). The building was renamed 30 July 1984 in commemoration of the 109th anniversary of the Army Chaplaincy. Chaplain (Major) Charles J. Watters, a Catholic Priest of Jersey City, N.J., was killed in action in Vietnam and posthumously awarded the Medal of Honor by President Nixon in 1969.[187]

The Chaplain Board, a field operating agency of the Chief of Chaplains, moved to Fort Monmouth in September 1979. It executed programs in support of various religious and moral activities of the Army and focused on meeting the changing needs of the Soldier. The board also assisted the Chief of Chaplains in developing concepts of ministry and professional guidelines for chaplains and religious activities.

U.S. ARMY INFORMATION SYSTEMS MANAGEMENT ACTIVITY (ISMA)/PROJECT MANAGER, DEFENSE COMMUNICATIONS SYSTEMS-ARMY

ISMA, located in Squier Hall (Building 283), was formerly the Communications Systems Agency (CSA) and was later assigned to the Army Information Systems Command (previously the Army Communications Command at Fort Huachuca, Arizona). The changes resulted in the establishment by the Army in mid-1984 of a staff agency and a major command to coordinate the modernization of the Army's information management, communications, command and control systems. Thus, the Army Communications Command at Fort Huachuca, the Army Computer Command at Fort Belvoir, and their associated agencies merged to form the new Information Systems Command (USAISC).

The Information Systems Management Activity (ISMA) at Fort Monmouth was a subordinate command of USAISC and a project management office of the Army Materiel Command.

The activity handled the acquisition and fielding of a wide variety of information and telecommunications systems in support of the worldwide Defense Communications System. In addition to undertaking projects for the Army, Navy and Air Force, the activity supported the State and Commerce Departments, the National Security Agency, the Federal Aviation Administration, and foreign allied governments in improving the modernizing their communications systems.[188]

513th MILITARY INTELLIGENCE GROUP

The 513th Military Intelligence Group activated at Fort Monmouth in September 1982, along with three subordinate units: the 201st, 202nd and 203rd Military Intelligence Battalions. All resided at Fort Monmouth except the 203rd, which was headquartered at Aberdeen Proving Ground, Maryland.[189]

Activation of the units (all assigned to the U.S. Army Intelligence and Security Command) resulted in an additional 375 military personnel at Fort Monmouth. Fully staffed, the 513th was expect-

ACTIVATION CEREMONY
30 SEPTEMBER 1982

ed to result in an increase of about $7 million in the post's annual military payroll.

Increasing military requirements to provide rapid and accurate intelligence support to military commanders responsible for planning and executing peacetime, contingency, and wartime operations influenced the activation of these units. The 513th was realigned to Fort Gordon, GA in the Spring of 1994.

MODERNIZING THE FORCE

The phrase "Force Modernization" characterized the 1980s based on technologies developed largely in the 1970s. The introduction of tactical ADP (Automated Data Processing) systems gave the American Soldier new battlefield capabilities no other Army possessed. CECOM also introduced new secure communications systems, including the Single Channel Ground and Air Radio System (SINCGARS) and Mobile Subscriber Equipment (MSE). MSE was the product of the largest single contract ever awarded for C-E equipment, $4.5 billion, and heralded a new way of doing business. MSE would provide users with a means of communicating throughout the battlefield, regardless of location, in either static or mobile situations. MSE Project Manager, Colonel John R. Power, who assumed control just one month after the award of the MSE

AN/PVS-6 Eye Safe Laser Rangefinder, October 1983

contract, stated of the program, "I wanted to be a project manager of a major Army Signal program…I got to be the Project Manager of the Army's most major Signal program."

CECOM had taken the lead in finding ways to shorten the acquisition cycle through procurement and adaptation of non-developmental items and in standardizing tactical computers and software. The SINCGARS provided Very High Frequency (VHF) Frequency Modulation (FM) combat net radio communication with Electronic Counter-Countermeasures, or frequency hopping, and digital data capability.

In 1983, research and development contracts were awarded for the All Source Analysis System (ASAS), an automated tactical intelligence system that would provide all source correlated intelligence to commanders at division, corps, and echelons above corps. ASAS' all source fusion network could be used to generate timely, accurate, and comprehensive understanding of enemy deployments, capabilities, vulnerabilities, and potential courses of action.

Maneuver Control System (MCS) contracts were also awarded in the 1980s. The MCS, a collection of computer equipment, provided battlefield information by collecting, processing, and displaying data generated within the air/land combat environment. Using this system, a commander could improve the time-liness of his or her decisions and allocate resources accordingly.

In 1988, the GUARDRAIL/Common Sensor (GR/CS) was fielded to Korea. GR/CS, a corps level airborne signals intelligence (SIGINT) collection/location system, would provide near real time SIGINT and targeting information to tactical commanders throughout the corps area with emphasis on Deep Battle and Follow on Forces Attack support.

Concurrently, CECOM embarked upon an extensive internal reorganization. The continuing budget challenge within the Federal Government acted as one motivating factor for this change. Budget challenges dictated reduced spending and a renewed search for more efficient ways of doing business. The changes included the creation of a Command, Control, Communications and Intelligence (C3I) Logistics and Readiness Center and the establishment of the U.S. Army Garrison Fort Monmouth. CECOM also assumed responsibility for Vint Hill Farms Station, Virginia, and its garrison (which previously belonged to the Intelligence and Security Command).

CECOM's parent command, the U.S. Army Materiel Development and Readiness Command (DARCOM), was re-designated the U.S. Army Materiel Command (AMC) on 1 August 1984. This had been its original designation from 1962 to 1976.

SINCGARS Radio

COLD WAR CONNECTION

The competitive nature of the Cold War (1945-1991) fueled creativity among scientists and inventors worldwide. In the race for the best electronics, Fort Monmouth played a crucial role in the American defense system. It was well understood that the future would favor those with superior electronic capability. If a nation could increase the efficiency of its communications network, shield that network from enemies, and at the same time expose the networks of others, the advantage could prove more valuable than that of entire armies.

In 1983, the world again inched towards nuclear war during Exercise Able Archer 83. Soviet air defenses mistook this NATO exercise, which simulated a nuclear-release, as fake cover for a NATO attack. The country consequently put its nuclear forces on high alert. "The world did not quite reach the edge of the nuclear abyss," recalls U.S. intelligence agent Oleg Gordievsky in an article by college professor Lawrence S. Wittner, "But during Able Archer 83 it had ...come frighteningly close."

To protect the United States from such disaster, President Ronald Reagan proposed the Strategic Defense Initiative (SDI) that same year. This military research program sought to develop an antiballistic missile (ABM) defense system. An Architectural Survey and Evaluation of Buildings completed at Fort Monmouth in 2006 revealed that research conducted at the Pulse Power Center (buildings 2707-2710, 2713) in the Charles Wood Area contributed to the development of the SDI, known colloquially as the "Star Wars" project.[190]

Pulse Power Center, 1994

Famous Firsts

1980: *CORADCOM engineers received U.S. Patent number 4,197,500 for their apparatus for automatically selecting the best telecommunications channel.*

The apparatus provided extremely reliable communications over difficult terrain under varying propagation conditions.[191]

1983: *CECOM led a DOD initiative to define and implement a standard programming language for all mission-critical defense systems.*

The result, Ada, became a military standard (MIL-STD-1815a) in January 1983. Subsequently adopted as an international standard, Ada has been successfully employed in a wide variety of Army and DoD systems and has gained favor as well in the commercial sector.[193]

1989: *Fort Monmouth developed the Multi-Sensor Target Acquisition System (MTAS) ...*

a millimeter wave radio that would provide modern heavy forces with an independent, all weather sensor capability to augment infrared (IR) and visible sensor inputs. The radar afforded 360-degree independent search and acquisition of multiple ground and helicopter targets.[194]

1981: *The first Tactical Fire Direction (TACFIRE) system was fielded.*

TACFIRE was a tactical ADP system that automated selected field artillery command and control functions to provide efficient management of fire support resources.[192]

The Gulf War

The United States launched air strikes against Iraq on 17 January 1991 in an attempt to liberate Kuwait. CECOM was responsible with equipping and sustaining the force with the communications and electronics equipment it needed to fight. This was not an easy task. Units arrived in theater with only the equipment they owned. While some units possessed newer equipment, most units had at least some incomplete or damaged systems. The Army, and CECOM in particular, had to fill these gaps either through accelerated fieldings of new equipment or by reissuing items in theater before the ground offensive.

CECOM's Emergency Operations Center (EOC) began operating twenty-four hours a day, seven days a week on 7 August 1990 to address the situation. Although several organizations within CECOM set up their own crisis management centers, the EOC served as CECOM's focal point for all actions relating to the crisis in the Middle East. Employees worked around the

The EOC in action: Steve Jumper (on phone) receives a request from the field as Sal Panduri checks the computer log, August 1990

clock in order to equip Soldiers with everything from radios and jammers to night vision and intelligence systems. From day one, CECOM worked to sustain the equipment out in the field and ensure any follow-on items arrived in theater mission-ready.

On the eve of the Ground War, CECOM had fifty-nine military, 103 civilian, and 122 contractor personnel in or on their way to the Kuwaiti Theater of Operations.

CECOM completed 1,318 fieldings between July 1990 and February 1991, many accelerated specifically to meet the requirements for Desert Shield/Desert Storm. For example, CECOM managed to issue the SINCGARS system to an entire brigade within one week in order to equip the 1st Cavalry Division with SINCGARS radios before its deployment. This included not only the radios themselves, but also the operator and maintenance support training needed to sustain them. CECOM would repeat this same accomplishment in theater three more times before it was all over.

CECOM also supported the war effort through the purchase of commodities: the consumables, repair parts, and replacement items that kept forces viable wherever they operated. This complex, time consuming process ordinarily involved item managers, contracting officers and other employees across several organizations and functional areas. Many of these administrative processes were temporarily suspended due to the immediate needs of forces deployed overseas. By the end of the crisis, CECOM processed close to 180,000 requisitions, shipped six million pieces of equipment worth over $1.1 billion (including four million batteries), initiated 456 urgent procurement work directives valued at $113 million, and procured a total of 10.8 million pieces of equipment worth $326 million.

CECOM also established a Communications Security (COMSEC) Management Office in Saudi Arabia that opened 15 November 1990. While most theaters traditionally had a communications command responsible for managing COMSEC issues, one had not been set up for Operation Desert Shield. The reserve unit ordinarily assigned to Central Command, or CENTCOM, was not deployed due to obsolete equipment. Consequently, the Army Theater COMSEC Management Office (TCMO) was a significant development. CECOM, recognizing the need for dedicated COMSEC support in Saudi Arabia, acquired the necessary authorizations, resources and space to set up at the Royal Saudi Air Force Base in Riyadh. TCMO came under the direct control of CENTCOM shortly after operations commenced and remained operational until May 1991.

CECOM made extensive use of Logistics Assistance Representatives (LAR) during Desert Shield/Storm. LAR were civilian employees (GS-11 through GS-13) who provided hands-on technical assistance when needed. These LAR deployed to Saudi Arabia along with the divisions. This made them among the first civilians to arrive in the war zone. CECOM had forty-eight LAR ready to deploy within seventy-two hours of receiving the full deployment alert for Operation Desert Shield. CECOM LAR proved invaluable in providing assistance whenever Soldiers in the field asked for help regarding their equipment. In the theater, most LAR lived in tents. They ate, slept, and worked with the soldiers they supported. They were largely self-sufficient, however, and they participated in all the life support activities of their units. One LAR held the record for filling sandbags at VII Corps Main.

LAR who arrived early had a particularly difficult job. They were supposed to help units off-load their equipment, inspect it, then help move it to holding or assignment areas. But units arrived in theater daily, and before one unit could be made fully operational, other units arrived requiring immediate assistance. Nevertheless, the initial contingent of CECOM LAR in country played a very large role in determining the communications-electronics infrastructure of the theater, to include satellite linkages, telephone and message switch networks, and arrangements for special repair activities.

The LAR contributed beyond their numbers to the exceptionally high readiness rates of CECOM equipment. They played a vital role in maintenance and supply and found themselves, in the process, doing things they did not usually do. For example: some units arrived without their stock of tools and repair parts. To keep them in operation until their supplies caught up with them, the LAR "cross-leveled," that is, arranged the transfer of assets from units that arrived in theater more amply endowed. Several LAR purchased parts locally with their own funds to keep CECOM systems running. Their daily routine typically included visits to the supply support activities and direct and general support units, as well as a walk through the 403d Transportation Battalion area, the central receiving compound for air cargo, in hopes of finding misplaced commodities.

Two CECOM LAR assisted the 35th Signal Brigade by coordinating the exchange of FM radios with a rear unit for immediate use in system control at Corps Main. They helped the 426th Signal Battalion obtain BA-5590 and BA-1372 batteries that CECOM provided through a push package to a local Special Repair Activity. They obtained from the 40th Signal Battalion five different parts that the 50th Signal Battalion needed to make two tactical satellite links with Corps Main fully mission capable. This involved 660 kilometers of round-trip travel. They assisted the 327th Signal Unit Maintenance Company with technical advice on its inability to tune the AN/GRC-106 radio to lower frequencies: the suspected cause of the problem was the discriminator limiter. They found a truck-load of unclaimed CECOM equipment at a forward logistics base and called to have a CECOM supply LAR arrange for pick up

and delivery to the appropriate units. They assisted the Washington State Army Reserve with the installation and operation of an AN/VSC-2 in a M-1009.

Several LAR were involved in extraordinary efforts during the final preparations for the Ground War. Four of them worked day and night in cold rain to complete the installation of forty-seven SINCGARS radios in 1st Cavalry Division vehicles just forty-eight hours prior to the beginning of the offensive. A Joint (Tri-Service) Tactical Communications (TRI-TAC) team got an AN/TTC-39 switch of the 93d Signal Brigade into operation just before its deployment into Iraq. A Division Artillery LAR, who had been in country less than twelve hours, replaced a Firefinder Radar for the Tiger Brigade: it was instrumental in the initial success of the Brigade's assault. In addition to the logistics support they provided to the Army and other allied forces, the LAR furnished CECOM Headquarters with up-to-date information on logistical problems and the readiness of CECOM-supported systems. Then, when the Ground War ended and U.S. forces began to redeploy, the LAR remained behind to support the various clean-up operations involved in extracting CECOM equipment from the theater.

LAR Instructing Antenna Installation, Saudi Arabia

Contractors also played a vital role in the Gulf War. Technical assistance from contractors became necessary in cases involving very recently developed systems on which the effects of the desert (such as the intense heat) were not yet fully understood. CECOM in many cases planned on developing a support capability within the organization but could not do so before the system was sent to the Gulf. In other cases, especially with older items, CECOM no longer had the ability to maintain them. Contractors provided the necessary support.

Batteries represented a huge challenge for CECOM during operations. Wartime demands surpassed peacetime stocks. Batteries quickly died, due largely to the intense desert heat. Unfortunately, nearly every item in CECOM's inventory required numerous batteries. Battery producers were instructed to work around the clock by the time the Air Campaign started in January 1991. This continued until the conclusion of the ground war in early March. Different pieces of equipment, such as radios and night vision devices, demanded different types of batteries. Maintaining stock and ensuring that the right equipment received the right battery became a logistical concern for CECOM. CECOM decided to push shipments into the theater to a single control point for distribution, rather than filling individual requisitions as they were received.

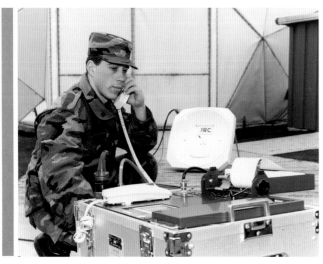

**CECOM personnel operating new equipment
that was to be deployed to Saudi Arabia**

When supplies of the BA-3517 battery were critically short, a CECOM production engineer developed a cable that allowed the M8 chemical alarm to use vehicular power sources. Federal Prison Industries fabricated 10,000 450-foot cables within five weeks; a CECOM industrial specialist supervised the operation; a Department of Defense civilian delivered connectors to the prison in his own automobile. In January 1991, shortly before the start of Desert Storm, engineers of the Center for C3 Systems designed a cable that permitted use of vehicle batteries to power Global Positioning System receivers. Volunteers from the Center and the Concurrent Engineering Directorate, working twelve to fourteen hours a day, assembled nearly 800 cables in five days, mostly in the Directorate's manufacturing technology facility. They shipped the first lot on 18 January, just two days after receiving the tasking and just fourteen hours after setting up their production line. In less than two weeks, responding to a request of the 11th Transportation Battalion, civilians of the Center for C3 Systems modified, tested, and shipped seventy-seven DH-132 combat crew helmets to permit wireless communication between the deck and bridge crews of the Lighter Air Cushioned Vehicle LACV-30. The battalion's supply officer compared the system to one used on Navy aircraft carriers: "[What] would have cost us about $1.8 million, cost us nothing." When CECOM software engineers discovered a bug in the firmware of the Quickfix Improved Communications Processor that corrupted the data exchanged with Trail-

blazer (a patently dangerous situation), they engineered and verified an emergency correction of the problem. One of them then traveled to Saudi Arabia to reprogram Quickfix circuits on site, using a reprogramming fixture developed by civilians of the CECOM Center for Software Engineering.

An employee of the Center for Signals Warfare (CSW), John B. Mitchell, developed a prototype signal analysis tool — a laptop computer and a standard spectrum co-processor card with appropriate software (generic and special purpose signal processing algorithms) - for the use in laboratory research projects. Seeing that the system, known as the Fast Response Intelligence Analyst Resource (FRIAR), could provide a plug-in/plug-out signal identification capability for tactical intercept systems, CSW technicians assembled ten systems using on-hand components and fielded them to the 24th Infantry, the 101st Airborne, and the 82d Airborne Divisions. FRIAR gave signal analysts in Saudi Arabia an easy-to-use automated capability for examining and identifying unknown signals and developing appropriate exploitation algorithms. The Air Force, Navy, and National Security Agency also deployed FRIAR.

Dr. Alan Tarbell, Chief of the EW/RSTA Center's Radar Division, determined that Firefinder, developed in the 1970s to detect artillery and mortar fire, had the range and could be modified to locate the launch sites also of Scud missiles. The Center had the modification, consisting of a tape cassette and a single sheet of instruction, in place well before the start of the Ground War. Addressing problems of helicopter safety in night operations, Edwin W. Wentworth III of the Center for Night Vision and Electro Optics (CNVEO) invented the Terrain Perception Enhancement Kit, which used infrared aiming lights to warn night-vision-goggled pilots of approaching dunes. Another CNVEO scientist, Henry C. (Budd) Croley, invented the Budd Lite, which troops on the ground found indispensable in night time operations for marking vehicles, landing zones, and roadways, and also for reading maps.

**Engineers demonstrate the cable tester they designed
and fabricated over the course of one weekend for use
by troops in Operation Desert Storm, February 1991**

Whether modifying software, procuring hardware, designing and producing special kinds of cables, accelerating fieldings, or rushing delivery to the theater, people at CECOM astonished themselves with the realization that, when circumstances dictated, they could do things in a fraction of the time they normally required. Many of the things they did on a quick reaction basis were very much beyond the ordinary. On the advice of the Navy, the Army had purchased Landing Craft Utility (LCU) for use in resupplying posts that had limited dockside facilities. On or about 10 December 1990, the contractor charged with outfitting the vessels put in an urgent call to CECOM for help: six of the boats, needed for operations in Southwest Asia and Panama, were to set sail on 21 December. The four-man crew CECOM sent to troubleshoot, test, repair, and certify the boats' communications systems had only nine days to get the water-crafts' systems in operation, and then to teach their first mates how to operate their teletypewriters, cryptology equipment, and radios. The boats departed for their destinations on schedule. Another example: one Friday afternoon in early February 1991, the Center for C3 Systems learned that soldiers in Saudi Arabia needed a way to test 26-pair cables. Over the weekend, Salvatore Romano and Martin Rosenzweig designed and built a hand-held test set that allowed soldiers to determine in well under a minute whether a cable was good or bad and which of its twenty-six pairs were open or shorted. The RDEC Prototype Fabrication Facility assembled twenty-six sets, using parts acquired in part from the local Radio Shack. Said Rosenzweig: "We don't mess around, as long as a Radio Shack is open."

The DoD, the Army, and CECOM learned many lessons during Operations Desert Shield and Desert Storm. Although the Gulf War was viewed as an overwhelming success for the nation, the experience demonstrated the undeniable need for enhanced communications and more integration on the battlefield, along with a better logistics infrastructure. These lessons became the impetus that shifted military strategy towards one that emphasized information dominance over brute force.[195]

The Monmouth Message

Vol. 48-No. 11   March 15, 1991

Home!

The Fort Monmouth community awaits the return of the 513th MI group, returning from Saudi Arabia, 6 April 1991

A Decade Of Realignment and Digitization - 1990s

The missions of CECOM and related Fort Monmouth organizations acquired enhanced significance in the 1990s when the Army Chief of Staff defined the Army's role in the new world order and identified requirements for decisive victory: to own the spectrum, to own the night, to know the enemy, and to digitize the battlefield.

Despite the important role it played in supporting these requirements, CECOM's worldwide civilian workforce fell from 7,375 to 6,501 during the 30 September 1990 to 30 September 1995 period, while its military strength dropped from 1,035 to 555. During the same time, the number of civilians assigned to all organizations at Fort Monmouth, including CECOM, fell from 7,732 to 6,385; the number of military fell from 1,826 to 761.

The Defense Management Review recommended consolidation of four AMC organizations that performed missions associated with Test, Measurement, and Diagnostic Equipment (TMDE). Implementation of this recommendation entailed the movement of the TMDE Product Manager and associated support personnel from Fort Monmouth to Huntsville, Alabama (Redstone Arsenal). In addition to the three military and twenty civilian positions of the PM Office, eighty-four civilian resident matrix support spaces were transferred from CECOM to the new Army TMDE Activity and fifteen civilian non-resident matrix support spaces were reassigned to the U.S. Army Missile Command (MICOM), for a total loss to Fort Monmouth of three military and 119 civilian personnel. The Department of the Army approved the transfer in February 1991, with an effective date of 14 June 1991. On that date, the three military personnel and two civilian technical personnel of PM TMDE transferred to the Army TMDE Activity at Huntsville. The remaining civilians of the TMDE core, who did not wish to transfer with their functions, found other jobs in CECOM, as did all collocated and non-collocated matrix support personnel. AMC formally established the Army TMDE Activity by Permanent Orders 41-2, dated 16 May 1991.[196]

Although the size of the force it supported decreased during this time, CECOM experienced little if any reduction in its workload. This situation challenged leadership first to find ways of reducing the civilian workforce without resorting to involuntary separations and then to accomplish its mission with the remaining personnel without sacrificing quality or service to the customer.

CECOM's leadership met the first of these challenges by imposing strict hiring freezes, offering incentives for voluntary

early retirement or separation, and reassigning employees from eliminated positions to vacant positions of higher priority. It addressed the second challenge, initially, through a large-scale reorganization that focused on vertical integration (the development of multi-functional mechanisms for the management of weapon systems from cradle to grave) and through the development of a workforce committed to the principles of Total Quality Management. Subsequently, the command focused its work on the objectives of the total force, as defined by Department of the Army, and on the "core competencies" of the Army Materiel Command, namely technology generation and application, acquisition excellence, and logistics power projection.[197]

CECOM managed half the Army's Advanced Technology Demonstrations (ATD) in 1994 and 1995 and participated in a large percentage of the balance. Of the seven ATDs the Army completed in FY95, CECOM managed five: Common Ground Station, Survivable Adaptive Systems, Radar Deception and Jamming, Multisensor Aided Targeting - Air, and Close-in Manportable Mine Detector. Achieved in collaboration with Program Executive Offices and users, these ATDs focused on speeding the insertion of emerging technologies into operational systems.

Altogether, CECOM owned nearly a quarter of all the Army's approved Science and Technology Objectives (STO). It managed sixteen of the thirty-five Advanced Concepts and Technology II programs the Army awarded in FY95. It also had the Army's most active Independent Research and Development (IR&D) program and one of its largest Small Business Innovative Research (SBIR) programs.

In administering its IR&D program, CECOM focused on outreach and publicity. In FY95, the command hosted eight Technology Interchange Meetings, representing forty-one IR&D projects valued at $21.9 million, and consecrated the program one Level II Advance Planning Briefing for Industry. IR&D successes that year included the Embedded GPS/Inertial System, which involved the integration of a Global Positioning System (GPS) receiver module with a ring laser gyro inertial navigation system in a single chassis, and the SPANet Asynchronous Transfer Mode (ATM) Switch. The latter, installed in five Army, Navy, and Air Force locations, were connected via satellite and terrestrial links to form a test network known as the Joint Advanced Demonstration Environment.

CECOM placed $10.3 million of "seed money" on seventy-three SBIR contracts in FY95. SBIR products delivered that year included the Soldier's Personal Adaptive Monitor - a miniature, low-power, helmet-mounted display for the dismounted soldier featuring a full 640x480 monocle that could be adapted to either eye and a VGA-quality resolution that permitted uncompromised display of computer graphics - and a Collaborative Scenario Generator for an Integrated Decision Aids Demonstration System. The latter enhanced methods of generating command and control scenarios to support tactical mission planning for law enforcement agencies and the military. A high-priority of the Battle Command Battle Laboratory for use at the Command and General Staff College, it was co-funded by the Office of National Drug Control Policy to support planning and operations in areas of high-density drug traffic.

The following list of eighteen ATDs and the thirty-four additional STOs that CECOM owned in 1995 illustrated the scope of the command's R&D mission. In the first list, ATDs are grouped according to the objectives of the Army's strategic plan. The remaining STOs occur in the order in which they were listed in the Army Science and Technology Master Plan.

CECOM ATDs (1995): Protect the Force, Advanced Image Intensification, Close-in Man Portable Mine Detector, Off Route Smart Mine Clearance, Remote Sentry, Bi-Static Radar for Weapons Location, Vehicle Mounted Mine Detector, Multispectral Countermeasures, Dominate Maneuver, Target Acquisition, Hunter Sensor Suite, Execute Precision Strike, Multisensor Aided Targeting – Air, Radar Deception and Jamming, Common Ground Station, Air/Land Enhanced Reconnaissance and Targeting, Win the Information War, Digital Battlefield Communications, Combined Arms Command and Control, Survivable Adaptive Systems, Battlespace Command

and Control, Project and Sustain Combat, and Total Distribution.

CECOM STOs (1995): Advanced Helicopter Pilotage, Joint Speakeasy - Multiband Multimode Radio, Improved Spectrum Efficiency Modeling and Simulation, Aviation Integration into the Digitized Battlefield, Range Extension, Communications Countermeasure Demonstration, Intelligence Fusion Demonstration (Completed in FY95), Orion (wide bandwidth SIGINT electronic support package on a short-range UAV), Tactical Intelligence Data Fusion, Multimission UAV Payload, Digital Communications Electronic Attack, Rapid Force Projection Initiative, Aerial Scout Sensors Integration, Electronic Integrated Sensor Suite for Air Defense, Low Cost, Low Observable Multispectral Technology, Mine Hunter/Killer, Tactical Electric Power Generation, Integrated Photonic Subsystem, Information Warfare On-the-Move, Networking and Protocols, Battle Planning, Advanced Optics and Display Applications, Modular High Density, High Performance Processor Technology, Soldier Individual Power Source, Electronic Warfare Processing Techniques (completed in FY95), Non-Communications Electronics Support Measures/Electronic Countermeasures Techniques, Diverse Wavelength Sources (completed in FY95), Advanced Electro-Optic and Infrared Countermeasures, Advanced Radio Frequency Countermeasures, Smart Focal Plane Arrays, Advanced Focal Plane Arrays, Multi-Wavelength Multi-Function Laser, Advanced Aided Target Recognition Processing and Algorithm Exploitation (completed in FY95) and Electronic Terrain Board.

The scope of the command's R&D mission is further reflected in the number and variety of specialized, world-class research facilities the RDEC operated in 1994 and 1995, as displayed in the following partial list: CECOM R&D Facilities, Command and Control Laboratory, C2 Concepts Laboratory, C2 Prototyping Laboratory, Interactive Speech Technology Laboratory, Navigation Laboratory, System Testbed for Avionics Research, Advanced C2 Decision Support Aids and Interoperability

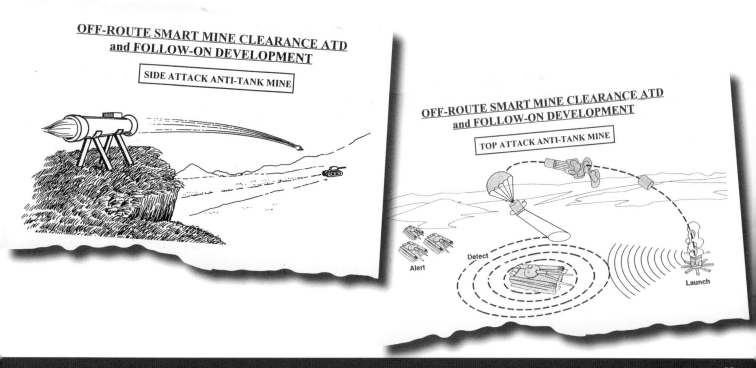

OFF-ROUTE SMART MINE CLEARANCE ATD and FOLLOW-ON DEVELOPMENT

SIDE ATTACK ANTI-TANK MINE

OFF-ROUTE SMART MINE CLEARANCE ATD and FOLLOW-ON DEVELOPMENT

TOP ATTACK ANTI-TANK MINE

Alert   Detect   Launch

Laboratory, Communications Systems Design Center, High Speed Communications Test Facility, MSE Support Facility, Modeling and Simulation Facility, Development Engineering Facility, Digital Integrated Laboratory, Software Prototyping and Integration Laboratory, Command, Control, and Communications Integration Laboratory, Tactical Data Fusion Laboratory, Simulation and Modeling Laboratory, Advanced Sensor Evaluation Facility, Electronic Warfare Survivability Integration Laboratory, Local Area Communications Integration Laboratory, Commercial Communications Technology Laboratory, Joint Advanced Demonstration Environment Testbed, Army Battle Command System, Interoperability Laboratory, Antenna Evaluation Facility, IEW Technology Assessment Center, Secure, Instrumented 5000-Meter Laser Test Range, Sensor Evaluation Test Range, Minelanes Laboratory, Physical Security Equipment Laboratory, Infrared Focal Plane Array Microfactory, Radar Target Measurement System (mobile), Interactive Speech Technology Laboratory, Avionics Validation and Test Facility, and Environmental Test Facility.

CECOM established the Commercial Communications Technology Laboratory to assess the military utility of commercial innovations and to ensure the timely, successful, and cost-effective insertion of commercial technology in military systems. Industry used the Laboratory to demonstrate to the Army various products and technologies, including (in 1995) Digital Trunked Land Mobile Radio, Direct Broadcast Satellite, ATM software for video teleconferencing and collaborative planning, Broadband Code Division Multiple Access, handheld FM radios, and ATM switches and multiplexers.

For early proof and testing of its airborne innovations, Team Fort Monmouth relied in large measure on the System Test Bed for Avionics Research (STAR, a uniquely configured UH-60 Blackhawk helicopter) and other facilities of the Command, Control and Systems Integration Directorate (C2SID) Airborne Engineering Evaluation Support Branch. This organization participated significantly in developing prototypes of the Multi-Sensor Aided Targeting - Air system, the Aided Pilotage System, the Personnel Locator System, the Soldier 911 system, the Doppler Embedded GPS Navigation System, the Radar Deception and Jamming ATD, and the Team Antenna.

In FY95, CECOM completed construction of a 20,000 square foot facility for its Space and Terrestrial Communications Directorate to serve as a Defense Satellite Communications Systems (DSCS) Operations Center and focal point for analysis and evaluation of DSCS network control. Also in FY95, CECOM completed a $1 million renovation of nearly 10,000 square feet of space in the Myer Center to accommodate Computer Engineering Laboratory Five.[198]

BASE REALIGNMENT AND CLOSURE (BRAC)

During the Cold War, the Soviet Union posed the primary threat to America's national security. U.S. military equipment, doctrine and training centered on effectively dealing with that threat. While the conclusion of the Cold War meant the threat was diminished, the infrastructure had not adjusted accordingly. This imbalance of forces and threats led many to believe that more bases existed than necessary. The first round of Base

Realignment and Closure (BRAC) consequently occurred in 1988. Fort Monmouth was impacted by 1991 with the decision to move the Electronics Technology and Devices Lab (ETDL) of the Army Research Lab out of Fort Monmouth to Adelphi, Maryland.

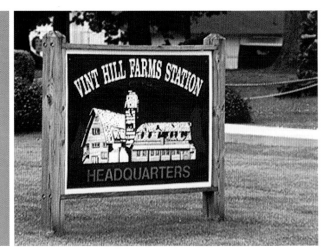

Vint Hill Farms Station in Virginia

During FY93 and into FY94, CECOM prepared detailed plans to implement three BRAC '93 decisions: realign activities at Fort Monmouth, disestablish the Belvoir RDEC, and close Vint Hill Farms Station (VHFS). The BRAC Division of the Program Analysis and Evaluation Directorate submitted final iterations of these three plans to AMC and Department of the Army on 29 October 1993. Legislation required the government to complete BRAC '93 actions by the end of FY98. However, in response to entreaties of the Clinton administration, the Department of Defense advanced the deadline to 30 September 1997.

In connection with the disestablishment of the Belvoir RDEC, BRAC '93 eliminated five business areas, moved five to the Tank-Automotive Command (TACOM), and realigned six "in place" to CECOM (Countermines, Low Cost Low Observables, Physical Security, Battlefield Deception, Electric Power, and Environmental Controls). By virtue of a memorandum of agreement between CECOM, TACOM, and the Aviation and Troop Command (ATCOM), CECOM took operational control of its six new business areas and their personnel on 1 February 1994. Department of Army formally completed the disestablishments and realignments on 30 September 1994, three years before the deadline.

The realignment of Fort Monmouth involved the relocation of the Chaplain Center and School from Main Post to Fort Jackson, South Carolina; the relocation of CECOM activities from the General Services Administration (GSA)-leased CECOM Office Building (COB) in Tinton Falls to Main Post; the closing and disposal of the Evans Area and the relocation of its occupants to Main Post and the Charles Wood Area; and the disposal of excess housing in Olmstead Gardens. These actions had no effect on manpower authorizations, but did entail migration of personnel. An unrelated action - the relocation of the 513th Military Intelligence Brigade to Fort Gordon (494 military and six civilian employees) - vacated space on the Main Post for CECOM administrative personnel in Tinton Falls.

The closing of Vint Hill Farms Station entailed the transfer to other locations of 762 civilian and 399 military positions and the elimination of 240 civilian positions. Organizations and personnel transferred to Fort Monmouth included the on-site components of CECOM's Intelligence and Electronic Warfare Directorate, C3I Acquisition Center, and Legal Office; on-site components of the Program Executive Office for Intelligence and Electronic Warfare, including the Program Manager for Signals Warfare; and all the Intelligence Material Management Center except the missions identified for transfer to Tobyhanna Army Depot (the wholesale supply, repair, and maintenance missions). The Intelligence and Security Command (INSCOM) was to transfer its VHFS entities -the Mission Support Activity and the Force Modernization Activity-to Fort Belvoir. Fort Meade was to receive functions and personnel of the Operational Security Evaluation Group. Other activities, such as the Technical Contract Management Office, the TMDE Support Center, and VHFS contingents of the Health Services Command were to be reassigned. Of the 240 positions slated for elimination, seven belonged to the Information Systems Command, ten to PEO IEW, eighty-six to CECOM, and the remainder, 137, to the VHFS Garrison. To meet the accelerated closing date (30 September 1997), CECOM entered into a memorandum of agreement with the Industrial Operations Command to effect the early transfer of IMMC's unique maintenance and repair mission to Tobyhanna and negotiated with the Corps of Engineers to accelerate the start of new construction for the Intelligence Materiel Management Center (IMMC), Intelligence, Electronic Warfare Directorate (IEWD), and PEO elements that were to move from Virginia to Fort Monmouth. The LRC took its first tentative step toward relocating the IMMC in late 1995 with the establishment of the Signal Intelligence Division (Provisional).

The closing of VHFS and the Evans Area and the realignment at Fort Monmouth involved expenditures, estimated at nearly $70 million, for construction and renovation of facilities. This included construction of a new Chaplain Center at Fort Jackson (Project 42280, $8 million), renovation of buildings in the 1200 Area for CECOM administrative and readiness organizations (Project 42708, $22 million), and renovation of Building 2700 (the Myer Center) for research and development organizations, including all the pieces of PEO IEW (Project 42683, $4 million). Projects 42708 and 42683 included "Information Management Area" (IMA) expenses of $6.5 million.

To accommodate IMMC activities at Fort Monmouth, the VHFS BRAC Implementation Plan called for the renovation of Building 1201 (Project 42681) to provide for a "restricted controlled" site with two Special Access Program areas, a Sensitive Compartmented Information (SCI) Facility, and class A vaults for SCI material. Renovations were to include elevator support and installation of an Intrusion Detection System. The plan also called for construction of a pre-engineered building for use in IMMC equipment fabrication programs. It estimated the total cost of construction for IMMC at $5.4 million.

For relocating the IEWD from VHFS and Evans to Main Post (Project 42682), the BRAC Implementation Plans stipulated construction of a building with SCI and Special Access Program areas, computer labs, common support areas, and functional areas for graphic arts and classified document destruction; a Guardrail V test bed using existing sixty-five foot high equipment towers; remote non-SCI laboratory facilities to include a high-bay shop; and a pre-engineered storage building. The estimated cost of construction, to include utilities, uninterruptible power supply, emergency generators, fire protection,

Chaplain Museum

an integrated access control system, site improvements, and perimeter fencing totaled $19.2 million.

Apart from the inconvenience some employees might experience in moving their place of work from one location to another within the established boundaries of the installation and apart from the congestion associated with the consolidation of the work force in a smaller geographic area, BRAC '93 had no immediately identifiable impact upon CECOM employees at Fort Monmouth. The displacements they might suffer with internal realignment were for them a small price to pay for release from the initial DoD proposal, although not adopted, which recommended realigning the command to Rock Island Arsenal, Il. CECOM civilian employees at Vint Hill Farms Station were not nearly as fortunate. To them, the prospect of being transplanted from the Elysian fields of "The Farm" to the urban milieu of the Jersey shore was, at least at first glance, every bit as distressing as the prospect of removal to Rock Island Arsenal had been to workers at Fort Monmouth. Even worse for many at Vint Hill Farms - garrison employees primarily, but also employees unwilling to relocate - was the prospect of being left without a job of any kind.

Rock Island Arsenal, Illinois

CECOM hoped to reduce the number of involuntary separations through attrition, spurred by the Voluntary Early Retirement Program/Voluntary Separation Incentive Program (VERA/VSIP) incentives and intensive out-placement assistance. For the latter, as detailed in the BRAC Implementation Plan, CECOM employed the services of the local Transition Assistance Office, the Department of Defense Civilian Assistance and Re-Employment and Priority Placement Programs, the Defense Outplacement Referral System, the Office of Personnel Management Displaced Employee Program, the Interagency Placement Assistance Program, and job fairs. There were also state programs, not identified in the plan, that assisted displaced workers. Involuntary separations - the "Reduction in Force"- would occur in FY97; the number of appropriated fund personnel separated involuntarily would depend, of course, on the success of the attrition efforts.

CECOM's BRAC managers expected many employees who were subject to transfer of function to accept reassignment to Fort Monmouth: an unusually large percentage of them responded favorably during a non-binding survey. However,

many of them had little effective choice: at a time of retrenchment, there were few opportunities in other agencies at comparable grades for employees with the highly technical, specialized skills of the VHFS work force. Transfer of function was a bitter pill for these employees to swallow and the command provided considerable assistance to them, to include, for example, assistance in exploring other employment options.

A "Transition Team" was created at Vint Hill Farms Station to assist affected employees, contractors, and family members in the adjustments associated with their "transition" to new jobs or new homes. The team, chaired by the Army Community Services Officer, had about forty members representing all the on-post organizations. During FY94 it organized an installation-wide survey in coordination with the Civilian Personnel Office and the Virginia Employment Commission Rapid Response Office to identify the transition needs of soldiers, civilians and contractors. It co-sponsored a large job fair and staged a relocation fair with subject-matter experts from Fort Monmouth, Fort Gordon, and Tobyhanna Army Depot. Though smaller than the fairs of earlier years, the Fourth Annual Job Fair (July 1994) emphasized opportunities for displaced workers in such fields as telecommunications, computers, engineering, logistics, equipment operation, and law enforcement. The Transition Team also orchestrated efforts to obtain for the work force the benefits of the state's Economic Dislocated Workers Adjustment Assistance (EDWAA) retraining program. A special effort at cooperation between the state and Department of Defense made Vint Hill employees eligible for early registration in this program, and on 2 June 1994 the Director of the Governor's Committee on Employment and Training personally informed them of the program's skills assessment, training, and placement services. The Transition Center had 39,000 client contacts in FY94, up 12,000 from FY93. Programs it offered in addition to relocation briefings included career planning services and workshops on "Financial Planning for Transition," "Workplace Stress," "Change Management," and "Focus on the Future."

The Civilian Personnel Office at Vint Hill, a component of the Garrison, offered other kinds of assistance. On 23 November 1993, it conducted the first Defense Outplacement Referral System (DORS) registration for Vint Hill's civilian employees. Subsequently, it organized a comprehensive one-on-one counseling program to address concerns and questions regarding transfer of function; outplacement through the Priority Placement Program, the DORS, and the Interagency Placement Program; separation benefits; continuation of health insurance; and re-employment eligibility. It framed a BRAC Personnel Strategy Plan in which it set a schedule for communicating BRAC-related information to the work force. And in September in anticipation of a reduction-in-force, it surveyed eligible employees to ascertain the number that would be swayed by voluntary separation incentive pay to accept early retirement.

CECOM had already begun the move from Tinton Falls. Between June and October 1993, the Command Group, the Public Affairs Office, and the Program Analysis and Evaluation Directorate moved out of the CECOM Office Building (COB) into Russel Hall. Though Russel Hall was but a way station pending relocation to a permanent home in Watters Hall (Building

1207), the Command Group deemed the move wise both as a sign of good faith and as a measure that could help the installation avert scrutiny in BRAC '95. Meanwhile, in the wake of the move to Russel Hall, BRAC Division personnel launched correspondence through AMC to effect a reduction in GSA rental charges for the space vacated in COB. Concurrently, realizing the impossibility of completing the Fort Monmouth realignment by September, 1997 if the Chaplain Center remained on post until its scheduled moving date (January 1997), CECOM prevailed upon the Department of the Army to advance the Chaplains' removal by a full year (to January 1996), as proposed by TRADOC.

By March 1996, CECOM had begun the construction and renovation of all the facilities required to accommodate incoming employees. Ground-breaking for the largest of new facilities - the $14 million, 90,000 square-foot complex to accommodate three hundred IEWD employees on Main Post, occurred in September, 1995. The LRC initiated the first BRAC-related relocation of personnel from Vint Hill to Fort Monmouth in summer 1995 with establishment of the SIGINT Division (Provisional). By the end of the fiscal year, seven employees had relocated to this new Division from Vint Hill's Intelligence Materiel Management Center. Meanwhile, in August 1995, the Assistant Secretary of Defense approved the transfer of 246 excess housing units in the Charles Wood Area (Olmstead Gardens) to the Department of the Navy.

The final decisions of BRAC '95 required the disestablishment of the Aviation and Troop Command (ATCOM) in St. Louis and the transfer of 178 of its personnel (including eight military) to Fort Monmouth. These were the people who were responsible for the acquisition and logistics support of aviation-related communications-electronics materiel. With the disestablishment of ATCOM, CECOM also acquired the Program Manger for Electric Power and the Weapon System Manager for Physical Security, both which remained in place, however, at Fort Belvoir.

DOD's BRAC '95 Joint Cross-Service Working Group on Laboratories, recognizing Fort Monmouth's quality and the achievements of Fort Monmouth activities, recommended Fort Monmouth as the site for DOD's Center for Command, Control, Communications, Computer, and Intelligence. This was the basis for the DOD recommendation to relocate Rome Laboratory, an Air Force organization, to Fort Monmouth and Hanscom AFB. However, the BRAC's final decision rejected this recommendation, notwithstanding a GAO report that recommended the relocation as being both a more efficient use of facilities and cheaper for the taxpayer. The Department of Defense also recommended relocation of the Military Traffic Management Command (MTMC) from Bayonne, New Jersey, to Fort Monmouth. The BRAC, however, gave the MTMC a choice of installations, including Fort Monmouth, onto which it might eventually relocate. The disposal of additional excess housing in the Howard Commons area of Fort Monmouth was also ordered.

It's over
CECOM building empty, end of an era

by Debbie Sheehan
Public Affairs Office

"When you have 5,000 people working together in one place, it's like a small town," said Mary Maurer, who held the position of CECOM building manager until last month. "It was a small town. It had everything; all the unique people you would encounter in a small town."

Maurer walked down a deserted hallway in the building and pointed to an area the size of a ball field. "This was the last area to be moved and funny thing, they were the first to be moved in." Fifty chairs sat in the middle of the room, work stations on the floor in pieces, odd pieces of furniture were here and there.

"I give the movers a lot of credit. They had a system where they would move things in a way that would allow people to leave work on a Friday and be in a new space and up and rolling on a Monday. It wasn't easy, things didn't always fit, they really cooperated and worked together," Maurer said.

Perhaps the biggest impact of the Base Re-alignment and Closure Act here was the decision to move out of the leased CECOM Building and relocate to the main post. "It was 725,000 square feet, built and designed for the Army. We were the sole tenant and, except for the cafeteria, leased the whole building," said Maurer. "When it was built, it was to consolidate and bring workers from Philadelphia and older buildings on post to one new "state of the art" facility.

"Here it was easy to relate with different activities. You would see people in hallways, in passing, they were accessible. If you had a problem, you could just walk around the building and talk to someone. If you needed a signature, you could just go to that office and get it."

She remembered all the high level visitors to the facility. "People could see that they were a part of something important. They would see the soldiers and generals on a daily basis," she said.

Maurer said that different commanding generals had different requests. "One was very tall and could see if there was dust on a cabinet or if a cigarette butt was stuck somewhere outside. Another would call when we had high level visitors coming and request that the dandelions be removed from the lawn.

"There was always something going on. This was never a boring job. I was always learning; laughing about something about the job. It was fun."

One thing that was not fun about the CECOM Building was the parking lot.

"There was only room for 3,000 cars and we had 5,000 working here. At one point we started towing cars," she said. "I would have to ride with the tow truck drivers and iden-

END OF AN ERA, continued on page 2

photo by Gregory Brower

Charlie Pruitt (left), foreman for the Base Realignment and Closing moves dictated by Congress, sees some of the last boxes leaving the CECOM Office Building before its recent closure.

CECOM expended a tremendous amount of effort during each round of BRAC in order to stay competitive while complying with the decisions reached by the commission. Unlike many other installations, the 93 and 95 BRAC rounds resulted in gains for Fort Monmouth, which meant additional work. The command met the logistical challenge to physically relocate so many employees while avoiding any interruption to its mission.[199]

RESHAPE II

During FY94, it became apparent that the anticipated Program Budget Guidance for FY96 would not support the existing command structure. The CECOM Executive Advisory Committee (CEAC) had already determined that the basic command structure was correct for accomplishing the mission: therefore, the Resource, Self Help, Affordability, Planning Effort (RESHAPE) II Task Force focused strategies to avoid involuntary personnel separations while achieving affordability in directly funded accounts with major shortfalls and reimbursable accounts with eroded customer bases.

In August 1994, MG Otto J. Guenther, CECOM Commander Jul 1992 - Jan 1995, informed the work force that funding reductions anticipated for FY96 required a new round of "reshaping" initiatives. His announcement indicated the need to cut the work force by another 500 - 600 employees. To lessen the potentially adverse impact of this cut, CECOM again requested approval for Voluntary Early Retirement Authority (VERA) and Voluntary Separation Incentive Pay (VSIP). Guenther made the announcement by video tape "to reduce rumors and provide employees with plenty of time to make their plans while continuing to accomplish their mission." To assess the interest of the work force in the VERA/VSIP incentives, the Personnel and Training Directorate distributed more than 1,900 survey questionnaires to employees who were eligible for optional or early retirement. At the end of FY94, one in seven CECOM employees (fifteen percent of the work force) was eligible for optional retirement.

Guenther's message to the employees is quoted, verbatim, below:

Last month, I commissioned a reshape team. I tasked the team to begin examining command-wide functions that could be reduced or eliminated and how future resources will be allocated to each Command Executive Advisory Council principal. I also tasked the team to develop strategies to reduce the impact of anticipated reductions and to develop timelines to meet reshaping challenges.

The challenge that we face is an issue of affordability in fiscal year 1996 and beyond. We are projecting shortfalls in both our direct and customer funding that will require reductions in our work force beyond what we expect from attrition.

Based on what we currently know, we would have to reduce our on-board strength by approximately five hundred to six hundred employees by the beginning of fiscal year 1996 to maintain an affordable work force through the end of the fiscal year.

Please keep in mind, these numbers are based on information we have today. Changes in the methods of calculating strength figures at the DOD level, as well as other unknown events, could cause these numbers to change.

There are several key reshaping events that are now taking place. Having prioritized our missions and functions and having allocated our resources, we will then develop a "new reshaping organizational concept." The next step is to request approval from AMC headquarters for VERA and VSIP. We will also be determining our options for approving VSIP applications if we're required to limit eligibility to only employees whose resignation or retirement would directly save an employee facing separation. The VERA/VSIP window would then open for a specific period of time; right now we project late this calendar year or early next year. Based on the results and other efforts to minimize the impact of the reductions, we would then be in a position to know if a RIF would be necessary.

During the short VERA/VSIP window, 19 December 1994 through 13 January 1995, CECOM received and accepted 587 applications - a sufficient number to avoid involuntary separations. MG Gerard P. Brohm, CECOM Commander Jan 1995 - Sep 1998, communicated the good news to CECOM employees by letter dated 27 January 1995. Thereupon, the task at hand was to "cross level" the work force: to reassign or promote employees from abolished positions to valid vacancies. The Acting Director of Personnel and Training assured employees that there would be no reductions in grade or pay as a result of cross-leveling.

Reshape II achieved its objectives partly through efficiency initiatives. The LRC centralized the Computer Aided Acquisition and Logistics Support (CALS) Office in headquarters; the Readiness Directorate brought in outside work to increase the number of reimbursable positions from 118 to 132; Materiel Management employed mergers to eliminate two branches; the IMMC realigned seven directorates into five divisions; and LMD eliminated one commodity branch and one support division branch. The RDEC had already accomplished a major restructuring (during Reshape I) in which it reduced fourteen "boxes" to six on the organization chart. Reshape in other organizations generally involved elimination of missions and/or reduction or transfer of services. The result was a reduction of the work force to an "affordable" 6,560 spaces.

Also as outcomes of Reshape II, the first-line supervisor to employee ratio increased from 1:10 to nearly 1:12 as of the end of FY95; the overall supervisor to employee ratio grew from 1:7 to 1:8.5; the clerical support ratio increased from 1:10.7 to almost 1:13; and the command came in under the AMC-specified FY97 "cap" for high grades. The VERA/VSIP window opened as scheduled in December 1995 and January 1995, and CECOM once again averted an involuntary "Reduction in Force."[200]

TEAM C4IEWS FORMS

Amidst the BRAC realignments directed by higher headquarters, CECOM began a strategic alignment of its own in 1993 when it formed TEAM C4IEWS (Command, Control, Communications, Computers, Intelligence, Electronic Warfare and Sensors). Several organizations comprised this partnership: CECOM, PEO C3S (Command, Control and Communications Systems), PEO EIS (Enterprise Information Systems), PEO IEW&S (Intelligence, Electronic Warfare and Sensors), ARL (Army Research Laboratory), and DISA (Defense Information Systems Agency). Although the names of some of these organizations changed through the years, their commitment to the partnership did not. The overarching goal of the partnership was best represented by its mission statement, which read: "We, the leaders of the above C4IEWS member organizations, commit to work together to support the vision of Fort Monmouth as a premier global Center of Excellence in developing and supporting superior C4IEWS systems and equipment as well as new architecture for strategic communications, automation and defense information infrastructure."

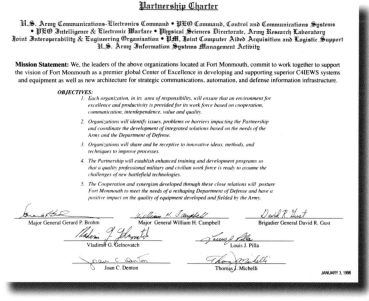

In essence, the signatory organizations agreed to look beyond organizational boundaries and work together to develop innovative integrated solutions for the Warfighter. The partnership managed to overcome organizational differences and better support the Soldier by formalizing their cooperation.

Two examples of this alliance were the Digitization of the Heavy Forces at Fort Hood, TX, a PEO C3T-led project supported by the rest of the partnership; and the creation of the Stryker Brigade Combat Team (SBCT) at Fort Lewis.[201]

Team C4IEWS would eventually be renamed Team C4ISR (Command, Control, Communications, Computers, Intelligence, Surveillances, and Reconnaissance).

TRAINING THE WORK FORCE

CECOM redoubled its efforts to promote the professional development of civilian employees in the 1990s. In FY93, the command increased its annual expenditures for training from $200 to $500 a person. In September 1992, General Guenther established the Senior Professional Development Committee, chaired by his civilian Deputy, to seek educational opportunities for CECOM employees and to oversee cross-training assignments for as many as fifteen employees a year in the GS-12 through GS-15 pay grades. In September 1993, the command inducted 150 employees, selected from a pool of more than 500 applicants, into its first Civilian Leader Development Program, a three-year training sequence that would qualify graduates to apply for AMC's new Senior Managers' Executive Development Program. The latter was designed to identify, through AMC-wide competition, outstanding employees with demonstrated executive abilities who were likely to become candidates for various senior-level and Senior Executive Service positions. Of the thirty-eight participants selected in 1993 for its first iteration, nine were from CECOM. Concurrently, in FY93, two Materiel Management Directorate employees were among the fifteen that Department of Army selected worldwide to participate in the Department of Defense Executive Leadership Development Program.

Every year civilians of the Research, Development and Engineering Center (RDEC) were selected for exceptional training opportunities. In FY89, for example, an RDEC computer scientist won an Army-sponsored research fellowship. An electronics engineer, a woman, was accepted for the Masters' Program in Electronic Warfare at the Naval Post Graduate School and was the first Army civilian of either gender to be admitted. Another woman with high potential enrolled for managerial training and developmental experience in Office of Personnel Management's (OPM) Women's Executive Leadership Program.

Members of the CECOM Civilian Leader Development Program exsecutive board were: George Mudd, Larry Smith, Richard E. Kelly, Ken Kociela, COL Norman K. Southerland, Thomas Sheehan, Ken Morgan, Eugene Bennett, and Wanda Nieves

Concurrently, CECOM worked with nearby college and university faculties to build training opportunities for existing and future employees. A close joint research relationship between computer scientists of the RDEC and a local college built the college's capabilities in software engineering to such an extent that it was able to establish a new PhD program in that discipline. Consequently, in FY89, the Army Materiel Command designated the RDEC to serve as mentor for thirteen software engineering interns. By FY91, at the instigation and with the help and encouragement of the RDEC, the same college established a curriculum leading to the MSEE-ME degree - the Master of Science in Electronic Engineering with concentration in Military Electronics. RDEC civilians also helped the University of Pennsylvania devise a Management of Technology program - a rigorous course of study to develop skills for creative leadership in the development of emerging technologies.

To foster the training needed by personnel of and applicants for the Army Acquisition Corps, the CECOM Commanding General appointed an Acquisition Manager and established an Executive Acquisition Council. The Council staged an exposition of educational opportunities to showcase the programs of local colleges and universities, including the MBA program offered on post by Monmouth College and the graduate-level courses in business and public administration that Farleigh Dickinson University offered, also on post. To qualify personnel of the acquisition work force for entry into the Acquisition Corps, the Council also initiated the development of a twenty-four credit certificate program at Brookdale Community College.

To compensate in some measure for the diminution of a military presence in the command, CECOM inaugurated at least three programs to give civilian employees knowledge of the military and experience with soldiers. Participants in two programs - SMERF (Subject Matter Experts Return to the Field) and DEFEWS (Design Engineers in Field Exercises with Soldiers) - lived, ate, slept, and worked with soldiers to acquire a more intimate understanding of their needs and wants. For civilians who had no opportunity to participate in these on-hands experiences, the Logistics and Maintenance Directorate developed a course of instruction on "Force Structure." It offered the course initially only to employees of the Logistics and Readiness Center but then, on popular demand, opened it to the entire Fort Monmouth/CECOM community under a "train the trainer" concept. Together, these programs helped keep the civilian work force focused on the CECOM Bottom Line - the Soldier.[202]

CECOM'S RESPONSIBILITY FOR INFORMATION TECHNOLOGY AND SUPPORT BROADENS

A single, integrated engineering organization was considered critical for coherent progress leading to the force of the future as information age technology began to blur the distinction among tactical, strategic, and sustaining base capabilities. To address the requirement, a Signal Organization and Mission Alignment (SOMA) study was conducted in order to determine the most efficient way to organize the Signal Corps' information management capabilities. All of the information management, acquisition, engineering, and procurement operations of the former Army Information Systems Command (ISC) were consequently assigned to CECOM. Through this reorganization, effective 1 October 1996, CECOM gained the Information Systems Engineering Command (ISEC) at Fort Huachuca, AZ. This added a total of nearly 1,600 civilian and 400 military personnel without relocation. Also as part of this reorganization, the Information Systems Management Agency

(ISMA), already located at Fort Monmouth, was realigned in place. ISMA began reporting to CECOM/AMC and not to ISC at Fort Huachuca. A year later, ISMA became part of the Systems Management Agency at Fort Monmouth (which would later be absorbed into PEO EIS).

In addition to these Army-directed organizational changes, the Army Materiel Command directed CECOM to take operational control and management oversight of the Army Missile Command's Logistics Systems Support Center (LSSC), St. Louis, MO, and the Industrial Operation's Command's Industrial Logistics Systems Center (ILSC), located in Letterkenny Army Depot, PA and Rock Island, IL. CECOM's operational control of the LSSC and ILSC did not involve any personnel relocation. CECOM furthermore acquired Software Development Centers at Fort Lee, VA, Fort Meade, MD, and the Information Systems Software Center (ISSC) at Fort Belvoir, VA.

Beyond the increase in personnel, these realignments represented a marked increase in mission. Prior to 1997, CECOM was focused primarily on the operations/tactical domain of the spectrum and the technologies, software, sensors and products needed within the battlespace. In 1997, however, CECOM gained responsibility for the infrastructure side of the spectrum. This meant the responsibility for executing IT infrastructure improvements across all Army posts, camps and stations. The realignments of 1997 gave CECOM responsibility for information technology across the full spectrum of operations, from the sustaining base to the battlespace.

Further 1997 reorganizations within the Army Materiel Command formally placed Tobyhanna Army Depot (TYAD) under the direct control of CECOM. Located in northeastern Pennsylvania, TYAD was the largest full-service communications-electronics maintenance facility in the Department of Defense with, at that time, more than 2,700 employees and fourteen forward operating locations located throughout the world.

Aerial view of Tobyhanna Army Depot

Tobyhanna, the DoD's recognized leader in electronics maintenance, was responsible for a wide array of products. The depot's primary specialties included engineering, maintenance and manufacturing services, systems integration, repair, overhaul, power projection and high tech training.

TYAD was on the winning side of many BRAC realignments. In addition to its work supporting Army communications-electronics systems, TYAD gained responsibility for depot-level maintenance on the guidance and control systems for the Maverick, Sparrow, and Sidewinder missiles used by the Navy and Air Force. The 1995 closure of the Sacramento Air Logistics Center shifted all Air Force ground communications equipment to TYAD. Forty percent of Tobyhanna's work supported the Air Force by 2002.[203]

TASK FORCE XXI, ADVANCED WARFIGHTING EXPERIMENT

The Task Force XXI Advanced Warfighting Experiment was the culmination of battlefield digitization efforts within the U.S. Army. A "real world" environment tested many of the systems conceived in the 1990s.

The brigade-sized task force consisted of two heavy battalions, one light infantry battalion, and a brigade support slice. Each of these exercises, held at the National Training Center at Fort Irwin, were designed not only to assess the technical aspects of these digitization efforts, but also to provide senior Army leaders with a sense of how these systems would perform in the hands of Soldiers actively engaged in combat operations.

The lessons learned from this experiment went a long way in determining the value of these systems in combat. While not every system involved met the goals of this exercise, the lessons learned from the experience proved invaluable in helping Army engineers and scientists better refine and improve these systems.[204]

The command's strategy in this effort involved partnerships with TRADOC, program executive offices, and industry; deployment of scientists and engineers to field new technical capabilities; emphasis on the application of technology (rather than its generation) to expedite the fielding of useful products; and establishment of "beta" sites with users to evaluate technologies and produce immediate improvements. On the road to Force XXI CECOM had or would have the assistance of two intermediaries: the "Battle Labs," with which the command field tested its concepts, and "Task Force XXI." Task Force XXI was a project to turn the 4th Infantry Division (formerly the 2d Armored Division) into a prototype of the Force XXI Army. As "Experimental Force" (EXFOR) for Force XXI, the 4th Infantry Division would test CECOM's architecture for the digital battlefield in brigade and division-level exercises that were to begin in 1997.

In the meantime, CECOM was making Task Force XXI possible by solving difficult technical problems on the spot (in "real time" in military parlance) and by supplying troops of the Battle Labs with technologically advanced systems - equipment and software - to supplement systems already in the field. These efforts would ultimately give the Army a basis for deciding what the force of the Twenty-First Century would look like, both technically and doctrinally. At the core of Force XXI and the experiments leading to it was what had come to be known in the Army as "digitization." Digitization - the Digital Battlefield concept - was (in the simplest terms) the use of computers and digital transmission technologies to link all an army's soldiers and equipment, giving commanders the ability to assess the disposition of friendly and enemy forces quickly with a glance at a flat-panel display, whether in headquarters or a forward command post. For CECOM, digitization - the military application of the "information revolution" - meant supplying

Task Force XXI, Advanced Warfighting Experiment, held at Fort Irwin, CA

the tools that would give soldiers the right information in the right place at the right time. Army Acquisition Executive Gilbert F. Decker spoke of digitization objectives in these terms:

The ability to dominate the battlefield or to conduct operations other than war efficiently will depend completely on having the pertinent information in the right hands at the right time. In military terms, this is often summarized as situation awareness.

It seems clear to warfighters and technologists alike that if commanders and decision makers at every echelon of the Army are completely aware of their total situation at all times, they will react with a course of action that will place them inside the opposing forces' decision cycle. Thus, with numerically inferior forces, one can achieve combat leverage dominance.

The Army's ambitious strategy to equip its forces with digital technology, in the words of one reporter, "catapulted electronics to the forefront of its plans and thrust CECOM into the limelight." Though the Department of the Army established an "Army Digitization Office" in the Pentagon to supply some Army-wide oversight in the implementation, it was clear from the outset that the nuts and bolts of the effort - responsibility for actualizing the concept - lay squarely upon CECOM and its RDEC; this, in turn, gave the command an increasingly prominent role in service debates on acquisition, training, and doctrine.

It also gave the command a rare opportunity. The products of CECOM "touched" every Army system. While the diversity of these products - and the complexity of the CECOM mission - was noteworthy, it was their integration and the resulting synergy that would have the greatest impact on the art of war in the early centuries of the third millennium. In conjunction with its partners in Team Fort Monmouth, the RDEC supplied products that transcended individual organizations, putting CECOM

in position to accomplish what no other organization in DOD could accomplish - that is, to integrate the Army. To this end, CECOM's RDEC wrote the Army's Technical Architecture - a set of "building code" standards to be employed in all new C4IEW systems - and accepted responsibility for functioning as the Army's System Engineer to enforce this architecture. Department of the Army (DISC4) required an architecture that was comprehensive; an architecture applicable to all weapon, soldier, and information systems; and an architecture that supported joint and combined operations. Recognizing CECOM's leadership role in achieving and maintaining this architecture, General Leon E. Salomon, AMC Commander, Feb 1994 - Mar 1996, named General Brohm AMC's lead for the Experimental Force (EXFOR) Coordination Cell that he established at Fort Hood to support the 4th Infantry Division with the fielding of all types of equipment and prototypes.

Army officials described the path to Force XXI during the Association of the United States Army conference at Fort Monmouth in May 1995. To achieve the goal, they said, the Army had to improve the inter-networking of current legacy systems while concurrently developing a future, fully digital Battlefield Information Transfer System (BITS). It already had contracts in place to equip the EXFOR with computers (the $1.5 billion Common Hardware/Software II contract with a GTE/Sun Microsystems team and the $240 million Appliqué contract with TRW). The "Appliqué," known formally as the Force XXI Battle Command Brigade and Below (FBCB2) System, consisted of a computer, software, a GPS receiver, and a communications interface. It would supply a digital link for weapon systems that had none (meaning most of the existing systems) and give troops digital maps with nearly real-time updates displaying the location of friendly and enemy forces and raising situational awareness, thereby, to "unprecedented levels." General Campbell, PEO

Helping hand
Army provides digitized training to Marines

by Debbie Sheehan
Public Affairs Office

The Marines landed, took their training and now they are on their way. Eleven Marines and two support contractors recently invaded Fort Monmouth for a six week training course in operating and maintaining the Enhanced Position Location and Reporting System (EPLRS) Network Control Station NCS(D)E.

The EPLRS is a highly sophisticated computer based tactical communications network that has been fielded within the Army and served as the Army's digital backbone during Task Force XXI.

The Marine Corps has ordered over 800 of the most recent versions of the EPLRS radio sets which can be used in either a man-pack configuration or installed in a vehicle to network with the NCS(D)E.

The Marines took their training in Building 2507, the shelter and integration facility within the Research, Development and Engineering Center's Command and Control Directorate (C2D) Systems Prototyping Division in the Charles Wood Area.

The joint training effort was put together by Project Manager Tactical Radio Communication Systems (PM TRCS), Venntronix Corporation and the Marine Corps' Systems Command, (MARCORSYSCOM) Quantico, Va.

This pilot course combined both operations and maintenance in the NCS(D)E. Previously, separate courses were provided for operators and maintainers.

"Both job skills are now performed by one individual," said Terry Harmon, lead instructor for the course and training manager for Venntronix Corporation, a CECOM contractor.

photo by Michael Allison

Eleven marines recently spent six weeks here training in maintaining tactical communications. Left to right are: Cpl. Roger Fasoldt, Cpl. Kyle Adams and Cpl. Ryan Willett.

"These are the first Department of Defense personnel to be formally trained on the downsized NCS, "said Harmon. "Our objective was to train these Marines to operate and maintain the NCS(D)E, and to develop proficiency through extensive hands on performance during the six -week course."

Marine Col. Robert Lampkin commanded the troops. "Unlike the Army, our systems are not digitized; only one division is working in this arena. Hopefully in the future, all will be identical; everyone will have the same host computers to operate world wide."

The Marines were training for a three month Limited User Evaluation (LUE) that is taking place at 29 Palms, Calif. this summer. Lt Col. JJ Spegele, the senior Marine here, monitored their progress. "These Marines are key to our efforts during this summer's initial use of EPLRS as the Marine Corps' communications system from regiment to battalion."

"The training material and technical manuals were completed in time for the course because we developed the dynamic logistics support package," Harmon said. "With the accelerated schedule, only through the joint cooperation of PM TRCS, C2D, MARCORSYSCOM, and our personnel were we able to bring together the training assets, equipment, and facilities required to make the training successful.'" "The students trained on the actual equipment that they will be using during the evaluation," said Lampkin, the Army's EPLRS Product Manager.

Marine Cpls. Roger Fasoldt, Kyle Adams and Ray Willett were students in the course. "Basically the unit gives a positive location and lets you know where to go or not to go," said Willette, who is stationed at Camp Pendleton,

Calif. "The first four weeks were devoted to learning system operations, the rest on how to maintain and repair the equipment."

Fasoldt said he found the atmosphere for the course was easy going. "But this is a high visibility, high pressure mission," he said.

Adams, also from Camp Pendleton, said that learning how to use the Net Control Station units was easy. "The test will come at 29 Palms when we have to field the equipment."

Fasoldt said there will be information in the field transmitted to a command center. Once the information is gathered it is passed through this wireless network. The Marines will deploy two Net Control Stations during the LUE.

"We will have one NCS up and a backup we can bring up if required, which is important to mission success," he said. " That unit is the same as the one in the control area, set up with a frame and batteries. It looks forty years old when you look at the box," Fasoldt said.

The Marine Corps will employ these systems in a variety of platforms to support a tactical commander's need for battlefield situational awareness and command and control.

"Not only will we be looking at helicopters and tanks, we will use this to help deter friendly fire," said Adams. "We will be the best qualified with the most time in on this equipment."

"Using the PM TRCS, EPLRS Logistics Object Oriented Database developed by Venntronix for the Net Control Stations, we adapted and supported the Marine Corps unique training and logistics requirements,' said Steve Layton, director of Operations Venntronix Corp.

These Marines will be working with Marine Corps units who will be the first digitized Marines. Spegele said, "The Marines, like the Army, are just beginning to employ EPLRS as a communication system."

photo by Michael Allison
Cpl. Frederick Stewart with the Network Control Sta-

for Command and Control Systems, thought that computer capacity would not be a problem, but that there might be a problem with the communication pipes that would be needed to network the computers. Moreover, Appliqué-equipped systems would not have the ability to link with state of the art systems such as the M1A2 Abrams Tank and the AH-64D Apache Longbow attack helicopter, both of which had been built as digital systems from the ground up. General Gust, PEO Communications Systems, spoke of the Army's strategy to address these problems. The Army had a program under way to seamlessly interconnect the "legacy" systems that it had acquired in recent decades at a cost of nearly $10 billion: the long-haul TRI-TAC tropospheric scatter system, MSE, SINCGARS, and Enhanced Position Location Reporting System (EPLRS). The SINCGARS radios, originally developed to handle voice rather than data, were to be upgraded to provide a throughput of 4,800 bits a second and would also come equipped with an internet controller card developed by ITT that would allow them to pass data to an EPLRS interface and to one another using standard Transmission Control Protocol/Internet Protocol (TCP/IP). EPLRS and SINCGARS would, in turn, pass data through a Tactical Multinet Gateway (TMG), a rugged version of a commercial router designed by GTE that would interface with TRI-TAC and MSE. Major General Joe Rigby, head of the Army Digitization Office, said that the Army planned to buy 750 Internet Controller Cards and seventeen TMGs for the experimental brigade. General Guenther, Director for Information Systems, Command, Control, Communications and Computers (DISC4), said that the fix provided by these two hardware enhancements would "vertically and horizontally integrate our legacy systems" - a remarkable accomplishment since these systems were not originally designed to dovetail in such a way to handle data. Because bandwidth requirements, especially for video teleconferencing and multimedia, would quickly exceed SINCGARS's capacity, CECOM was planning to release a solicitation for a Near-Term Digital Radio that would provide a throughput of at least 144 kilobits per second. CECOM expected bids from Hazeltine (which already had a 180 kbps digital radio), Hughes, and ITT. For long-haul communications between the United States and forward-deployed forces, said Gust, "We have to start thinking T-1 circuits." CECOM intended to provide these circuits initially through Lockheed Martin's MILSTAR satellite system with battlefield connectivity provided by mobile, HMMWV-mounted terminals developed by Raytheon and Rockwell as prototypes for the MILSTAR Advanced Terminal. Beyond that, according to Rigby, the Army wanted to rely more on commercial technologies to satisfy its high-bandwidth needs: this would include deployable cellular systems, personal communications systems, wideband data radios, and commercial satellites.

For the RDEC, navigating the path to Force XXI involved the following objectives: (1) Capitalize on the "explosion" in information technology to harness the potential of automation and digital communications for use in near-term applications and in the ultimate digitized Army; (2) build a flexible architecture and infrastructure for all Army systems; (3) continue developing new technologies for C4IEWS, combat identification, targeting systems, and software-intensive systems; (4) be the "technical bridge" between Battle Labs, basic research, early

technology, PEO/PM programs, and industry; (5) integrate military and commercial technology; (6) expedite the insertion of technology into existing products; (7) work with the user to integrate technology with doctrine; and (8) profit from technology to increase the lethality of U.S. weapon systems, protect the force, and exploit the vulnerabilities of opposing forces.

In working toward these objectives, the RDEC found two tools especially valuable in 1994 and 1995: the Digital Integrated Laboratory and the Advanced and Battle Lab Warfighting Experiments.

THE DIGITAL INTEGRATED LABORATORY

In 1994, building upon its earlier successes in use of the Army Interoperability Network to test and perfect techniques for interconnecting Army and Joint Service Systems, CECOM established the Digital Integrated Laboratory (DIL) specifically to support development of the Force XXI architecture.

Craig Criss, chief, Software Prototyping Laboratory, with MG Guenther at the ribbon cutting ceremony opening the Software Prototyping and Integration Laboratory

Army policy mandated use of the DIL to develop, maintain, improve, and certify interoperability between and among the C4IEW hardware and software systems that were to be employed in Task Force XXI and follow-on division and corps-level warfighting experiments. The linkages that existed between DIL, the Battle Laboratories, contractors, and others and the ability to reconfigure the DIL readily to replicate any existing or evolving C4IEW environment allowed CECOM to perform rapid prototyping in each case at the outset of the certification process. For each system it certified, the DIL undertook in all its activities, including modeling and simulation, to eliminate as many potential problems as possible before committing the system to an exercise. One reporter likened the process to a "grueling virtual boot camp"- "a crucible in which the Army burned off applications that fell short of interoperability and ease of use requirements." Moreover, in providing an electronic link between the user, the contractor, and the materiel developer, the DIL gave the real user - not the schools, but soldiers in the field - the ability to provide input in real time to both the material developer and the contractor. The goal, said RDEC Director Robert Giordano at a 1995 Advance Planning Briefing for Industry, was to link developers and soldiers by "bringing the lab into the field and the soldier into the lab."

In subjecting systems to an electronic gauntlet that simulated combat, the DIL allowed soldiers to get a feel for a prototype system, spot its weaknesses, and suggest changes well before the Army committed the system to the battle experiments of the EXFOR. "This," said General Brohm, "produces tremendous leverage in terms of moving quickly and efficiently into the future without the fits and starts of the past." With the DIL, Brohm noted, the Army moved beyond arguing about what direction it would take and, instead, provided sufficient discipline to the development process to allow leaders to make confident decisions, incrementally, about what should be added to the capability packages of its hardware and software. "Discussions now focus on how objectives will be achieved," he said, "not on what the objectives are."

One of the DIL's heaviest near-term users was TRW, Inc., contractor for the Force XXI Appliqué. TRW subjected software for the Appliqué to extensive integration and interoperability testing in the DIL, with a "user jury" at Fort Knox evaluating incremental software releases and working directly with TRW and its subcontractors to make improvements.

CECOM officials demonstrated some of the other DIL-tested capabilities in March and May 1995 during live demonstrations in Washington and Fort Monmouth. These included:

Dismounted Soldier System: A pencil-sized camera mounted on the helmet of a soldier walking down a street in Haiti picked up digital video "footage." The images were downloaded through the soldier's radio to an antenna unit of a nearby HMMWV, then relayed to a field Tactical Operations Center and on through a Secure Telephone Unit III connection to command headquarters. At headquarters, commanders were able to freeze, play back, or zoom in on the images and transmit instructions back to the soldier while the camera was still rolling.

Electronic Mail: An e-mail message from a soldier in Georgia, generated on a ruggedized laptop, was relayed through half a dozen digital platforms, including an ATM switch, using a wireless internet Protocol. Before the message flashed on a screen at Fort Monmouth, it passed through several networked PCs, a HMMWV-mounted work station, a SINCGARS radio, and a surrogate satellite communications terminal on the roof of the Myer Center.

Prototype Aviation Mission Planning System: This system linked digital elevation data with digitized, two dimensional colored maps to create realistic three-dimensional "landscapes" through which a pilot, flying a work station, could rehearse a bombing run. Using a mouse, the pilot designated a target on the map, displayed on one screen, then "flew over it" in the simulated landscape produced on another screen.

Common Ground Station: In this test, fusing intelligence data and imagery from multiple sources, Synthetic Aperture Radar images of enemy troops moving under cloud cover were overlaid on digital terrain maps and fused with precise position data collected from the surveillance aircraft. This supplied the information needed to send in an unmanned aerial vehicle for a closer look with a digital camera, whose live images aided commanders in planning an ambush.

Digital Radio Communications: The DIL tested the reliability of digital radio links between tanks on the move by placing the vehicles in a simulated mountain environment. When a tank moved behind a simulated "ridge" (for example), it automatically lost its line of sight connection.

The DIL also provided a forum in which companies could show the Army their wares. And, according to Larry Denis, operations manager of ITT's Defense and Electronics Office in Tinton Falls, it provided industry with a venue in which to test new or evolving commercial systems within the Army's technical architecture or conduct early evaluations of military systems under development. According to Colonel Robert Shively, chief of Special Projects Office for digitization, most of the testing conducted by the DIL involved real equipment, rather than modeling and simulation. However, the DIL employed modeling to analyze information obtained from exercises - such as, for example, data on the volume, type, and destinations of traffic and the timeliness of transmissions - to create a picture of Force XXI requirements. A subset of the DIL, the Software Prototyping and Integration Laboratory (SPIL), opened in October 1994 with a core staff of eight experts and several AMC software interns to provide the acquisition and development communities a quick-reaction, low-cost software development capability.

THE BATTLE LABS AND WARFIGHTING EXPERIMENTS

Department of the Army established the Battle Labs in coordination with the Training and Doctrine Command to test and refine technology and doctrine for the digital battlefield. In line with its commitment to its ultimate customer, the soldier, CECOM's RDEC was the first in AMC to establish and maintain a permanent field presence at all the Battle Labs. CECOM consecrated twenty-one positions to this role: eleven for permanently assigned personnel and ten for personnel on rotations ranging from ninety to 120 days. In FY94, the command developed and implemented Memoranda of Agreement with the Battle Lab directors to legitimize its presence and specify its support to the Battle Lab missions.

CECOM field tested its new technologies in warfighting experiments, most but not all of which were staged by the Battle Labs. In FY95, CECOM supplied concepts and hardware for testing and evaluation in Prairie Warrior (Combined Arms Center), Warrior Focus (Infantry School), Focused Dispatch (Armor Center), Unified Endeavor, and Joint Warfighter Interoperability Demonstration (JWID) 95. In more than one of these exercises, CECOM and its PEO partners had more than fifty people in the field, living with the "troops" and participating in the evaluation of equipment and doctrine. For Force XXI and Synthetic Theater of War initiatives, CECOM produced an interface between the Brigade/Battalion Battle Simulation and various equipment simulators. This permitted simulation of virtual combat elements in order to improve training and support evaluation of proposed battlefield systems and organizations while, at the same time, conserving resources. In Focused Dispatch, CECOM supplied a systems architecture design and engineering support that allowed soldiers to test their warfighting capabilities simultaneously against live and "virtual" forces with fully realistic, terrain-dependent tank-to-tank, tank-to-simulator, and simulator-to-simulator communications.

Some CECOM-supplied technology or another played a role in every such experiment, but especially in Warrior Focus, JWID 95, and Unified Endeavor, where CECOM applied the advanced communications technologies that it developed in the Adaptive Systems and Digital Battlefield Communications Advanced Technology Demonstrations. The "mother" of all warfighting experiments, the brigade and division-level exercises of Task Force XXI, would further evaluate technologies tested on a smaller scale and refined in the earlier Battle Lab exercises. Other CECOM contributions to Task Force XXI included:

The Hunter Sensor Surrogate (HS2). A lightweight, deployable and survivable vehicle platform with an advanced low-observable, long-range targeting capability, the HS2 combined second generation thermal imaging, day-time television, eye-safe laser rangefinders, embedded aided target recognition, and image compression and transfer technology.

Direct Broadcast Satellite (DBS). CECOM acquired three DBS terminals for Task Force XXI to support rapid dissemination of data on the battlefield. JWID '95 sponsored demonstrations of the Global Broadcast Service/DBS for each of seven receiver sites - Fort Gordon, Fort Huachuca, Fort Leavenworth, Fort Hood, Fort Lee, Fort Monmouth, and the Naval Research Laboratory - using Common Ground Station Data. The demonstration showed users the technology's battlefield capabilities.

Personal Communications System (PCS). Adaptation of commercial cellular telephone technology for dismounted infantry use, connecting through MSE.

Surrogate Digital Radio (SDR). Use of commercial technology to provide a wide-band "Data Hauler" for seamless transmission of information to forward-deployed units.

Message Standard. The RDEC aggressively pursued the development of the MIL-STD-188-220 Information Standard for military networks, for which it received joint service approval in record time (six months). The variable message format of this standard had the flexibility to meet the operational situations of the moment while restricting content to the required information only.

ADVANCED TECHNOLOGY DEMONSTRATIONS AND TECHNICAL ACHIEVEMENTS

The U.S. military would have a large array of sensors on the battlefield of the Twenty-First Century. Deployed aboard manned and unmanned platforms, as well as in orbiting satellites, these sensors would provide commanders with imagery, signals intelligence, telemetry, and other information crucial to maneuver, the performance of weapon systems, and protection of the force. But to exploit the capabilities of its sensors fully - to make sense of the abundance of information they provided - the U.S. military depended on high-speed microprocessors, advanced electronic packaging, and new algorithms that allowed commanders to correlate information from various sources, sift it, and view it in immediately useful form. This manipulation of sensor information was knows as "sensor fusion."

One Army program that exploited sensor fusion (and correlation) was the Hunter Sensor Suite. It was the "hunter" part of the larger Rapid Force Projection Initiative (RFPI), which focused on moving targeting data quickly between "hunters" and "killers" to protect light forces from armored attack. Hunter would seek targeting data from a variety of sensors. It would have advanced Forward Looking Infrared (FLIR) sensors, two television cameras, a laser rangefinder, and a seeking module mounted on a stabilized platform aboard a HMMWV. The operator or operators in the vehicle would have access to as many as three displays: one for acoustics information, a second for TV and FLIR information, and a third for a remote sentry. The suite was intended to serve initially as a decision aid for the driver. Ultimately, the data gleaned would help the operator select targets and send information over Combat Net Radio (CNR) to a tactical operations center. The initiative would also look into linking sensors directly to weapons (the "killers"). Technology for the program included infrared focal plane arrays, high density integrated processors, and standard advanced detector assembly packaging. Particularly challenging was the use of acoustic sensors to identify ground targets. Historically, most work on acoustic sensors focused on identifying and classifying air-based targets: Identification of ground-based targets required engineers to filter out noise from wind, engines, wheels, and other environmental sources. Image compression was also a challenge for which the Army worked with industry to examine existing and emerging algorithms, such as wavelets. Perhaps the most difficult aspect of the program, however, was the joining of different technologies that had previously been demonstrated only alone or in different applications. The need for an interoperable reconnaissance architecture pressed home during the War for Kuwait, when U.S. forces deployed an assortment of sensors and other systems that could not communicate, contributing to the Military's inability to get intelligence quickly to those who needed it most. CECOM prepared a prototype of the system, the Hunter Sensor Surrogate, for Task Force XXI.

The Common Ground Station (CGS) ATD demonstrated transmittal of responsive, timely, correlated multi-sensor intelligence data to brigade commanders "on the move." Significant achievements of the demonstration included integration of the Army combat information process with the DOD Operational Support Office Global Broadcast System; proof of the ability to distribute multi-media data base information to low-cost, widely distributed remote sites; and proof of the benefits of using facilities in a sanctuary site for providing nearly real-time intelligence data to remote locations. Technological breakthroughs included an expanded, automated, and integrated capability for receiving, processing, correlating, displaying, and disseminating intelligence and targeting data using Tactical Exploitation of National Capabilities, Enhanced Tactical Radar Correlation, Unmanned Aerial Vehicle video, the Synthetic Aperture Radar Target Recognition and Location System, Second Generation Forward-Looking Infrared, the Joint Surveillance Target Attack Radar System (JSTARS), and Firefinder products. The ATD integrated the CGS processing capability with an advanced multibeam, multiband phased array antenna system that provided simultaneous reception of information from multiple sources. The CGS also featured advanced work stations and software and an advanced, multimedia distributed data base for filtering, storing, and processing information. Transferred or transferable products of the ATD included the antenna (to the Digital Battlefield Communications ATD for satellite communications on-the-move), Software [supplied to the JSTARS Project Manager for the Ground Station Module program and to the Advanced Research Projects Agency for the Battlefield Awareness and Data Dissemination Advanced Concept Technology Demonstration (ACTD)], and information correlation algorithms (to the Intelligence community for use in Task Force XXI). CECOM demonstrated CGS capabilities in JWID 95.

In the Advanced ATR Processing and Algorithm Exploitation program, CECOM integrated signal processing prototypes of various kinds to achieve nearly real-time gauging of target acquisition data from multiple sensors (e.g. second generation thermal sensor, millimeter wave radar, laser radar, etc.). The aim was the development of technologies that would enhance both the lethality and survivability of future weapon systems. CECOM tested the prototype hardware using the Multi-Sensor Aided Targeting - Air algorithm and passed the proven technology on to the Electronic Integrated Sensor Suite program, the Hunter Sensor Suite ATD, and the Navy Rolling Airframe Missile program.

The Multi-Sensor Aided Targeting - Air (MSAT-Air) ATD demonstrated the economical fusion of second generation forward-looking infrared and millimeter wave radar in an airborne, automated target acquisition suite (Apache Longbow). CECOM conducted the demonstration under simulated battlefield conditions in a fully-operational flying test bed emulation of the RAH-66 target acquisition system. It included feature-level fusion algorithms, real-time processing, and man-in-the-loop evaluation, and it proved the technology's ability both to increase the lethality of Army weapon systems and shorten target search times (thereby increasing aircrew survivability). Essentially, the integration of infrared and radar technologies gave pilots visual displays of targets and terrain not otherwise seen. The program's software algorithms transitioned to the Target Acquisition and the Air/Land Enhanced Reconnaissance and Targeting ATDs, where they were to be upgraded for use in on-the-move operations. The Rotorcraft Pilot's Associate ATD incorporated some of the MSAT-Air's system design concepts, and CECOM surrendered the prototype system (aircraft, processor, FLIR, and radar) for use in the Commanche program.

In FY94, the RDEC developed a state-of-the-art ultra wideband receiver/downconverter for communications intercept that delivered five times the performance with one-hundredth the size and one thirtieth the weight of previously available technology.

It was a single VXI-based circuit card covering the 10 MHz to 40 GHz range. The RDEC also developed a single VXI-based circuit card that provided 4 GHz of instantaneous bandwidth at a rate of sixteen gigasamples a second with a 10 KHz repetition rate. These products were to be used in the IEW family of common sensors, included the Ground Based Common Sensor, Advanced Quickfix, and Guardrail.

In March 1995, the RDEC fielded a Guardrail/Common Sensor system together with a Ground Tethered Satellite Relay and all the required support equipment. Valued at $758 million, this was the RDEC's largest fielding ever to a single battalion. The system detected enemy radar and radio signals, traced the signals to their sources, and relayed the information to tactical commanders on the battlefield almost instantaneously.

Two of the five ATDs CECOM completed in FY95 focused on techniques for protecting the force: the Close-in Manportable Mine Detector ATD and the Radar Deception and Jamming ATD.

The Close-in Manportable Mine Detector (CIMMD) ATD demonstrated the reliability of various technologies and combinations thereof in detecting both metallic and non-metallic anti-personnel and anti-tank mines. The demonstration evaluated various brassboard technologies against major mine types under varying environmental and operational conditions. It included evaluation of infrared and ground penetrating radar technologies both independently and in combination, and it proved the ability of existing technologies to detect more than eighty percent of the non-metallic anti-tank mines and more than half the non-metallic anti-personnel mines. Brassboard technologies developed for the program included separated aperture balanced bridge microwave sensors, thermal imaging, and synthetic pulse ground-penetrating radar. This ATD led to evaluation of selected prototype hardware components for immediate acquisition, use of the infrared camera in conjunction with the AN/PSS-12 mine detector in a Dismounted Battlespace Battle Lab Warfighting Experiment (Warrior Focus), and transfer of technologies to the Hand-Held Standoff Mine Detection Sensor Program.

The Radar Deception and Jamming (RD&J) ATD employed a sensor suite to provide a knowledge-based management system for effective use of an aircraft's on-board countermeasures. Successfully tested in flight at the Electronic Combat Range, Eglin Air Force Base, the RD&J demonstrated integration of aircraft survivability equipment, situational awareness, non-cooperative target recognition, precision direction finding and location, target cueing of weapon systems, countermeasures control, and resource management. Key subsystem products of the program included the Sensor Fusion Processor, Advanced Threat Radar Jammer hardware, the User Data Module, a Digital Terrain Map System, an Aircraft Avionics Integrated Navigation System, a Data Collection and System Control Display, and simulations of a C3I data link, weapons systems, and an expendable dispenser. Integration of multi-sensor data fusion, automated tactical situation assessment, and response management with next-generation defensive systems and existing aircraft avionics produced a flyable, integrated Electronic Warfare system with a significantly increased probability of survival. All the ATD's software transitioned to the Advanced Threat Radar

Jammer program. In addition, the ATD established a baseline for future integration of the Advanced Threat Countermeasure System and the AN/AVR-2A Laser Warning Receiver.

Radar Deception and Jamming (RD&J) ATD

In FY94, also to protect the force, the RDEC, using signature management technology, developed a survivable high-mobility camouflage skirt that was an effective signature suppressor and was capable of withstanding at least fifty miles of off-road maneuvering. During that time the Center also built and delivered the world's very first scanned electron beam x-ray source for use in the Photon Backscatter Imaging Mine Detector.

The Combined Arms Command and Control ATD was a "capstone" information technology program addressing a broad range of Army C3I digitization issues. It worked at developing both the Force XXI and the Task Force XXI systems architectures, which specified the networking and distributed data base technologies required to support battlefield synchronization, horizontal and vertical integration, and situational awareness.

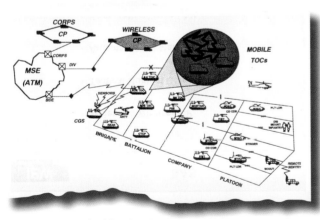

Survivable Adaptive Systems (SAS) ATD

The Survivable Adaptive Systems (SAS) ATD demonstrated the ability of high-capacity communications network technologies to support the multimedia (voice, data, imagery) information requirements of the future battlefield, including on-the-move communications. The technologies demonstrated involved wireless local area networks (LAN) of both narrow and wide bandwidths, a tactical fiber-optic LAN, network

management, gateways, and security. Break-throughs of the program included use of commercial standards and protocols in automated tools for configuring and managing complex tactical internetworks; use of a tactical multinet gateway to link older, "legacy" communication networks using commercial protocols and products; use of wideband packet data networks in several frequency bands for high-capacity data communication on the move; and use of a secure packet radio and software to disseminate intelligence data during on-the-move operations. SAS products applied to other programs included the Tactical Multinet Gateway and the Automated Network Management capability, transferred to Task Force XXI and the Joint Tactical Communications Systems Project Manager (PM JTACS); the Surrogate Digital Radio, to Task Force XXI; the AN/VRC-99 wideband packet radio and the wireless network capability, to the Signals Warfare Project Manager; and a dual-use wireless LAN, to the Program Executive Office for Command, Control, and Communications Systems.

In April 1995 during Proteus 95/Unified Endeavor, the 2d Armored Division deployed for a simulated warfighting exercise using RDEC-supplied Asynchronous Transfer Mode (ATM) technology, integrated into the Division's Mobile Subscriber Equipment (MSE) network, to provide the commander with a unique video teleconferencing and multi-media capability. The Proteus wide area network provided seamless communications from deployed brigades to division rear, corps, and into the National Capital Region. For the first time in an Army exercise, the division commander, using desktop video teleconferencing, could play an integral role in collaborative planning with his forward division elements. During his August 1995 visit to Fort Monmouth, Army Chief of Staff General Reimer learned first-hand of the benefits of the system when CECOM engineers linked him via the ATM/MSE system directly to Fort Hood and the Commander of the 2d Armored Division.

CECOM also deployed Trojan SPIRIT II to the National Training Center for Proteus 95, where the system employed ATM to transmit "hypermedia" from fixed strategic sites to maneuver brigades. Subsequently, at the direction of the Army Chief of Staff, CECOM integrated Trojan with the Theater Missile Defense Tactical Operations Center for a demonstration at the Army Commanders' Conference. Grouped in a Defense Intelligence Support Element configuration, a Trojan SPIRIT II system at the Army War College received intelligence data, imagery, and motion video from the Analysis Control Element at Fort Hood. Lieutenant General Menoher, Deputy Chief of Staff for Intelligence, was reportedly "exuberant in his praise for both the concept and the execution." During FY95, in the very year of the system's acquisition, CECOM carried out twenty-three DA-directed fieldings of the Trojan SPIRIT II to Forces Command, USARPAC, USAREUR, and the U.S. Army in Korea.

One of the technology break-throughs of FY95 was the development of a database-to-database interface that eliminated the need for text-based message formats. Using a "clearing house" data base that contained a superset of all applicable data types, the interface coordinated data elements between different systems in a local or wide area network, thereby producing the standardization inherent in message formats. This reduced processing overhead and bandwidth while focusing on pertinent, time-critical information to provide commanders and analysts the data they needed, when they needed it, in a usable format.

Land Warrior Demonstration, 1996

CECOM researchers were responsible for developing the individual soldier's computer and radio that were to be the heart of the initial Land Warrior System. The Army expected to field about 5,000 of these first generation systems by 1999, beginning with the 82d Airborne Division. The second generation system was to incorporate the CECOM-developed, helmet-mounted "pencil" camera, in addition to a wide view-field image intensifier and a card-based radio/computer. CECOM put some of its digital technology to the test in Haiti in the spring of 1995. Soldiers from the 2d Armored Cavalry Regiment, equipped with commercial still and video cameras as well as the helmet-mounted "pencil camera" captured full-color digital images that were transmitted from the theater to the Pentagon nearly instantaneously via SINCGARS radio and commercial telephone.

To produce more affordable, more capable software for the Army, CECOM engineered the shift from stovepipe software systems (i.e., systems designed for specific applications) to generic, layered "architecture-based" systems, using off-the-shelf products and open system standards. The resulting demonstration project, Software Technology for Adaptable, Reliable Systems (STARS), improved software quality, reduced support costs by forty-nine percent, and abridged the time spent in software development and remaintenance. The STARS project was one of twenty-four, selected from a field of 325 applicants, to receive the FY95 Federal Technology Leadership Award.

The Army's Common Operating Environment (ACOE) was an integrated architecture with standard, modular system and application support layers. The common computing infrastructure enabled disparate C2 systems to communicate and gave commanders the ability to tailor software applications to meet individual needs. The Army required that all new systems, as well as system modifications, use the ACOE. The ACOE requirement was one of a host of steps the Army took to get its C2 effort - the Army Battle Command Systems (formerly the Army Tactical Command and Control System) back on track. Two ABCS components in particular -- the Maneuver Control System (MCS) and the All Source Analysis System (ASAS) highlighted the Army's attempt to use commercial technology, standards, and substantial input from users to rejuvenate the ABCS.

Among other strikes against it, MCS had come under fire for not meeting user needs. The Army banked on software developed by the Battle Command Lab in tandem with Mystech (Falls Church, VA) to turn the program around. The new software development project to enhance the MCS, built upon technology developed for ASAS. Named "Phoenix," the new MCS software system featured a Windows-based user interface, tear-away menus, an icon builder, and the ability to run multiple applications simultaneously. The voice-activated system also provided common scaleable map displays, enemy and friendly force tracking, and desktop video teleconferencing. The Army put Phoenix's C2 capability to the test in Warrior Focus and Prairie Warrior. Upgrading of the ASAS to its Block II configuration also involved user input and technology experiments in the EXFOR, which gave components of a prototype ASAS software system, known as Warlord, a "good shakedown" in working together with more mature software.[205]

TECHNOLOGY APPLIED TO LOGISTICS

The Logistics Anchor Desk (LAD), a CECOM-engineered Joint Advanced Concept Technology Demonstration (ACTD), linked computer workstations in a network that extended from Ramstein, Germany, to Fort Leavenworth, Kansas, to give soldiers an early glimpse of terrain in Bosnia and help them, thereby, plan the movement of supplies across an unfamiliar landscape. In essence, LAD allowed logisticians to do a "paper" reconnaissance well before the arrival of the first troops. The data itself was not new: the maps, for example, were all derived from other sources. What was new was the integration on a single platform of information and programs from many agencies and the dissemination of this information across the network to everyone who needed it. In addition to coordinating current data, the LAD could project deployment activities weeks into the future. Features included the Mapping and Analysis Tool for Transportation, video teleconferencing, and a Knowledge-Based Logistics Planning Shell (KBLPS) that did "what-if" analyses. USAREUR logistics coordinators used another LAD feature, Time-Phased Force Deployment Data, extensively during the deployment of troops to Bosnia. This utility allowed them to create and display a near term (twenty-day) projection (a pictorial representation) of future deployment activities.

Typically, planners used two LAD terminals side-by-side: one for tracking current data, the other for playing out the hypothetical analyses. To illustrate some of the system's applications: By clicking on a series of icons, planners initiated a search to identify the number of kinds of Army tactical (mobile) bridges in Europe; they overlaid the results of the search on a map of Western Europe to show exactly where the bridges were. Hypothetically, at least, with links to Army supply centers and real-time updating of logistics information, LAD could provide "total asset visibility" of equipment in storage, in transit, and in the field. USAREUR used the KBLPS feature to plan the construction of base camps in Kaposvar, Hungary, a major staging point for the 1st Armored Division. According to Colonel Jim Paige, Director of the LAD ACTD:

> USAREUR was trying to figure out how many people they had to feed or sleep in base camps, knowing 10,000 people would be clothed, housed, and fed over 'X' amount of days. When you run the number through KBLPS, you don't have 10,000 people at any one time, you actually have 6,000. You don't have to buy as many beds.

That knowledge, by preliminary estimates, saved the Army more than $1 million. Meanwhile, at the Atlantic Command in Norfolk, Virginia, LAD operators used the KBLPS to track petroleum, oil, and lubricant usage in Bosnia in order to estimate the fuel requirements of allied forces by comparing them to similarly sized and equipped forces of the U.S.

CECOM initiated development of the LAD in 1994 as part of the Total Distribution ATD. The program became a Joint Advanced Concept Technology Demonstration when other services voiced an interest in its capabilities. CECOM fielded the LAD two years ahead of schedule to support operations in Bosnia. The LAD matured into a command and control system more quickly than expected. Logisticians installed a LAD workstation at Kaposvar on 15 March 1996 and were on track then to install yet another, on 28 March, at Task Force Eagle in Tuzla. Meanwhile, in real-world non-Bosnian applications, government employees employed LAD, following the crash of a charter airliner off the coast of the Dominican Republic, to support rescue planning by generating maps that depicted nearby air and seaports. And when a heavy storm severely damaged a prison security wall in Barbados, operators of the Atlantic Command employed a LAD feature, called the Global Logistics Awareness Display, to locate supplies of concertina wire available for immediate shipment.

In another "logistics power projection" initiative, CECOM engineers conceptualized and designed a state-of-the-art system for rapid distribution of software to the field in combat and other operational environments. Such a capability proved especially valuable to users in remote locations. Using this breakthrough technology, known as the Rapid Open Architecture Distribution System (ROADS), the Army could securely transmit a new software version electronically to a field site, where it could be replicated, distributed, and installed in on-site computers.[206]

QUALITY IMPROVEMENT

The Logistics and Readiness Center (LRC) competed for and won the President's Quality Improvement Prototype (QIP) Award for 1996. Winning the competition was in itself a remarkable feat. The LRC application was one of eighteen in the Department of the Army. A panel of twenty ranked these eighteen applications and forwarded the top six to the Office of Personnel Management (OPM) for the national competition. OPM selected ten finalist organizations from a field of thirty applicants representing all sectors of the government. It then subjected each finalist to on-site scrutiny by a team of quality experts from both the public and private sectors. This team thoroughly examined LRC operations, conducted in-depth interviews with all members of the Center's Executive Quality Council and randomly selected employees, viewed team meetings and demonstrations of LRC processes, and researched the LRC's back-up documentation. It judged the finalists not against one another, but in comparison to a pre-established standard of excellence. Although the OPM judges usually selected multiple winners - they had the authority to select as many as two for the top prize and as many as six for runner-up honors - for 1996 they chose only the CECOM LRC.

Quality Week at Gibbs Hall, 19 October 1995

Created in 1988, the award program applied the same stringent standards used in the private-sector for the prestigious Malcolm Baldridge National Quality Award. The intent was to recognize world-class organizations whose improvement efforts resulted in more effective use of the tax dollar and the delivery of higher quality products and services. The program promoted awareness of quality issues, spurred implementation of quality management practices in federal agencies, and gave these agencies working models by which to assess their own Total Quality Management (TQM) progress. According to Tony Stevens, the LRC Quality Coordinator who spearheaded CECOM's application effort, "Winning the QIP puts us in a very distinguished group: since the award's creation, only three other Army organization have won, and none of those was a first-time submitter as we were."

The good news came in a surprise announcement by General Brohm during the farewell luncheon for Colonel Norman Southerland, who was Director of the LRC when it submitted its application. Brohm received the award on 5 June 1996, during the Ninth Annual Conference on Federal Quality. "I am extremely proud of this award," he said: "It recognizes the superb management and work force as well as the technical excellence and outstanding customer service of our LRC organization. It is a fitting and well-deserved recognition of what the LRC and our entire command is contributing to the Army and our national defense."

COL Jack Dempsey, Victor Ferlise, Anthony LaPlaca, Kevin Carroll and Frank Fiorilli with the President's Quality Achievement Award, September 1997.

The effort the LRC put into winning the President's Award for Quality (PAQ) was indicative of the kinds of things the Center routinely did to achieve its major objectives. The quality competition involved intensive effort at all levels by members of the LRC Executive Quality Council, multi-functional teams, process action teams, and the LRC work force at large. Seven focus groups, one for each criterion of the competition, collected source data and drafted input. A "champion" from the Executive Quality Council led each group and made sure that all the required resources and support were on hand. Members of the focus groups, representing all the Center's directorates, collaborated to give the application a center-wide perspective. Each group also had a facilitator who led meetings, gathered input from members and coordinated with the PAQ application manager. On learning that it was among the ten finalists, the LRC established a new team of "ambassadors" to accommodate the competition's site examiners. Members of this team arranged for facilities, set up interviews, assembled back-up data, and served as guides. Then, having won the award, the LRC set up teams to develop a case study, a workshop, a videotape, and display booth for presentation at the forthcoming quality conference.

CUSTOMER SERVICE

The QIP Award recognized CECOM for the application of practices and procedures that made it the Army's most efficient, most responsive supplier of equipment and logistics support services. The superiority of CECOM's performance in this regard was reflected in key indices. CECOM's "stock availability" rate for FY95 was 91.3 percent. Stock availability rates for the other four AMC logistics support organizations ranged from 82.1 percent for the Aviation and Troop Command to 90 percent for the Missile Command. The availability of critical "non-mission-capable" supplies in CECOM was 94.4 percent. In other AMC organizations, it ranged from 79.9 percent for

the Aviation and Troop Command to 92 percent for the Armament and Chemical Acquisition and Logistics Agency (ACALA). CECOM back-orders as a percentage of demand were just 1.3 percent in FY95. Other AMC organizations had backorder rates ranging from 2.4 percent (Tank-Automotive and Armaments Command) to 4.3 percent (Aviation and Troop Command). CECOM also had the lowest rates in AMC for Materiel Release Denials. Depots issued Materiel Release Denials when equipment tagged for release to users was missing components, was unavailable, or was damaged. CECOM's rate for FY95 was 0.8 percent. The rate in other AMC organization ranged from 0.9 percent (ACALA) to 1.7 percent (Aviation and Troop Command). These indices were a direct reflection of a command's proficiency in forecasting demand, purchasing stock, and keeping the depot inventory accurate.

Cost was another measure of efficiency. Each quarter, AMC computed the logistical operations support cost per item for each commodity command. For CECOM in FY95, the cost per item was $2,297. The per-item costs of other AMC organizations ranged from $4,187 for ACALA to $8,512 for the Tank-Automotive and Armaments Command. While some of CECOM's advantage in this index could be attributed to "economies of scale" - CECOM managed fully half of all the stock-numbered items in AMC - it also reflected the LRC's aggressive search for ways to cover expenses, boost sales of products and services, and reduce operating costs.

Reduction of operating costs was more difficult in this era of Army retrenchment during which deactivating units turned back large quantities of equipment. To keep escalating inventory costs under control, CECOM took strong measures to "right-size" requirements and dispose of obsolete stock. CECOM managed to reduce the value of its depot inventory during this time from $2.9 billion in 1993 to $2.2 billion in 1995. Concurrently, the command reduced the value of "stock due in beyond requirements" from $24 million in 1993 to $12 million in 1995. When the amount of stock due in together with the amount on hand exceeded what the LRC thought it needed to meet the demand, the LRC had in effect bought too much or had bought it too soon, and this was costly to the government. The fifty percent ($12 million) reduction in the value of the excess stock reflected the LRC's aggressive approach to acquiring the right amount of the right equipment at the right time. This approach included innovative "just in time" acquisition strategies such a direct vendor delivery, packaged buys, omnibus contracting, and delegation of ordering authority to item managers.

Customer satisfaction was partly a product of the LRC's ability to deliver everything the customer ordered and to deliver it all at reasonable cost. But the timeliness of the delivery was also a factor. In this, once again, CECOM was an AMC leader. In partnership with the depots, the command shipped 89.4% of available stock on time in 1995, compared to 85.5 percent for the Missile Command (next best case) and 77.8 percent for the Tank-Automotive and Armaments Command (worst case). Reductions in the time it took the LRC to assemble provisioning requirements and procurement data packages and award contracts helped with the customer satisfaction index, as did a number of other specific improvement, as follows:

From its surveys, the LRC learned in 1992 that customers

were unhappy having to wait eight hours or so for responses to inquiries on the status of their requisitions. By 1995, with upgraded automation, LRC employees could answer questions about requisitions and deliveries while the customer was still on the phone.

With upgraded computer equipment, the Security Assistance Management Directorate was able to reduce acquisition processing times for foreign military sales by thirty percent. This shortened the lead time for deliveries and boosted total foreign sales to more than $1.6 billion in 1995, a ten percent increase over the previous year.

In formal partnership with Tobyhanna Army Depot, the LRC introduced into the depot a teaming concept that effectively reduced the time for preparing high-priority ships from 3.5 days to less than a day. Concurrently, it reduced the time for preparing routine shipments by more than two-thirds.[207]

TEAMING FOR CHANGE

In FY96, the LRC revamped its structure totally to embody the teaming concept from the ground up, aiming ultimately to provide customers with "one stop" service while reducing staffing and coordination times for many processes previously shared by different directorates. A strategy to guide the move from a traditional hierarchical structure to a structure based on functionally integrated teams - "LRC '97: Teaming for Change" - emerged from numerous executive and work force planning sessions. It depicted the integration of the directorates of Materiel Management, Logistics and Maintenance, Systems Management, Product Integrity and Production Engineering, and Intelligence Materiel Management to constitute multi-disciplinary teams in three weapon system directorates and one consolidated Directorate for Logistics Engineering Operations. The schedule for implementing this reorganization purposely coincided with the BRAC-related relocation of the LRC from the CECOM Office Building in Tinton Falls to Main Post Fort Monmouth. The precept for forming integrated weapon system teams was to focus operations on specific customer groups and commodities, creating multi-functional environments in which workers of different disciplines would interact freely to ensure the highest level of customer satisfaction.

The LRC tested its teaming and empowerment concepts first at the Communications Security Logistics Activity (CSLA) in Fort Huachuca, Arizona. The center's first director, James Skurka, set up the experiment there in 1990 to determine how effective wide-scale deployment of functionally integrated, self-managing teams could be. The CSLA teams were empowered to perform functions traditionally reserved for management: budgeting, training, setting of priorities, scheduling of work, appraising, rewarding, measuring, and disciplining. The test was a resounding success: by 1995 CSLA was a nationally recognized pioneer in the federal use of self-directed teams. It was a nominee for the Quality Improvement Prototype Award in both 1994 and 1995; in three consecutive years, its personnel addressed the Federal Quality Institute's Annual Conference on Quality. Its director, Richard Dion, won the 1994 John W. Macy Award for excellence in the leadership of Army civilians.

The LRC initiated testing of the concept at Fort Monmouth in July 1995 with the formation of four prototype teams: Power Sources, Sensors, SINCGARS, and ASAS/Combat Terrain Information System. These teams, chartered under the direction of Tony LaPlaca, the center's Associate Director, would provide "lessons learned" to help prepare the LRC for full scale reorganization in FY96 and FY97. The teams consisted of workers in all the pertinent LRC disciplines: item managers, quality assurance specialists, engineers, Integrated Logistics Support (ILS) managers, provisioners, technical writers, and catalogers. With the intent of breaking new ground in ways of conducting business, the prototype teams paid big dividends:

The entire workforce of the Command, Control, Communications, Computers and Intelligence Logistics Readiness Center teamed to plan for changes in the center beginning in 1997

they improved awareness of program needs, shortened mail routes, abridged formal staffing and documentation requirements, increased responsiveness to customers, and heightened job satisfaction.

CIPO: A FOCUS ON INTEROPERABILITY

Section 912 of the fiscal year 1998 Defense Authorization Act included several requirements pertaining to acquisition. In an April 1998 report to Congress responding to some of those requirements, the Secretary of Defense noted that "joint operations have been hindered by the inability of forces to share critical information at the rate and at the locations demanded by modern warfare." To attack this problem, the Secretary directed the creation of a study group to examine ways to establish a joint command, control and communication integrated system development process, advance command, control and communication integration and interoperability between the services and achieve efficiencies across the developmental process leading to reduced costs of acquisition, support and operations. The study group included CECOM, the Air Force Electronic Systems Center (ESC) and the Navy's Space and Naval Warfare Systems Command (SPAWAR). The study group established a Joint Command and Control Integration/Interoperability Group consisting of the commanders of CECOM, ESC and SPAWAR, as well as three CINC (Commander in Chief) Interoperability Program Offices (CIPO), each comprised of CECOM, ESC and SPAWAR personnel. One CIPO was located at Fort Monmouth, one with ESC at Hanscom Air Force Base, and one with SPAWAR in San Diego, CA. The CIPOs were to assist in making old technologies/systems more interoperable, to ensure new technologies were "born joint," and to enhance the capabilities of the CINC of the nine unified commands. Priorities for the CIPO included increasing situational awareness to fight as a coalition force, and reducing fratricide. The CIPO at Fort Monmouth was disbanded in June 2006.[208]

REVOLUTIONIZING MILITARY LOGISTICS

CECOM provided the logistical support for virtually all electronics-related items in the U.S. Army inventory. By the late 1990s, this organization, and every other tasked with a similar mission for other commodities, relied on a computer system that was over thirty years old. In response, the Army Materiel Command, in conjunction with CECOM and private industry, established the Wholesale Logistics Modernization Program (WLMP). The program would later become simply known as LMP.

LMP sought to modernize the Army's logistics system and use the same computer-based tools as private industry in order to create a better supply system. LMP addressed requests in an almost real-time environment instead of running batches of requisitions. This dramatically improved CECOM's responsiveness to customer needs.

This modern, enterprise-based program additionally allowed logisticians to obtain information and insight far beyond that indicated by traditional printed reports. This meant improved analysis of different types of data and resulted in enhanced decisions on the part of the logistician. This meant decreased time for field units to order and receive the items they required and decreased CECOM time and money in providing this service to the Army.

LMP IN TIME			
1997	1999	2000	2003
LMP begins with request for proposals to industry	$800M FFP contract awarded to CSC	SAP selected as ERP package	Goes live: C-E LCMC and Tobyhanna Army Depot
2004	March 2005	September 2005	March 2006
DoD Award for Seamless Information Technology	SAP Customer Competency Center certification	Clinger-Cohen Act (CCA) compliance	PEO EIS assumes LMP leadership

While LMP represented a revolutionary improvement over the previous system, one of the most interesting aspects of this project was the innovative nature of the contract with the Computer Sciences Corporation (CSC). AMC created a strategic alliance with private industry and purchased ten years worth of the service instead of buying the system outright. Additionally, all of the data existing in the old system was successfully transferred into LMP. Each government employee whose job was negatively affected received a "soft landing" that offered a $15,000 signing bonus and a three-year contract with CSC, among other benefits.

Another Army process-reengineering initiative, known as Single Stock Fund (SSF), targeted the purchase of Secondary Items (replacement assemblies, repair parts and consumables). Before SSF, there was a distinct separation between the wholesale and retail level, and a very complicated purchasing and procurement arrangement. With few exceptions, items that left the wholesale area would subsequently "disappear" into the retail system. Field units would be given all the parts they wanted free of charge. This created little incentive to limit inventory. An "iron mountain" of spare parts was stockpiled throughout the Army.

Under SSF, the distinctions between the wholesale and retail level logistics structures greatly diminished. AMC obtained visibility into the assets of the Directorate of Logistics at every post, camp, and station, and paid for the repairs on every item. This simplified the management and funding processes and increased the visibility of assets, since the wholesale level could see what the retail level had in stock. Consequently, logisticians could not transfer items from the wholesale to the retail unit, but from unit to unit if necessary. This created additional flexibility and improved efficiency.

SSF ultimately enabled the Command to see more of its inventories, manage them more intelligently, and capture costs with greater clarity.[209]

STREAMLINING THE CONTRACTING PROCESS

CECOM and the Army recognized the need to speed the process by which they acquired and delivered the best technology available to Soldiers. CECOM's efforts in acquisition reform successfully allowed the command to use commercially available products and software and adapt them where necessary to meet the needs of the Soldiers. In instances where products and software were not available off-the-shelf, CECOM developed the new technologies needed to enhance overall capability.

The Interagency Interactive Business Opportunities Page (IBOP) launched on 14 May 1999. Supporting all U.S. commands, Army leaders saw the IBOP as an innovative and easy way to expedite the process of passing solicitation and contract information to and from potential bidders.

IBOP was a significant step towards implementing a totally paperless and more efficient contracting environment. Designed to capture the entire solicitation process from posting draft documents to electronic signature of contracts, IBOP revolutionized the business and provided a main point of information dissemination regarding solicitations for DoD. Furthermore, IBOP was successfully exported to other federal agencies. The Department of Energy, Department of State and all of its embassies world wide, Department of Commerce, U.S. Special Operations Command, Army Forces Command, and U.S. Navy SPAWAR all used it.

IBOP was one example of CECOM's efforts to leverage technology growth and modern commercial software applications to accomplish the rapid contracting solutions demanded at the time. Reverse auctioning, which compelled sellers to bid down through vibrant competition with other sellers, was another.[210]

Acquisition reform was as much a priority for General Brohm as it had been for his predecessor, General Guenther, and it remained a priority for General Guenther in his role as Army Director for Information Systems, Command, Control, Communications and Computers (DISC4): "We're looking at a rapid action process," he said in 1995, "where we identify key things we want to buy, they're brought to the board, decided on, and funded."

Guenther had been carrying the gauntlet of reform for the Army at least since the beginning of 1994. During the first week of January, as Commander of CECOM, he directed Victor Ferlise, his deputy, "to take a fresh look at procurement." Ferlise, in turn, invited the directors of the command's three centers, the Chief Counsel, the Director of Program Analysis and Evaluation, and the Deputy Director of Materiel Management to gather for a "brainstorming session." This group, which subsequently included representatives of the PEO community, constituted what came to be known as the "Fort Monmouth Acquisition Re-engineering Team."

There was at the time a sense of history in the making. "Our upcoming meeting on acquisition reform is of critical interest to both me and the CG," wrote Ferlise on 11 January 1994:

The CG sees this contribution as the pinnacle of his career as an acquisition officer. I believe we have a unique opportunity to advance revolutionary concepts in acquisition reform. Our credibility with higher headquarters has never been higher, and I am sure this will facilitate acceptance of our ideas.

Ferlise asked the invitees to come to the meeting prepared: they were each to meet with their best people to develop a one-page paper addressing the possibilities of reform in their own areas of expertise and, if they had not already done so, they were each to read a book by Michael Hammer and James Champy, *Re-engineering the Corporation.*

CECOM and the State Depoartment sign a Memorandum of Agreement allowing the sharing of technology, May 1998

The session, held as scheduled on 18 January, produced a number of ideas:

Teaming on major procurements through an integrated team approach: Identify players up front and work together early in the requirements generation and definition process through the post award. The team would utilize concurrent engineering (parallel) principles and eliminate the current sequential "picket fence" process. Teaming would establish responsibility, reduce administrative (non-value added) functions, provide flexibility, and streamline/shorten the overall process. Teaming would also eliminate reviews - except for a senior level review.

Empowerment: Item managers should have the authority to buy spares.

Information Technology: Use advanced automation technology. Use wide area networks and the Electronic Bulletin Board, along with face to face contacts. Review the current acquisition data base. Permit all players, including PMs and contractors, access to data bases.

Flexible Contracts: Increase the use of flexible contracts (ID/IQ, RQR)

Additionally, the meeting called for a re-examination of some existing practices - the role of the competition advocate in the acquisition process, for example, and the policy of having Legal Office branch chiefs sign off on acquisition summary sheets to the commanding general. The participants also briefly addressed the need to involve the community, through support to the Battle Labs, in the definition of requirements. The session closed with taskings to each of the participants and plans for a follow-up meeting in February.

Guenther shared some of these ideas with leaders of the Signal Center on 24 January 1994 at a home-on-home conference in Fort Gordon. He spoke first of the process improvement initiatives that were already in place at CECOM: Advanced Planning Briefings to Industry, omnibus contracting, professional development programs, the Commanding General's semi-annual conference with Chief Executive Officers, the Electronic Bulletin Board System, one-on-one sessions with prospective contractors, and the appointment of a non-developmental item (NDI) advocate. He then spoke of the challenge that faced the Army, quoting an October 1993 statement of John Deutch, Under Secretary of Defense (Acquisition and Technology): "The process is in terrible shape. It's not the result of past people having been either stupid or dishonest. It's the result of an infinite number of regulations." Even so, said Guenther, the Army could not wait for the "Top" to change the laws: to meet the challenge, CECOM and its AMC counterparts had to re-engineer the process from the bottom up. Finally, he spoke of CECOM's response to the challenge: the constitution of the Acquisition Re-engineering Team and some of the ideas it brought to the table in its first meeting. In this he placed particular emphasis on use of an integrated team approach for major procurements.

The idea of an "integrated team approach" had been around for awhile. A TQM expert who sat in on the 18 January proceedings noted:

> We could benefit from studying the lessons learned from our own [Communications Security Logistics Activity] at Fort Huachuca, which long ago formed teams organized around commodities and made them self-directed. We may not want to be that aggressive initially, but they [CSLA] can tell [us] about things like phasing in the concept [and] the types of people most likely to succeed in this environment.

And the SATCOM Project Manager employed an "integrated team approach" in the Special Project Office he established in October 1993 to shepherd CECOM's first "Pacer Procurement." The concept was not altogether new; nor were most of the other ideas broached in the Acquisition Re-engineering Team's first meeting. Nevertheless, in the words of the TQM expert (Thomas Cameron), their implementation in most instances meant "a definite paradigm shift."

CECOM proved itself adept at shifting. An independent survey, performed toward the end of 1995 at the request of the Army Vice Chief of Staff, concluded that of all the Army's contracting organizations CECOM's was the most efficient. By all appearances, CECOM also led the Army in streamlining procurement services and implementing innovative contracting methods. Edward Elgart, Director of the CECOM Acquisition Center said he also believed that CECOM was the DOD leader. From the end of 1993 to the end of 1995, building on such initiatives as electronic contracting, omnibus contracting, pacer acquisitions, oral presentations, and use of ordering officers for Indefinite Delivery/Indefinite Quantity (IDIQ) contracts, CECOM cut the average cycle time for procurement by about thirty-six percent.

The Electronic Bulletin Board System (EBBS) permitted instantaneous communication between CECOM and more than 7,000 contractors; its use produced continuous, measurable reductions in the time required to place new contracts. CECOM launched its first paperless procurement in April 1992, when it used the EBBS to release solicitations for the engineering and manufacturing development of the Single Channel Anti-Jam Manpack Terminal (SCAMP) and the Secure Mobile Anti-Jam Reliable Tactical Terminal (SMART-T). At the time, the EBBS had multiple "islands" operating with many different phone numbers and passwords; the second phase system, which came on line in mid-1994, had a single phone number and a single password and could accommodate concurrently as many as sixty-four users, twenty-two hours a day. As of that time (about May 1994), the Acquisition Center had processed eighteen to twenty additional solicitations through the EBBS. Its goal was to employ the system for all solicitations of more than $25,000. Other AMC organizations, following CECOM's, also issued electronic solicitations. Nevertheless, as of the end of February 1996, CECOM was the only command in the Army that used electronic solicitations for all (one hundred percent) of its mission procurements, and with successful receipt of a contractor's encrypted proposal, CECOM had already taken the next step in journey toward totally paper-free contracting.

"Omnibus Contracting" produced a less costly, less time-intensive approach to soliciting, awarding, and administering service contracts. Having identified related functions in each of three domains - Business and Information Systems, Logistics and Readiness, and Research, Development, and Engineering - a team of procurement specialists in the C3I Acquisition Center consolidated requirements so as to reduce the total number of service contracts required in CECOM, thereby reducing overhead and improving the efficiency of operations. The Center awarded the Computer Systems Development Corporation first contract under the Omnibus umbrella on 17 September 1993. This contract, for an estimated $24 million, provided CECOM's Corporate Information Directorate and its customers with support services for workplace automation and telecommunications. On 30 September 1993, the Center awarded the second Omnibus service contract to ARINC to support the Army Interoperability Network. Advantages of Omnibus Contracting included standardization of contractual documents (uniformity promoted clarity and elimination of ambiguity), reduction of contractor overhead expenses (fixed costs could be spread over a larger base), and reduction of government overhead costs (fewer contracts to administer). In addition to saving the government money, Omnibus contracting enabled users to obtain needed services in a matter of days with fewer resources than would be required to acquire the same services through separate contracts in a fully competitive acquisition environment. To promote uniformity and efficiency, the Acquisition Center established a single, centralized Omnibus Contracting Team. Before the inception of omnibus contracting, the Acquisition Center had jurisdiction for more than 125 major service contracts. With omnibus contracting, the Center expected by the end of FY96 to reduce this total to twenty-three (four valued at about $95 million for business information systems; five, at $200 million for logistics board readiness; and fourteen, at $1.4 billion for research and development).

The "packaged buy" concept was a similarly conceived innovation used in the procurement of spares and repair parts in which items of similar product technology and manufacturing processes were lumped together for acquisition from a single supplier through a single indefinite delivery/indefinite quantity type of contract.

With Pacer procurements, Fort Monmouth pioneered the use of what would later be known as Integrated Product Teams (IPT). Smart, experienced people with authority to make decisions participated in all phases of the acquisition process, from the preparation of planning documents, system specifications, and the statement of work through the final selection of a winning contractor. With such people, CECOM could accomplish a major procurement in less than a hundred days. CECOM applied the Pacer concept initially in FY94 in two high-visibility acquisitions: the Tri-Band Super High Frequency Tactical Satellite Terminal (TRIBAND) and the Tactical Endurance Synthetic Aperture Radar (TESAR).

The Army needed the TRIBAND for its newly formed Power Projection Command, Control, and Communications (Power PAC3) Company. With TRIBAND, the Power PAC3 Company would support Army forces headquarters and liaison teams with critical beyond line-of-sight communications, quickly deployed for joint and combined task force operations. The Project Manager for Satellite Communications (PM SATCOM) received word of the requirement by telephone on 22 October 1993: the Signal Center wanted six prototype terminals on or before 30 September 1994. The Deputy Chief of Staff affirmed the requirement by memorandum to the DISC4, dated 7 January 1994.

When the Procurement Administrative Lead Time (PALT) in CECOM averaged 234 days, meeting the required delivery date, just 344 days from the initial telephonic request, required extraordinary measures. To speed procurement, the Department of the Army directed the PM to procure the TRIBAND terminals on a non-developmental item (NDI) basis. The six prototypes were to use existing technology, and the winning bidder was to provide initial spares, depot-level maintenance, and field support for the five-year life of the system. Even so, the Acquisition Center's first projected date for award of the contract was 30 June 1994. When the PEO insisted on an award on or before 31 March, the Acquisition Center and other elements of the CECOM matrix rallied to arrive at an award on 24 March, seven days ahead of schedule and more than three months ahead of the Acquisition Center's initial estimate.

As a first step in achieving this feat, on 25 October 1993, the PM established a Special Project Office to shepherd the acquisition. Staffing of the office began with PM personnel and embedded matrix support components, but expanded quickly to include representatives of CECOM functional elements, including the RDEC, the LRC, the Legal Office, and the Acquisition Center. This team received high-level management support not only from the PM, the PEO, and CECOM, but also from the Signal Center and Army Staff. More significantly, rather than being motivated by functional requirements, the team focused on a goal: to deliver the product by 30 September. To this end, it adopted a "zero-based" requirements philosophy by virtue of which the team pushed matrix support elements to justify

all the requirements they wanted included in the Request for Proposal (RFP) and other acquisition documents. Team leaders had the power to limit requirements, even to the point of discharging intransigent proponents. According to the PM's after action report:

> The team enforced a "no business as usual" policy. ... Each time the bureaucratic process demanded an action, a document, or a deliverable that didn't make sense, the requester was asked why. "Boiler plate" input to the specification or Statement Of Work provided by the matrix was drastically tailored to fit the program. Leaves were delayed or canceled, duty hours extended to include nights and weekends.

The team also worked hard to make industry a player in the process. Frequent communication and industry input during the formative stages of the procurement data package helped ensure timely release of the solicitation. The PM posted each revision of the specification and the statement of work to the Electronic Bulletin Board and responded to industry comment on these revisions usually within twenty-four hours. To further streamline the acquisition process, the team received authority to waive requirements, including requirements for internal review. During the pre-solicitation phase, it waived the requirement for an Operational Requirements Document. During the solicitation phase, it waived requirements as well for the Board of Solicitation Review, the Proposal Evaluation Adequacy Review, and the Contract Review Board. The presence on the team of functional representatives with authority to approve actions obviated the need for time-consuming committee reviews. It also shortened the time required for preparing essential documents. It took only five days to write the Source Selection Evaluation Plan, for example, because fully empowered representatives of all participating offices were part of the process. They could make changes "on the fly" without having to wait for approvals. Concurrently, the Product Integrity and Production Engineering Directorate established a "Superteam" to bypass the formal requirements of the Systems Data Review Board by working informally with the appropriate functional experts. The Superteam was also instrumental in reducing from forty-one to eleven the number of items in the Contract Data Requirements List. To accomplish this, it actively challenged the necessity of each item. To accelerate the assessment of performance risks (the Performance Risk Assessment (Analysis) Group (PRAG) process), the Special Project Office, using the Electronic Bulletin Board, asked bidders to submit draft PRAG information one month before the due date for submitting formal proposals. This request, dated 28 December, had a suspense date of 10 January, two days before the 12 January release of the solicitation. The Acquisition Center issued the Request for Proposal as a "paperless solicitation" on the Electronic Bulletin Board, just twenty-four hours after receiving the PM's procurement data package.

Anticipating the formal source selection process, the TRIBAND team informed potential bidders that they were to be evaluated on "key discriminators," that logistics would not be evaluated (instead, bidders were to be given a ceiling price for initial spares and maintenance support), and that technical proposals were to be limited in length to one hundred pages. As a result, both the source selection plan and the proposals were more precise, and there were fewer "Items for Negotiation" arising from the proposals. Typically, a source selection generated between 250 and 300 items for negotiation (IFN) for each bidder, the resolution of which required two to three days of face-to-face negotiation. The IFN for the eight bidders in the TRIBAND acquisition ranged from twenty-five to forty-nine. The government team eliminated face-to-face negotiation and, so, cut at least three weeks from the program's PALT. It saved another two weeks (approximately) when it eliminated the Initial Competitive Range Determination that normally preceded release of the IFNs. When it sent out model contracts with its request for best and final offers, the Acquisition Center saved the two or three days of administrative time usually associated with the signing of the contract after notification of the winner.

With all this, CECOM managed to award the TRIBAND contract in just 154 days. Just seventy-two days elapsed between the issue of the solicitation and the award, representing a seventy percent reduction in the command's usual PALT. During the debriefing of unsuccessful bidders, 5 April 1994, there was a general consensus among them that CECOM had conducted the acquisition "very competently and professionally" and that the short acquisition cycle had been "very beneficial since bid and proposal costs were reduced."

The TESAR was the Army component of the joint Unmanned Aerial Vehicle (UAV) program. The TESAR consisted of both ground and airborne subsystems, of which the latter made up the bulk of the UAV payload. PEO IEW managed Army participation, but the CECOM RDEC ran the TESAR Source Selection Evaluation Board. The urgency facing CECOM and the PEO in the TESAR acquisition stemmed from the need to supply the Navy's platform and data link contractors with the form, fit, and electrical/mechanical interface information they needed to meet their development schedules. Following the first publication of the procurement synopsis in Commerce Business Daily on 3 December 1993, CECOM had just ninety-eight days to award the TESAR contract. A contributing and/or complicating factor was the designation of the program as an Advanced Concept Technology Demonstration (ACTD). ACTD candidates had to have existing or maturing technologies with enough "off-the-shelf" stuff to yield a "fieldable brassboard" that had been examined in detail for doctrinal and tactical exploitation. As opposed to Advanced Technology Demonstrations, which were driven by the maturity of the underlying technologies, ACTDs were driven by their impact on military capability. User involvement was, therefore, critical to their success. Alternatively, ACTD designation brought with it the ability to abridge the early stages of the acquisition cycle, both by bypassing the concept exploration and definition phase and by the potential for abbreviating the engineering and manufacturing development phase. It also brought to the program strong support from the Department of Defense, assurance that senior leadership in the Army, CECOM, and Fort Monmouth would apply the resources the program needed to achieve its objectives on time, and freedom from the need to fulfill such formal requirements as the Operational Requirements Document, the Test and Evaluation Master Plan, and the Army Systems Acquisition Review Committee process.

In local implementation, in addition to waiving the Operational Requirements Document and the Test and Evaluation Master Plan, the Product Manager exercised his authority to waive the requirement for a formal Systems Data Review Board. Instead, knowledgeable experts in the CECOM matrix performed informal reviews of various technical data elements, paring about

The ITT and the US Army team gather for the signing of the Advanced Threat Radar Jammer (ATRJ) contract, July 1994

sixty days from the program's pre-solicitation phase. Management also waived the requirement for review by the Senior Board of Solicitation and confined deliberations of the Source Selection Advisory Council to a single meeting.

From its inception, assumed to be the 16 August 1993 memorandum from the Assistant Secretary of the Army (RDA), to the release of the draft solicitation on 3 December, the acquisition took 109 days (vs. the usual time of one year). PALT, from the release of the solicitation to the award of the contract on 9 March 1994, took just ninety-six days (vs. six months). A key contributing factor was the early appointment of a Source Selection Evaluation Board, constituted of people who were familiar both with the source selection process and the technical and programmatic aspects of the TESAR. In just six weeks, senior members of this Board prepared the program's statement of work, specification, acquisition plan, Justification and Approval, and Source Selection Evaluation Plan. This team also made good use of the Electronic Bulletin Board to speed communication with prospective bidders.

Although very different in kind - the TESAR was an Advanced Concept Technology Demonstration while the TRIBAND was primarily a systems integration effort - the two programs had much in common. They both had backing at the highest levels in Department of the Army and above and full support of their PEO and the CECOM Commander. Program managers for the two systems both had authority over a wide range of activities, as a result of which they were able to shorten the time their management chains spent in review and oversight. Early on, the managers of both programs recruited smart, experienced people for their teams and retained them through all phases of the acquisition, thereby ensuring continuity of corporate knowledge and easing the transition from one phase to the next: People who prepared the acquisition strategy, system specification, statement of work, source selection plan, and other such acquisition documents were major players, as well, in the source selection process. In both programs, the government teams developed and maintained close ties with industry. They met with industry prior to the release of their RFPs and encouraged prospective bidders to provide comment on and input to drafts of this document and the system specification. In communicating with industry, they both made extensive use of the Electronic Bulletin Board (EBB). In all their activities, the Pacer acquisition teams examined the processes they employed in order to abridge the time needed for their execution. In both, they exercised their authority wherever feasible to eliminate unnecessary requirements and time-consuming internal reviews.

Pacer (Integrated Product) Teams worked well in selected high-profile acquisition programs. To employ some of the Pacer processes in other, more traditionally structured acquisition programs, CECOM established a Lead Time Reduction Red Team. This team, organized in April 1995, implemented strategies for which it claimed a 102-day reduction in acquisition lead time between December 1994 and November 1995. It drew the preponderant proportion of its members from the Materiel Management Directorate but also had representatives from the Product Integrity and Production Engineering Directorate and the Acquisition Center.

The Red Team's usual tool was the weapon systems reviews it conducted with item managers and their matrix support elements. The team reviewed the programs of forty-six weapon systems during its first seven months, analyzing data bases and acquisition strategies to identify ways of abridging cycle time. Reduction initiatives included, where appropriate, elimination or reduction of testing and inspection requirements; use of aggressive delivery schedules; maximum use of commercial specifications and standards; and the application of Value Engineering. The team also developed an on-line lead-time tracking system and a lead-time reduction bulletin board to keep the work force apprised of its actions and objectives. It estimated savings of $1.6 to $1.7 million for each day it cut from the Command's average administrative and procurement lead time.

Edward Elgart, director, CECOM Acquisition Center, addresses members of the National Contract Management Association

Secretary of Defense, William J. Perry's June 1994 call for acquisition reform stipulated the use of industry standards and performance-based specifications in future procurements. His objective in eliminating specifications that were unique to the military and requirements that added no value was to free defense acquisition from constraints that made the development and procurement of materiel both difficult and expensive. This, in turn, accelerated use of state-of-the-art commercial technology in U.S. weapon systems and promoted integration of the military and commercial industrial bases. The Army issued its plan for implementing Perry's initiatives in November 1994.

Implementation of the Army plan at Fort Monmouth entailed a "bottom up" review of all the specifications and standards used in the acquisition of military C-E equipment, with a view toward determining which of these documents were to be abolished, re-engineered, or retained. Disposition options included cancellation, inactivation pending development of a new design, replacement by a non-government or commercial standard, and conversion to a performance-based specification.

Local implementation also involved the organization of a multi-disciplinary team, known initially as the "Standardization Improvement Working Group," then as the Standardization Program Team (SPT). Charter members of the team included one action officer from each of three organizations - the Program Executive Offices for Intelligence and Electronic Warfare (PEO IEW), the Program Executive Office for Command, Control, and Communications Systems (PEO C3S), and the CECOM LRC. Associate members represented the LRC, the RDEC, the Safety Office, and the Acquisition Center. Kenneth Brockel, in his capacity as Principal Assistant for Specification and Standards Reform, functioned as senior advisor to the SPT. Fort Monmouth Standards Executives - Bennet Hart (PEO C3S), Edward Bair (PEO IEW), and Colonel Norman K. Southerland (LRC) - approved the SPT's Master Action Plan on 5 February 1996.

This plan outlined a strategy for accomplishing specification and standards reform through education, cultural change, an overhauling of the standards process, imposition of methods for identifying and eliminating excessive contract requirements, and the development of new management tools. With the concurrence of the Standards Executives, the SPT turned its attention to preparing a goal-oriented five year business plan, to enlarging the existing internal specification and standards reform training program, and to creating an informational "Home Page" for the program on the Internet. The SPT also initiated development of a local "roadshow" to introduce elements of the Master Action Plan to the work place, academia, and industry.

It was more difficult to implement acquisition innovations than to imagine them. Implementation depended upon the existence of a responsive, well-trained work force. Knowing this, Elgart and his managers allocated fourteen percent of the Acquisition Center's non-payroll budget to professional development. They encouraged employees to continue their education and reimbursed them for expenses incurred in taking graduate and undergraduate courses. They implemented a cross-training/developmental assignment program in coordination with other components of Team Fort Monmouth.

As of February 1996, CECOM had inducted more than forty-five employees into this program, giving them opportunities to hone technical skills and improve their understanding of customer needs and processes. Developmental assignments, typically 120 days in length, met the needs of both the organization and the individual, as illustrated in the following examples:

An Acquisition Center branch chief trained in a program management position in a Program Executive Office. The branch chief brought back precious insight into the customer's support needs and viewpoints.

A procurement contracting officer trained in New York in Defense Contract Management Area Operations. He returned to CECOM with a better understanding of the relationships among pre- and post-award activities.

A legal advisor trained in the Materiel Management Directorate. He gained a better understanding of materiel management processes, and materiel management personnel gained appreciation for the legal aspects of their actions.

Another special program gave entry-level Army officers (captains and majors), who were well trained to be managers, the technical skills and experience base they needed to perform well in the Army acquisition arena. The program featured individually tailored training plans, developmental assignments, cross-functional experiences, and mentoring.

One measure of performance - perhaps the key measure of performance for the Acquisition Center - was timeliness of service. To measure performance, the Acquisition Center initiated a requirement that all its buying branches use Statistical Process Control charts for tracking various types of procurement in terms of Procurement Administrative Lead Time (PALT) and Administrative Lead Time (ALT). With Automated Procurement System (APS) 1-A Sheets, PALT and ALT counted the days for achieving each milestone in a procurement action and, so, identified the total amount of time required to achieve the action. Comparing actual performance to the baselines established for each milestone helped the Acquisition Center identify problem areas and areas of substandard performance. Performance anomalies invoked the preparation of corrective action plans that identified the reasons for the de-

Red Team puts CECOM in 'black'

by Helen Roche and Joyce Moffat
Directorate of Materiel Management

Strategies, reviews and actions initiated by a CECOM Red Team have reduced acquisition lead time by 102 days since December 1994, saving millions of dollars.

The actions are a response to an executive order in which the Secretary of Defense challenged all Department of Defense (DoD) agencies to reduce cycle times.

The Army Materiel Command (AMC) set goals for the Materiel Support Commands (MSCs) and instituted across-the-board reductions to lead times. They also formed an AMC Lead Time Reduction Process Action Team (PAT).

CECOM established a Lead Time Reduction Red Team this Spring, made up of volunteers from within CECOM, with technical expertise in different areas.

They are a multi-functional, integrated product team comprised of Anna Marie Van Brunt, Gary Webber, Dan Roddy and Sandra Castro, from the Directorate for Materiel Management (DMM), Len Jacques from Product Integrity and Production Engineering Directorate and Pete Kasper and Linda Cooper from the Acquisition Center. The team welcomed its newest member, Paula Fren, from DMM last month, who will be a player for a four month period.

The teams goals were to streamline the acquisition process to minimize Administration Lead Time (ALT); to pursue and institute strategies to reduce existing and future lead times; to insure data base accuracy; and to insure metrics accurately capture lead time activity and are visible throughout the entire process.

Members of the Red Team established to streamline the acquisition process are (from left): Dan Roddy, Bill Riehl, Len Jaques, Paula Fren, Linda Cooper, Gary Webber and Sandra Castro. Members of the team not available for the photo are: Anna-Marie Van Brunt, Pete Kasper and Hilary Ruske.

photo by Dave Brackmann

One of the actions initiated by the Lead Time Reduction Red Team was to have weapon systems reviews with the item managers/matrix. To date, they have reviewed 46 weapon systems, which focused on data base and acquisition strategy analyses.

One day of ALT reduced saves $1.6 million, and one day of PLT reduced saves $1.7 million.

Other PLT reduction initiatives include, where appropriate, the elimination of pre-production First Article test (FAT); minimiz-

ing initial production FAT; the reduction to test/inspection to minimum levels; analysis of "driver" systems for best possible PLTs; utilization of aggressive delivery schedules; and commercial specs/standards to maximum, and the use of Value Engineering (VE) to save money.

As a result of the Red Team's efforts in these areas and in conjunction with the workforce actions, CECOM has exceeded

TEAM, continued on page 8

form training program, and to creating an informational "Home viation or delay and, so, helped managers devise appropriate solutions including, for example, redistribution of resources and/or retraining of employees.

Training and other forms of acclimation produced results. Said Elgart (February 1996): "Our work force is creative, responsible, risk-taking, and accountable for owning their piece of mission achievement and customer satisfaction." The motto of the Acquisition Center, "Setting the PACE2," focused all its employees on a "Proactive Approach to Contracting Excellence through People, Automation, Continuous Process Improvement, and Education." With all this, the Acquisition Center set for itself the ambitious, but not unreachable goal of becoming a Centralized Defense Acquisition Center, whose employees could successfully award and manage contracts for everything from batteries to bombers.[211]

THE ORDERING OFFICER PROGRAM

CECOM instituted its Ordering Officer Program in FY94, once again setting a precedent in the Department of the Army. In coordination with the LRC's Material Management Directorate, the Acquisition Center authorized seventeen item managers in the LRC to place delivery orders on high-volume, pre-priced contracts without the intermediation of any Contracting Officer. The Ordering Officer Program thereby established the item manager as the focal point for receiving and validating requirements, obtaining funding approval, and executing delivery. Wherever implemented, the program reduced administrative lead time, increased control over stockage levels, and freed Contracting Officers to work in other essential acquisition activities. Colleen Preston, Deputy Under Secretary of Defense for Acquisition Reform, witnessed the first signing and execution of a delivery order by an Ordering Officer in June 1994. CECOM planned to employ the program in the future in all suitable indefinite delivery, indefinite quantity, pre-priced contracts.

One such contract was the contract CECOM awarded UNICOR in May 1994 for range quantities of 103 different items. In addition to the incorporation of the ordering officer concept, the contract benefited the command in the sense that it placed upon it no other obligation than to buy what it bought on the first delivery order (thirty-nine products valued at $4.2 million), but allowed it to buy up to the maximum quantity as many times as it wished during the life of the contract. CECOM awarded the second phase of this effort, for production of fifteen products valued at $1.9 million, in September 1994. Meanwhile, UNICOR assigned a full-time representative to CECOM to effect coordination between the command and the various federal prison facilities, to supply updated status reports, to troubleshoot problems that might affect delivery of quality products, and to expedite deliveries required by exigencies.

With pre-priced IDIQ contracts, Ordering Officers in the LRC saved time and money in the procurement of sustainment materiel, and by ordering just what the Army needed just in time, they were able to reduce both the size of their depot inventories and the expense of storing these inventories.[212]

COMMERCIAL ACTIVITIES (CA) STUDIES

The commercial activities study process placed government operations in direct competition with private industry to determine which provided the best service at the lowest cost. The winner was subsequently assigned the task. Garrison operations comprised the primary functions under review at many domestic installations. The last CA study done at Fort Monmouth had been completed in 1982.

During the mid-nineties, the Army once again utilized CA studies as one of many tools available to achieve greater efficiency and effectiveness. CECOM announced three studies to Congress on 26 February 1999: the Information Mission Area (IMA) study at Fort Monmouth, the Fort Monmouth Garrison (FMG) Base Operations (BASOPS) Directorate of Logistics (cataloging) study, and the Tobyhanna BASOPS study of information technology and public works.

The final recommendations from these studies became available in 2002. The period of performance for all three areas was for one year, with four one-year options to follow. The in-house cost estimate prevailed for both the TYAD and IMA studies. The annual cost savings per year were estimated at $2.9 million and $8.9 million, respectively. The Fort Monmouth Garrison Study resulted in a win for a contractor, which would result in an approximate annual savings of $1.4 million.[213]

THE BALKANS

When the first U.S. troops deployed to Bosnia for Joint Endeavor, CECOM had already been involved for more than two and a half years in planning for and supporting U.S. military operations in the Balkans. It participated in Operation Able Sentry, prepared estimates of the communications equipment needed to rearm Bosnian Muslim forces, and participated in advanced planning for large-force operations in both Bosnia and Croatia. From time to time, CECOM support involved the deployment of personnel, for example:

Responding to an AMC FAST request, RDEC supplied Task Force Able Sentry a long-range infrared surveillance capability for various observations posts along the Macedonian-Serbian border; it responded throughout the year to urgent requests for the technical and logistical support required to keep this equipment ready and operating. In May 1994, for example, CECOM dispatched two Night Vision personnel to install 2x extenders on several AN/TAS-6 night observation devices. The two men, one government employee and one contractor, arrived in Germany to assemble the devices on 25 May and moved on to Macedonia to install them on 1 June.

From 18 - 25 September 1994, personnel from the CECOM Safety Office paid a safety assistance visit to U.S. Army units in Macedonia. They surveyed ten of eleven U.S.-manned observation posts to identify improvements in lightning protection for their communications-electronics equipment. The two safety experts also provided the personnel who were there with general instruction in safety precautions during installation and use of electrical equipment. The AMC Logistics Assistance Office requested this on-site assistance in response to lightning accidents that had already rendered inoperative several of the communications systems used in these eleven observation posts.

Later, the command participated in preparations for Operation Daring Lion (the details of which were classified), and in October 1995 it participated in a NATO operation known as "Forceful Presence." At that juncture (the end of October 1995) the Software Engineering Directorate had prepared itself to deliver digitized maps of Bosnia; the Intelligence/Electronic Warfare Directorate was prepared to deploy Mini-RES, a miniature SIGINT system in two small packages that used FRI-AR IV software, and was also prepared to support the Signals Warfare Project Manager in the deployment of the Guardrail/ Common Sensor SIGINT system, along with a satellite remote relay; the Space and Terrestrial Communications Directorate (S&TC) was ready to deploy the AV-2095 self-steering antennae needed for satellite communications from a moving vehicle; it was also prepared to develop HF frequency assignments for Bosnia based on ionospheric measurements collected on a daily basis by the National Institute of Geophysics in Rome, Italy; the LRC's Readiness Directorate had deployed the Log Anchor Desk (in coordination with S&TC and the Command, Control, and Systems Integration Directorate) and was prepared to deploy two LSE C2 flyaway packages and as many as twenty-four Logistics Assistance Representatives; GTE had accelerated the fielding of and training for a new MSE software release; the LRC was working with the Global Positioning System (GPS) Project Manager to accelerate the fielding of 3,622 enhanced Precision Lightweight GPS Receivers (PLGR) to selected USAREUR elements and was overseeing installation of 2,200 SINCGARS radios in vehicles of the 1st Armored Division; LRC's Systems Management Directorate had initiated plans to swap 110 AN/PPX-3B interrogator sets for the less reliable AN/PPX-3A interrogator sets in V Corps; to accomplish this latter action, Tobyhanna Army Depot and the LRC's Logistics and Maintenance Directorate were preparing to dispatch personnel to

Europe; the Acquisition Center was negotiating with TRW and Electronics and Space, Inc., to extend contractor support for Guardrail/Common Sensor System 4 and its Advanced Quicklook equipment.

In its role as Executive Agent for Tactical Switched Systems, CECOM engaged in a variety of actions supporting Joint Endeavor. For example: To resolve compatibility problems between the AN/TYC-39 message switch and the British Ptarmigan store-and-forward switch, an RDEC team adapted an AUTODIN communications controller to provide a "handshake" between the two systems, thereby enabling communication between U.S. and British troops. In early January CECOM hosted two days of interoperability testing and training for the 55th Signal Company, a combat camera unit that would deploy soon to the theater to acquire digital images from the battlefield, compress the images, and transmit them back to strategic command sites "in near-real time." According to reports of this action:

The test strings involved a Macintosh computer transmitting files through STU-IIIs with tactical Digital Non-secure Voice Terminal adapters, KY-68 Digital Secure Voice Terminals, and CA-67 Tactical Terminal Adapters to tactical switching assets, back through a complimentary adapter (STU, 68, or 67) and terminated with another Macintosh acting as the receiver. The exercise provided the team critical tactical equipment experience prior to an anticipated Bosnian deployment.

Meanwhile, as Executive Agent, CECOM was receiving numerous requests regarding tactical applications of Motorola's Network Encryption System (NES) which, though primarily a strategic device, was being used increasingly to solve tactical data separation issues such as, for example, the use of a common communications backbone to carry both U.S. and Coalition data. CECOM tested several configurations using the NES in a tactical test bed to develop information for eventual communication to the joint community and incorporation into the Integrated Tactical/Strategic Data Network Quick Fix User's Manual. Also as Executive Agent, CECOM began collecting information in response to many requests from the theater for information on the interoperability of U.S. tactical systems with various commercial switching systems. By the end of March 1996, the Software Engineering Directorate had developed a new version of software for the TTC-39D Circuit Switch to fix a "toggling" problem in its interface with a KN-4100 DSN switch in Heidelberg and other problems identified by the Directorate's European Software Support Office.

The ubiquity of mines in Bosnia spurred the development of new mine detection and destruction technologies for which the RDEC's Night Vision and Electronic Sensors Directorate took the lead. Even before the deployment to Bosnia, the Director-

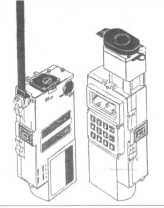

Published for the military and civilian personnel of Fort Monmouth, NJ

June 16, 1995

CECOM radio aided aviator rescue

by Henry Kearney and Cleo Zizos
Public Affairs Office

Last week's rescue of Air Force Capt. Scott O'Grady in the Bosnian pine forest had a CECOM and Fort Monmouth connection.

The PRC-112 radio which allowed the aviator to give his location to search teams flying overhead was procured by CECOM which manages distribution of the radios for use by Army, Navy and Air Force aviators.

The PRC-112s have two frequency capabilities which, when turned on, can be detected by aircraft equipment seeking to locate personnel on the ground, said Jim Mueller, an electronics engineer in the CECOM Command, Control, Communications, Computers and Intelligence Readiness Center (LRC) here, who manages procurement, fielding and logistics support of the radios. Mueller provides functional support to the Program Executive Office for Aviation Systems, St. Louis, Mo., as Assistant Project Manager for Special Mission Equipment.

Approximately 16,000 of the PRC-112 radios have now been distributed to Army, Navy and Air Force units. One of the major advantages of the radios over PRC-90 radios

which were used during Operation Desert Storm is their capability to allow detection by friendly forces only, Mueller explained.

"A PRC-112 is one aboard every aircraft platform," he said, "but we had to wait until Capt. O'Grady was debriefed to confirm that he had used our radio to aid his rescue. It's a great feeling to know that the radio worked the way it was supposed to work to save a life."

Mueller credited Tim Ryder and Rich Gunsaulis of the Research, Development and Engineering Center, who served as technical project leaders for the system for several years, as well as a host of other LRC personnel who have supported the system. ⊙

Radio Set AN\PRC-112 and Program Loader KY-913\PRC-112.

ate evaluated several countermine proposals and prepared to field to V Corps's 16th Engineer Battalion four field-expedient countermine systems - ground-penetrating radar that could be mounted on the front of a vehicle to detect buried, on-road, anti-personnel and anti-vehicular mines, hopefully before vehicle contact. Additionally, for use in Bosnia the RDEC delivered a prototype, remote controlled, vehicle-mounted mine detection system that consisted of a metal detector, an infrared camera, and the ability to transmit video to a control vehicle. The RDEC also fabricated and delivered vehicle mine protection kits for HMMWVs and five-ton vehicles, magnetic mine countermeasures equipment to overcome magnetic-influence mines, and a launched grapple hook for infantry use in clearing trip-wire mines.

Radio communications in the theater presented other unique problems for which CECOM supplied solutions. For example, in early March 1996 the Space and Terrestrial Communications Directorate responded to an inquiry on the availability of equipment in the Army inventory that could be used as a radio relay to extend the range of the VRC-12, PRC-77, and PRC-126 VHF FM radios in Bosnia: the ruggedness of the terrain there limited the range of these line-of-sight systems. Engineers of the Directorate supplied instructions detailing the equipment and procedures needed for configuring both the VRC-12 and PRC-77 to accomplish the radio relay/retransmission function. They also noted that two PRC-126 radios could be used for relay if interconnected by a specially fabricated cable. Though this capability was not documented in the radio's manual and the cable was not in the Army inventory, S&TC personnel had the knowledge and ability to fabricate such a cable in-house on receipt of parts.

Operations in Bosnia also afforded the command opportunities to test new capabilities. Team Fort Monmouth deployed the Shortstop "electronic umbrella" in early February 1996. The Army initiated development of this system for use in Desert Storm, but that operation ended before the system could be deployed. Shortstop protected forces from enemy fire by creating an electronic field that essentially tricked the proximity fuses of incoming mortar and artillery shells to explode prematurely (i.e., several hundred meters away from their intended targets). In US-based field tests, the prototype system knocked out more than 5,000 incoming shells - artillery and mortar rounds fired single and in barrages - before they reached their targets, achieving a one hundred percent effectiveness rate. The system worked by first reading the setting of a proximity fuse, which determined how far above ground the incoming shell was supposed to detonate, and then by sending out signals to the fuse to make it think it had already reached its target. According to staff writers of *Defense News*, "The Army's efforts to field Shortstop quickly shows a growing recognition that electronics will play a greater role on the future battlefield, especially with the trend for militaries to make dumb weapons like mortars smarter with electronic fuses."

CECOM was also instrumental in protecting airborne "targets." Thus, for example, when pilots operating in the Balkans reported an unidentified "threat," personnel of the CECOM Software Engineering Directorate developed an "unambiguous Threat Parameter List" for emitters programmed in the AN/APR-39's Bosnia Mission Data Set. This list allowed pilots and Quickfix ELINT technicians to identify and locate the threat.

By the end of February 1996, in support of the peacekeeping mission in Bosnia, the CECOM Acquisition Center had

Soldiers from the 754th EOD reunite with their families after returning from Bosnia, December 1997

awarded contracts valued at more than $16.5 million for state-of-the-art communications-electronics equipment and the field, engineering, and depot support of CECOM-supplied materiel. The fill-rate for Joint Endeavor requisitions approached ninety-

> *The slow movement is really taking its toll. The number of personnel here has doubled the amount that was previously planned. The LAR find it frustrating because none of the maintenance shops (organization or direct support) are set up and have the available space to accommodate them. This has had no effect on their performance and it shows how dedicated they are to the program and the mission accomplishment.*

six percent.

As of 18 January 1996, CECOM had eighteen of its LAR in the field supporting units engaged in Joint Endeavor. The Logistics Assistance Office (21st TAACOM - Forward) reported on 4 January 1996:

In addition, as of 18 January, CECOM had five GTE employees in Tuzla and one in Kaposvar to support MSE and another six, employed by various contractors, to support IEW systems in theater. Altogether as of that date, there were two military, twenty-seven civilians, and twenty-one contractors in Europe to support Joint Endeavor. By 4 April, the number of GTE personnel in Tuzla had grown to seven, and there were twelve CECOM LAR in Bosnia, four in Croatia, and two in Hungary. Lack of maintenance shops was no longer a problem at that date, but LAR in theater were constrained by convoy and vehicle restrictions to which contractors were not subjected. With a "protecting the turf" kind of pride their work, LAR complained that on several occasions, contractors had gone directly to LAR-supported units before they themselves could get there to provide assistance.[214]

In response to the capture of American Soldiers in Macedonia, the Balkan Digitization Initiative (BDI) was developed. It focused on installing a real-time vehicle tracking system designed to provide commanders with the precise location of any vehicle on patrol. The BDI, also known as "Blue Force Tracking," was a cooperative effort between U.S. Army Europe, the Program Executive Officer for Command, Control and Communications Systems (PEO C3S), the CECOM Logistics and Readiness Center, Tobyhanna Army Depot and TRW, Inc.

This program was considered critical to the entire effort in the Balkans since U.S. forces continuously patrolled areas where there was serious potential for conflict. In order to provide adequate protection and ensure mission success, complete situational awareness was critical to ensure that any problem encountered could be successfully handled. This system, along with all of the associated command and control mechanisms, was designed, built, and installed in seventy M1114 up-armored Humvees in less than seven months.[215]

THE WAR ON DRUGS

CECOM participated in every peace-keeping and humanitarian operation undertaken by the United States Army between February 1994 and April 1996. The CECOM Logistics Assistance Representatives were involved in all of them, supporting troops in their preparations for deployment, accompanying them to the field, and doing whatever they had to do there to keep their C-E equipment and systems operating. CECOM Item Managers filled requisitions for deployed and deploying units, often on an urgency basis, at an average rate well in excess of ninety percent. CECOM scientists and engineers devised effective solutions to technical and logistical problems encountered in the field.

To ensure visibility and maintain control over projects assigned by the AMC Counter-Drug Support Office, CECOM delegated "single focal point" responsibility to the Program Analysis and Evaluation Directorate (Program Development Division). Personnel charged with executing this responsibility reviewed the projects received from national, state, county, and local counter-drug activities to ensure their assignment to and their timely execution by appropriate organizations within the command. Additionally, the division represented the command at quarterly in-process reviews at AMC and acted as its liaison in drug-control relations with Department of Defense, the Office of National Drug Control Policy, and various law enforcement agencies. During FY94, CECOM lent counter-drug organizations equipment valued at $7.1 million.

There were in FY94, six on-going counter-drug programs involving CECOM: the Container Inspection System (state of art technology to generate high-resolution images of organic materials), the Counter-Narcotics Command and Management System (secure voice and data network to tie embassies in South and Central America to the Drug Enforcement Agency,

The CECOM LCMC supports counter drug efforts in many South American countries through functional support agreements and embedded matrix support

the State Department, Department of Defense, and Southern Command), the Mobile Examination Demolition Van, Theater Gull (counter-narcotics correlation centers, remote sites, and intelligence networks in Louisiana, Mississippi, and Alabama), evaluation of cellular phone intercept and direction finding equipment, and development of a prototype Mobile Non-Intrusive Inspection System.

In second quarter FY94, CECOM issued a draft "CECOM Drug Interdiction Support Handbook" containing information on the products, services, and capabilities available in CECOM to law enforcement agencies and procedures for obtaining them.

The RDEC (IEWD) continued its service as Technical Agent for the DARPA-sponsored non-intrusive Counter-Drug Technology Development Program. This program sought to identify and develop technologies that could be used to inspect cargo loads as large as full-size eighteen-wheel tractor trailers. During FY96, the RDEC worked on eight separate technologies, including the Portable Vapor Detector, the Portable Particle/Vapor Detector, the Portable Organic Vapor Detector, Wireless Communications Network, and various X-Ray techniques.[216]

As part of an interagency agreement between the Army and Treasury Departments, IEWD supplied program management and acquisition support to the U.S. Customs Service in the procurement of contraband detection equipment. Procurements in FY95 included six mobile and eight pallet-sized backscatter X-Ray systems and eight Mobile Support Systems. The latter contained all the tools needed for opening sealed containers. IEWD also supported the Immigration and Naturalization Service in the development of the Transportable Observation Platform vehicle, a truck-mounted forward-looking infrared sensor system using commercial hardware and IEWD software.

The RDEC engineered the Translation/Transcription Support System (T2S2) to facilitate communication of wiretap audio data between remote sites and the Utah National Guard linguist complement at Camp Williams, Utah. The data, collected by personnel of the Drug Enforcement Agency using dialed number recorders, was to be stored at four collection sites on Digital Temporary Storage Recorders (DTSR), then transmitted to Camp Williams for transcription and return, in text form, to the collection sites. During FY96, the RDEC implemented the first phase of this system, which involved installation of DSTRs, processing workstations, and T1 communications and routing equipment at Camp Williams and remote sites in New York, Florida, Texas, and California.[217]

On 4 August 2005, President Bush reaffirmed his pledge to Columbia to help rid the country of drug trafficking. Columbia was the major supplier of cocaine in the U.S. The command's support to national counter drug efforts continues into the twenty-first century under a functional support agreement with the National Drug Intelligence Center as well as embedded matrix support to the Drug Enforcement Administration.[218]

HURRICANES ANDREW AND INIKI

Hurricane Andrew, a category five hurricane, struck Florida and Louisiana in August 1992 causing sixty-five deaths and over $26 billion in damage. On 1 September 1992, five LAR

Local soldiers among many to aid victims

by Cleo Zizos/Capt. William DuPont
Public Affairs Staff

A team of three satellite terminal operators were among Fort Monmouth personnel deployed to Florida in support of Hurricane Andrew relief efforts.

Staff. Sgt. David A. Blanchard, Spec. Matthew W. Korb and Spec. Ramon B. Postel, who left here for Florida City on Aug. 30, have set up satellite communications equipment that provides access to the International Maritime Satellite, and are operating the standard "A" voice satellite terminals. Normally, Blanchard, Korb and Postel serve as satellite technicians in the Space Systems Directorate of the CECOM Research, Development and Engineering Center here.

"The terminals consist of a one-meter satellite antenna, transmit and receive capability for a touchtone phone and a 9600 BAUD modem," said Herb Groener, chief of the directorate's development research division and supervisor of the INMARSAT team.

An entire terminal is transported in two each suitcase-sized carrying cases. Blanchard is the primary trainer of all U.S. Army units worldwide that have the INMARSAT. The INMARSAT terminals were first used to establish a communications network for Desert Storm. Blanchard was requested, by name, to deploy to Florida to establish and train users of this terminal. The INMARSAT terminal is a commercially manufactured item.

The terminals access the INMARSAT satellite positioned over the Atlantic Ocean. That signal is then relayed to a large satellite terminal in Southbury, Conn. From the Southbury terminal, which is owned and operated by COMSAT GENERAL, calls are connected to commercial ground systems in New York City.

Seven contractor and CECOM civilian logistic assistance representatives (LARs) from Fort Bragg, N.C. also deployed to South Florida to provided technical support for the Mobile Suscriber Equipment communications systems.

Capt. Eric J. Schaertl, executive officer for Fort Monmouth Garrison, and Maj. Craig A. Peterson, a comunications-electronics officer in the Readiness Directorate's Logistics Management Office have been part of the Logistics Support Group since Aug. 31 and Sept. 2, respectively. They serve as operations officers in the Army Humanitarian Depot at Miami International Airport staffed by the Army Materiel Command, CECOMS's parent command. They are Fort Monmouth's direct link to the disaster area.

The CECOM Emergency Operations Center at Fort Monmouth, which prepares for full-scale operations upon imminent threat, began full scale operations immediately upon learning of the anticipated severity of Hurricane Andrew, according to an EOC official. ☉

deployed with elements of the XVIII Airborne Corps to South Florida, where they provided telephone and radio support to initial Hurricane Andrew relief efforts and were instrumental in planning a communications network to accommodate the military units that arrived thereafter. Among other things, they identified sources of failure in the avionics systems of OH-58C and UH-60 aircraft; established and maintained satellite and multichannel line-of-sight links for the Florida National Guard and active component forces, assisted units in setting up and operating Mobile Subscriber Equipment, maintained message centers for Joint Task Force Andrew, and proffered technical advice in the installation of automatic and semi-automatic switches for critical telephone circuits.

Hurricane Iniki, a category four hurricane, struck the Hawaiian island of Kauai on 11 September 1992, causing six deaths and $1.8 billion in damage. In the aftermath of Hurricane Iniki, CECOM LAR deployed to Kauai, where they helped install tactical satellite terminals to restore communications with Oahu. They also supported the installation and operation of a Mobile Subscriber Equipment network that grew, during the crisis, to support more than 1,200 subscribers (operations centers, fire stations, clinics, hospitals, and food distribution centers). This network reputedly carried more than 120,000 calls from 14 to 21 September.[219]

EARTHQUAKE AT NORTHRIDGE

On 17 January 1994, a 6.7 earthquake struck the Los Angeles area, leaving fifty-seven people dead and over 11,000 injured. The CECOM Command and Control Center received the Defense Department's activation order for supporting victims of the Northridge earthquake on 18 January 1994. The Center monitored the operation until it closed on 14 February. During this time, CECOM supplied four VRC-12 radios for the operation and developed information on government contractors that were affected by the disaster.[220]

OPERATION RESTORE HOPE

In response to the breakdown of civil order in Somalia in 1992, the U.S. led an effort, Operation Restore Hope, to secure southern Somalia for the conduct of humanitarian operations. In FY94, the RDEC evaluated both fielded military and commercial infrared sensors in terrain and conditions similar to Somalia's Mogadishu bypass main supply route. It determined that a number of three- to five-micron infrared cameras could reliably detect buried mines and mine clues. A RDEC test team traveled to Somalia where they demonstrated the camera and trained route clearing teams to use them. Meanwhile, responding to an urgent request in support of Operation Restore Hope, the RDEC designed, modeled, tested, produced, and deployed (all within less than sixty days) mine blast and direct fire protection kits for five-ton trucks. The kits dramatically improved crew survivability from mine blasts under the front wheels and cab of the vehicle. Also in support of operations in Somalia, as expressed by urgent requests from TRADOC and Department of Army headquarters, the RDEC supplied the 10th Mountain Division with four self-steering UHF satellite antenna systems. The RDEC designed and fabricated kits for installing these antennae on the division's vehicles.

One CECOM LAR was on site from 18 December 1992 through 15 February 1993, routinely working sixteen hours a day, seven days a week, in this austere, hostile environment. As a volunteer member of the advance Logistics Assistance Office team this LAR immediately confronted the fact that no operational Army or Marine Corps maintenance capability existed in the theater to support CECOM equipment. He pooled available Army and Air Force TMDE, obtained approval to use Air Force facilities, and established a repair shop in Baledogle. Having fabricated cables that permitted the use of operational AM and FM radios as test sets, he worked eighteen to twenty hours a day providing Direct, General, and Depot level maintenance for CECOM equipment until Army and Marine forces established on-ground support mechanisms. Though he had no formal training on TACSAT equipment, he helped operators identify and repair a faulty wave guide in an AN/TSC-93B terminal that provided the only means of communication be-

tween Baledogle and Mogadishu. While in Baledogle, he also completed repairs to FM radio sets of the Canadian Army, AN/GRC-193 AM radio sets of the Marine Corps, and INMARSAT terminals of the Air Force and the 10th Mountain Division. Having helped to establish the AMC Logistics Assistance Office in Mogadishu, he developed and implemented a plan for evacuating defective MSE equipment to the GTE Regional Support Center in Germany. While in Mogadishu, he provided Marine Corps technicians with technical assistance and training that enabled them to complete FM radio repairs that were beyond the scope of their usual activity. He established a data communications system using INMARSAT-C terminals to link remote outposts to the Division Support Command. He completed the repair of malfunctioning computer systems in several units, helped a team from Fort Lee rectify software problems that impeded the transmission of requisitions to the U.S. via satellite, and helped an Aviation and Troop Command LAR locate electrical problems that burned out the transformers of a refrigeration unit.[221]

RWANDA

A civil war broke out in Rwanda in 1994 resulting in the genocide of hundreds of thousands of its citizens. On 22 July 1994, responding to an AMC directive, the Telecommunications Center resumed twenty-four hour a day operations to support the Rwanda relief efforts. However, on 29 July, in view of low volume and limited staff, AMC authorized the CECOM Telecommunications Center to return to its usual operation schedule.

Nevertheless, as of 5 August 1994, the CECOM Command and Control Center was working twenty-four hours a day to provide proactive support for the humanitarian relief operation. CECOM had already deployed to East Africa six of its employees - a legal officer, an INMARSAT/ADP NCO, and four Logistics Assistance Representatives - and had processed

CW3 Berry returning from Somalia, Bldg 1204W, 3 February 1994

for deployment two additional civilians, both with experience in supply. The command had received about fifteen requisitions for the operation, mostly for batteries, which it processed "off-line" to preclude shipping delays. During the following week, CECOM initiated action to supply steerable antennae and installation kits for the operation's M998 HMMWVs and to ascertain the availability of MST-20 radio sets. The SATCOM Project Office shipped the antennae and mounting kits on 17 August.[222]

OPERATION UPHOLD DEMOCRACY

The elected government of Haiti was overthrown by a military coup in 1994. CECOM support of U.S. forces that were to be committed to Operation Uphold Democracy began in early September 1994 with preparations to support forces that were poised for the forced entry in Haiti. One of the major tasks in this effort was the identification of personnel and equipment that were to be deployed to Haiti as part of the AMC Logistics Support Element (LSE). For the LSE, CECOM's Readiness Directorate built a deployable communications system, known as the LSE C2 Flyaway Package. This system, an assemblage of commercial and existing military equipment, arrived in theater with its three-man installation crew on 18 October 1994. Fully deployed in less than three days, it provided the Joint Logistics Support Command with service equivalent to that of a full-fledged commercial telephone system, and more. Captain Paul Fitzpatrick, who deployed to Haiti from CECOM as a public affairs officer, described the installation and operation of the system for readers of the Monmouth Message. His account is replicated below.

The CECOM Command and Control Center activated to provide round-the-clock support to U.S. and allied forces deployed for the operation at 1545h, 15 September 1994. At about 1900h, 16 September, the Center received a request from Fort Drum (10th Mountain Division) for five steerable antennae

and mounting kits. With the cooperation of PM SATCOM, the Space and Terrestrial Communications Directorate, and Transportation, the Center delivered this material to Fort Drum at 1100h, 18 September.

Fort soldiers deployed MILSTAR system in Haiti

by Capt. Paul Fitzpatrick
Public Affairs

Operations Restore and Uphold Democracy have been host to a number of military firsts. The United States military is continuing to improve ways to conduct successful operations using the best people and equipment in the world. Fort Monmouth continues to contribute both.

Two Fort Monmouth soldiers deployed the first ground based MILSTAR system to Haiti last month to provide secure, communications between Joint Task Force Headquarters (JTF) in Haiti and the United States Atlantic Command (USACOM) at Norfolk, Va.

MILSTAR (Military Strategic and Portable Relay) is a multi- service communications program operating at Extremely High Frequency (EHF) Super High Frequency (SHF) and Ultra High Frequency (UHF). The system provides worldwide, two-way, anti- jam, survivable, secure, voice, teletype and data communications in support of battlefield operations.

"Our primary mission is to provide back up communications to the commanding general and the International Police Monitor for a direct line to USACOM", Staff Sgt. Walter R. Padilla said.

Padilla is the noncommissioned officer in charge (NCOIC) of EHF Research Facility for Space and Terrestrial Communications Tactical Support Branch, Fort Monmouth. He deployed the ground based MILSTAR system on very short notice with Sgt. Kevin V. Thomas, Assistant NCOIC of EHF.

Padilla and Thomas deployed to Haiti Sept. 29, with their MILSTAR equipment. Their equipment is considered very lightweight and compact with three terminals, connecting hardware and satellite receiver/transmitter dish weighing in at less than 400 pounds.

"We set up the equipment and were operating the same day we arrived", Thomas said.

The system is user friendly and very dependable in harsh environments according to the operators.

"We had a 98 percent effectiveness rate at this point which is better than any communications system I've seen down here", Padilla said.

The MILSTAR program began in the early 1980s as a satellite communications system designed for protracted nuclear war fighting missions and operations according to Walter D. Dietz, chief Technical Management Division, Program Manager MILSTAR.

The program was restructured in 1991 to focus on tactical users.

The first satellite was launched in Feb. 1994, with additional satellites to be added in the future. The MILSTAR program is managed by the Milsatcom Joint Program Office (MJPO) Los Angeles Air Force Base, commanded by Air Force Brig. Gen. Leonard F. Kwiatkowski. The Army Program Manager at Fort Monmouth is Col. William F. Jaissle.

The MILSTAR equipment deployed to Haiti is not even fielded equipment, which meant a tactical conditions shake down of

advanced research and development equipment.

Padilla and Thomas were sent on temporary duty (TDY) on Sept. 24, to Lincoln Labs in Lexington Mass. to pick up two of their three communication terminals, Advanced Single Channel Anti-jam Man Portable (ASCAMP), and enhance their training on the system so that they would be self sufficient in the field.

Lincoln Labs is an MIT operated Federally Funded Research and Development Center (FFRDC) located on Hanscom Air Force Base and has been a leader in MILSTAR system design since its inception.

The ASCAMPs went straight from the lab into field operation.

"The reliability of these advanced R and D models was very good", Padilla said.

The team operated 24 hours a day for over a month providing dependable Tri-service, secure communications. They were released from their mission on Oct. 30, and scheduled for redeployment. At present there are enough secure communications systems now in place to release MILSTAR assets.

Dietz credits the success of the MILSTAR deployment to team effort.

"In an extremely short time, a small team of soldiers and scientists provided a successful demonstration of a reliable communications system under field conditions", Deitz said. "The mission demonstrated the value of secure, anti-jam, EHF, man- portable terminals for tactical, tri-service communications and was the result of outstanding soldier, FFRDC and government cooperation". ○

Staff Sgt. Walter R. Padilla (left) and Sgt. Kevin V. Thomas with one of the MILSTAR (Military Strategic and Portable Relay) terminals that were deployed in Haiti recently during Operation Restore & Uphold Democracy. The ground based MILSTAR system linked the Joint Task Force Headquarters there and the United States Atlantic Command at Norfolk, Va.

On 29 September 1994, two soldiers of the RDEC's Space and Terrestrial Communications Directorate -- Staff Sergeant Walter R. Padilla and Sergeant Kevin V. Thomas -- deployed with the first MILSTAR ground stations to be installed in a tactical theater. Weighing in at less that four hundred pounds in total, the three Single Channel Anti-Jam Man Portable terminals were operating with ninety-eight percent effectiveness the very day of their arrival in theater, providing secure communications between Joint Task Force Headquarters in Haiti and Headquarters of the United States Atlantic Command (USACOM) in Norfolk, Virginia. The equipment, which was still in development, provided dependable tri-service communications twenty-four hours a day for more than a month until released from their mission with the arrival and installation of older, inventory communications systems.

The total number of personnel from CECOM and the Communications Systems Program Executive Office deployed to Haiti peaked in October 1994 at seventeen. As of the end of December, CECOM and other Fort Monmouth organizations had ten personnel deployed for the operation -- eight in Haiti and two at Fort Drum. During January, in conjunction with the transfer of the Haiti mission from the 10th Mountain Division to the 24th Infantry Division (Light), CECOM dispatched three additional LAR and one contractor (a GTE employee) to the theater.

Between 21 September 1994 and 31 January 1995, CECOM received 2,812 requisitions for equipment, supplies, and repair parts to support US forces in Haiti. It filled 2,571 (91%) of these requisitions.[223]

OPERATION VIGILANT WARRIOR

The next support challenge came quickly when Iraq started moving large formations of combat forces towards Kuwait in early October 1994. This threat to Kuwait caused the President to direct the rapid movement of selected U.S. forces into Southwest Asia (SWA).

CECOM's initial reaction to this rapid reinforcement of SWA came quickly. By early October, CECOM had six individuals identified and preparing for movement to SWA. Five of these individuals were Logistics Assistance Representatives (LAR) and one was going to be the Deputy Commander of the AMC LSE in SWA. Between 12-15 October, an additional seven individuals were identified and placed on-call for possible deployment to SWA.

The initial movement of CECOM personnel to SWA began on 17 October when three civilian and two military employees deployed from the AMC mobilization site, Aberdeen Proving Ground (APG), to Kuwait and Saudi Arabia. Their initial mission was to assist in the off-loading of equipment located on the prepositioned ships. Additionally, one civilian and one military employee were called forward to APG for final preparation for deployment to SWA. Additional personnel were also placed on-call for possible deployment.

One critical mission given to a Fort Monmouth organization, PEO COMM, was to conduct a retrofit evaluation of the Single Channel Ground and Airborne Radio Systems (SINCGARS) that were on the prepositioned equipment.

October 1994 ended with CECOM having nine individuals deployed or preparing for deployment at APG, and ten individuals on-call. Additionally, the SINCGARS assessment team had been called forward to APG to prepare for deployment to SWA.

November began with CECOM and Fort Monmouth preparing a special SINCGARS installation contractor team for deployment to SWA. This team would follow the assessment team that had arrived in Kuwait on 7 November. It consisted of two military personnel from CECOM, one individual from ITT, and fourteen individuals from Nations Inc. This team deployed to APG for final preparation on 3 November and started to deploy to Kuwait on 7 November.

As quickly as the crisis situation in SWA came up, it was resolved. In early December the SINCGARS installation team had redeployed from Saudi Arabia and Kuwait. They had installed 346 of the 492 installation kits and radios shipped to Saudi Arabia on vehicles that would be put back in storage on the prepositioned ships. One problem that was identified during the preparation of the prepositioned equipment was that the automotive type batteries were dead. CECOM, as the AMC battery manager, directed the Tank-Automotive and Armament Command to send an inspector to Saudi Arabia to determine what the problem was and how to correct it.

One of CECOM's primary missions was to provide communications, electronics, and intelligence equipment and related supplies and repair parts to U.S. forces. Between 11 October and 7 December, CECOM received 1,147 requisitions of which 1,072 were filled for a ninety-three percent fill rate.[224]

Y2K

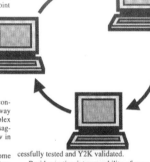

Exercise finds tactical systems Y2K compliant, compatible

by Cleo Zizos
Public Affairs Office

The Army Program Executive Office for Command, Control and Communications Systems (PEO C3S) and CECOM here achieved major successes in assessing Y2K compliance and compatibility of military tactical switched communications systems in a month-long multi-service interoperability exercise which ended two weeks ago.

Major players in the Joint Users Switch Exercise (JUSE 99-Y2K) were the Joint Chiefs of Staff; the Commander-in-Chief (CINC) U.S. Atlantic Command; the PEO C3S as the Defense Department's Executive Agent for Tactical Switched Systems/ Joint Network Management; and the CECOM Software Engineering Center and CECOM Space and Terrestrial Communications Directorate.

The exercise was conducted in a Y2K environment, testing tactical to strategic as well as strategic to tactical interfaces.

Months in planning, it involved Active, Guard and Reserve units from the continental United States and from as far away as Japan and Guam employing complex transmission, switching and voice messaging systems representative of those now in the field..

The units, participating from either home stations or deployed locations, were tied together with satellite communications, forming the backbone of the network.

a Joint Task Force.

That architecture, which was validated in a Y2K environment, included five different types of satellite terminals, 14 different voice switches, and eight different message switches.

The strategic to tactical messaging capability of the Defense Message System (DMS) (the replacement for the old AUTO-DIN system) with exercise facilities at Fort Detrick, Md., Fort Monmouth, and other interfaces to participating sites also was successfully tested and Y2K validated.

Besides testing interoperability of message and voice switching software, JUSE 99-Y2K rigorously evaluated the impact of

Reflecting the importance of network management to battlefield situational awareness, the Joint Defense Information Infrastructure Control System-Deployed underwent a JUSE 99-Y2K operational evaluation as it managed the non-secure e-mail system set up for the exercise.

The Joint Network Management System, a prototype system being developed by the Army for joint use, underwent a similar evaluation as it managed secure e-mail.

Critical Y2K dates against which systems and software were tested in the Y2K exercise environment were Sept. 8 to 9, 1999; Dec. 31 to Jan. 1, 2000; Feb. 28 to 29, 2000; Feb. 29 to March 1, 2000; and Dec. 30, 2000 to Jan.1, 2001.

The JUSE annual exercises began in 1996 as a Defense-wide global endeavor to assess concepts, interfacing schemes, untried network arrangements, and new technologies and operational approaches. Participation has grown since its inception to include communications units of allies such as the United Kingdom.

In the JUSE 99-Y2K eight U.S. CINCs participated.

Besides the CINC U.S. Atlantic Command as the lead, other CINCs involved were the CINC Pacific Command, CINC Southern Command, CINC Strategic Command, CINC Space Command, CINC Special Operations Command, CINC Cen-

Y2K compliance became a significant issue at CECOM, just as it did at many other technology-dependent organizations. Compliance represented the single largest IT project ever undertaken for the Army Materiel Command in general, and for CECOM in particular. An estimated $45 million was spent on project management costs alone during the four years of the project's lifecycle.

CECOM's role in this project was threefold. CECOM was not only responsible for ensuring that CECOM-managed tactical systems complied; it also supported AMC Headquarters in their overall implementation efforts and ensured the IT infrastructure at every AMC installation was Y2K compliant. Items such as telephone switches, traffic lights, and even refrigeration units had to be identified, inventoried and corrected before 31 December 1999. Following the completion of a comprehensive IT inventory, organizations had to decide whether or not to reengineer, retire, or replace every item that was not compliant. Only after all that had been accomplished could programmers and software engineers begin work on addressing the specific compliance issues in each piece of software.

Over 1.3 million items were inventoried and assessed during the Y2K project. Over 986,000 were corrected for potential problems.

Compliance efforts dramatically intensified as the Millennium quickly approached. Despite the numerous technical and managerial challenges ~~sociated with Y2K. CECOM clocks switched over to the new~~ associated with millennium without incident.

1991: CECOM's Henry C. (Budd) Croley invented a device known as the "Budd Lite" for the 24th Infantry Division.

The Budd Lite consisted of two infrared light emitting diodes mounted in a rubber holder with circuitry to make them blink. When attached to a nine-volt battery, the lights, which were detectable only with night vision devices, would blink for hours. While Concurrent Engineering Directorate (CID) engineers belittled the device, troops on the ground found them to be indispensable not only for their intended purpose of marking airfields, but also for marking vehicles, for unit identification, for light by which to read maps, and for marking lanes in fields.[226]

1994: CECOM built the world's very first scanned electron beam x-ray source for use in the Photon Backscatter Imaging Mine Detector.

The scanning rate of this device was approximately twice the rate of medical CT scanners.[229]

1994: CECOM personnel designed, built, and intgrated advanced electro-optic sensors - a second generation Forward Looking Infrared Radar (FLIR), an eye-safe laser rangefinder, a high resolution daylight television camera, and a binocular television display - into the M1 gunner's primary sight.

CECOM demonstrated the system in a concept evaluation program for the Armor School at Fort Knox. In a Combat Systems Test Activity demonstration at Aberdeen Proving Grounds in late winter 1994, a group of distinguished tank gunners and the CECOM senior leadership used the second-generation fire control system to find and "destroy" thirty of thirty stationary and moving targets. They reported that they were "very impressed" with the results of the quick-reaction in-house program that had produced the assemblage and with "the ability of the thermal system to recognize targets at long range, leading to precision shooting." This first installation of a second generation FLIR capability provided the user an early prototype for operational evaluation.[228]

1994: In January 1994, CECOM evaluated the first Laser Countermeasure System (LCMS) at the laser test range (Fort A. P. Hill).

The evaluation measured the system's effectiveness against direct view optics in daylight and against various night vision devices at night. The system met operational requirements for nighttime use and exceeded those of daytime operations. Individual soldiers could use it to find and disrupt threat optical and electro-optical surveillance devices. To improve the targeting accuracy of the LCMS, CECOM developed the High-Intensity Targeting System (HITS), a simple hand jitter and atmospheric turbulence compensation. The Project Manager tested the HITS successfully and incorporated it into the LCMS objective brassboard.

this project, no significant problems were as-

1994: Pfc James Pontius was the first person to speak on a voice circuit on the Milstar Developmental Flight Satellite during a test at Fort Monmouth on 21 February 1994.

BG David R. Gust, PEO Communications, became the first general officer to communicate over the satellite when he called the Milstar support facility at Fort Monmouth over the Air Force terminal. The Milstar satellite was launched in February 1994 to provide worldwide secure, survivable, highly jam resistant communications; satellite to satellite communication; autonomous operation; the ability to reposition to meet theater requirements, and the ability to provide direct support to mobile forces.[230]

by Cleo Zizos
Public Affairs Staff

Going to work on most days is often uneventful. But for one private first class here, a recent day at the office was like being the millionth customer at the local supermarket, when cameras flash and balloons fly.

"Well, not exactly," said Pfc. James Pontius, who was the first soldier to speak on a voice circuit over the Milstar Developmental Flight Satellite during a test on Feb. 21.

"With our test facility engaged in the tri-service over-the-satellite tests, it was time for scheduled voice tests to begin," explained Otto Eickmeyer, chief of the Tactical Support Branch of CECOM's Space and Terrestrial Communications Directorate.

"It could have been any one of my team mates, I just happened to be on duty when our turn came," added Pontius, whose name has since been inscribed on an impressive plaque specifically designed to commemorate the event.

Even so, the Milstar project manager and deputy project manager, Col. William Jaissle and L. Scott Sharp, believed the event warranted special recognition and arranged for the names of Pontius and team member, Spc. Harry Fuller, who was working with him, to appear in a place of honor on the plaque.

As a matter of fact, the names of the seven military operating team members also appear on the 14" x 21" plaque.

Meanwhile, Brig. Gen. David R. Gust, Program Executive Officer for Communications, who at the time was visiting the Raytheon Corp. Milstar facility in Sudbury, Mass., another of the testing sites, became the first general officer to communicate over the satellite, when he "called" the Milstar support facility at Fort Monmouth over the Air Force terminal.

Fort Monmouth is the only Army test site among 11 sites -two Navy and eight Air Force -- which are participating in the testing.

With the first Milstar satellite already successfully in orbit since Feb. 7, all that

photo by Russ Meseroll

Members of the Tactical Support Branch of CECOM's Space and Terrestrial Communications Directorate man the two terminals that are being used to test the Milstar Developmental Flight Satellite. They are, from left, Sgt. William Padilla, Pfc. James Pontius, Spc. Harry Fuller, Sgt. Sheila McClinton, Spc. Mark Lockwood, Spc. Douglas Bram, Sgt. Kevin Thomas and Spc. Kim Pontius.

remained was to test it, Eickmeyer said. Testing began on Feb. 21.

Over the past year, William Presho, the branch test director, had worked with support contractors, Booz Allen & Hamilton's Lino Gonzalez and Logicon's Frank Staber and Marc Fath to write the Army stand-alone portion of the test plan and procedure. Warren Pinney, a branch engineer, coordinated the procedure with the Milstar program manager.

Preliminary tests had been designed to verify the technical content and accuracy of the testing procedures and to anticipate any operational problems that might arise during the formal testing of the Milstar satellite, -- and had worked as expected.

The two terminals, which have been used to test the satellite since its orbit, will continue to be manned by the seven-member team, throughout occasional 10 and 11-hour days and some Saturdays. The team members-- Sgt. Walter Padilla, the noncommis-

sioned officer in charge; Sgt. Sheila McClinton, Sgt. Christopher Benson, Spc. Mark Lockwood, Sgt. Kevin Thomas, Spc. Douglas Bram, and Sgt. Kim Pontius -- will continue the testing through June.

Two team members, Staff Sgt. Carlos Febles and Spc. Ted Neary, serve as the Army's coordinators at Falcon Air Force Base, Colo., which is managing the testing.

"We knew it would work," Pontius said. "All our preparation pointed to it." ✪

The 21st Century

JOINT CONTINGENCY FORCE ADVANCED WARFIGHTING EXPERIMENT

One of the more significant lessons relearned from the Gulf War was the need for C4ISR interoperability between all of the armed services as well as foreign allies. The ever-increasing trend towards coalition warfare meant the U.S. Army would have to fight alongside units not just from different branches of the military, but also from different nations. Various types of communications systems were required to work in concert with one another in order to fight successfully in such an environment. CECOM participated in this effort.

A Joint Contingency Force Advanced Warfighting Experiment (JCF AWE) was held at Fort Polk, LA in September 2000 in order to establish how the digitization of light forces would increase lethality, survivability and operational tempo. These AWE initiatives played a vital role in Army transformation since they allowed leaders to more accurately determine just how well these systems worked. This endeavor heavily involved CECOM and Team C4IEWS because many of the new systems relied extensively on advanced communications and electronic components.

The En-route Mission Planning and Rehearsal System (EM-PRS) was one of the most interesting and significant systems developed by CECOM and tested during this exercise. This new system, installed on a modified cargo aircraft, allowed Soldiers to maintain situational awareness while in the air. Based primarily on a suite of enhanced communications equipment and onboard computers, EMPRS allowed embarked Soldiers to remain in constant contact with joint forces. It also provided a template for airborne Soldiers not just to change any aspect of their upcoming operation but to "rehearse it" and determine how likely these alterations would be to affect the success of the mission. This capability was especially critical on the mod-

ern battlefield due to the constantly changing nature of warfare. EMPRS was praised as the "crown jewel" of the exercise.

On 20 July 2001, Major General William H. Russ succeeded Major General Nabors as the Commanding General of CE-COM.

SEPTEMBER 11, 2001

At 0800 on 11 September 2001 a group of volunteers assembled at the Fort Monmouth Expo Theater to participate in a three-day force protection exercise involving law enforcement agencies and emergency personnel at all levels, from Fort Monmouth firefighters to the NJ State Police. The exercise included simulating a biochemical terrorist attack at Fort Monmouth and studying the emergency response that would take place after such an attack.

The group pretended to be horrified and upset when, just after 0900, the director of the exercise informed them that a plane believed to have been hijacked by terrorists had crashed into one of World Trade Center towers. The volunteers assumed this was part of the simulation. They quickly realized, however, that this was not so. The three-day exercise that took months to plan was cancelled in a matter of minutes. Volunteers were instructed to return to their offices and stay there.

Employees assembled wherever they could find a TV or a radio. They listened in horror along with the rest of the world as the first direct attack on American soil since Pearl Harbor occurred.

The Emergency Operations Center sprang into operation twenty-four hours a day, seven days a week. Fort Monmouth quickly realized it did not have enough manpower to monitor access to the post as required by the new threat level. Gates closed and access was limited to a few main roads. Employees volunteered, even on weekends and after duty hours, to help check identification cards at the gates. A Visitor Control Center was initiated to process visitors. Reserve Soldiers were activated to augment security on post.

Rescue Workers using infrared cameras attached to PVC pipe to search through the rubble at the World Trade Center site

While force protection measures were being upgraded and put in place, CECOM was tapped to help with the World Trade Center (WTC) site rescue effort. Fort Monmouth technologies helped rescue and recovery workers in a variety of ways. The world's smallest infrared camera, developed by CECOM and attached to PVC pipe, was used for finding and searching through voids in the rubble. A laser doppler vibrometer was also used to judge the structural integrity of the buildings. Electronic listening devices detected distress calls to 911 made from cellular phones. Additionally, hyperspectral flyovers monitored and controlled recovery operations from the air.

CECOM deployed a quick reaction task force to the Pentagon to install a communications infrastructure for 4,500 displaced workers. CECOM teamed with the Pentagon renovation office to provide engineering and integration support to renovate the Pentagon's command and control infrastructure in support of the Pentagon rebuild (Phoenix Project).[232]

HOMELAND SECURITY

The nation placed an unprecedented emphasis on Homeland Security (HLS) as a result of the 9/11 attacks. CECOM experienced first-hand the need for better communications, more integrated response plans and quicker response times. Given the nature of CECOM's mission and its close proximity to New York City, CECOM was in a unique position to help with future HLS efforts. HLS was one of CECOM's top initiatives in the months after 9/11.

Governor James E. McGreevey designated Fort Monmouth as the New Jersey Center for Defense Technologies and Security Readiness on November 10, 2003. McGreevey emphasized Fort Monmouth's important role in providing technologies to help fight the Global War on Terror. "The United States will never have a shortage of brave Americans who are willing to go into harm's way, but it has been our technological superiority that has made our Armed Forces the most advanced

and superior military force in the history of warfare. Much of today's modern weaponry depends upon software and electronics developed and supported at Fort Monmouth," said McGreevey.[233]

OPERATION NOBLE EAGLE

President George W. Bush announced the mobilization of reserves for homeland defense on 15 September 2001 in response to the terrorist attacks of September 11. The initial call up of reserve forces was for homeland defense only. It was later expanded to include reserve forces for Operation Enduring Freedom. Operation Noble Eagle began in October 2001 for Fort Monmouth with the arrival of Bravo Company, First Battalion, 181st Infantry Regiment from Boston, MA. The company's mission was to protect the Fort Monmouth community, its facilities, and personnel stationed on post. Bravo Company performed ID checking at the gates twenty-four hours a day, seven days a week; randomly searched vehicles; and conducted building and perimeter security. The 181st returned home in September 2002 and was replaced by Bravo Company, 104th Infantry Regiment, a National Guard Unit from Greenville, MA. The 50th Combat Support Battalion, Detachment One from West Orange New Jersey, replaced Bravo Company in June 2003. In addition to their regular duties, the Reserve forces became integrally involved with Fort Monmouth and the surrounding communities, paying visits to local veterans homes and schools and explaining life in the military and life as a Soldier.[234] Access control was then contracted to the Wackenhut-Alutiiq Corporation in early 2004 and then to TW & Company in February 2007.

Bravo Company, 104th Infantry Regiment, Greenville, MA

OPERATION ENDURING FREEDOM

One of the United States' initial responses to the attacks of September 11th was seizure of financial assets and disruption of the fundraising network of terrorist groups. Initial deployments then began to Southwest Asia and Afghanistan. On 20 September President Bush announced the start of the War on Terror and demanded that the Taliban in Afghanistan hand over all Al Qaeda terrorists living in their country or share their fate.

Operation Enduring Freedom commenced on 7 October 2001.

B-1, B-2 and B-52 bombers, F-14 and F/A 18 fighters enacted air and land strikes. Tomahawk cruise missiles launched from U.S. and British ships and submarines. Fort Monmouth's preparations for Operation Enduring Freedom began in the weeks following September 11th as all centers prepared to supply equipment and fulfill emergency requisitions. Fort Monmouth also deployed military, civilians and contractors. The highest demand items initially requisitioned included Lithium batteries, Firefinder radars, and night vision equipment. Batteries remained in short supply during Operation Enduring Freedom (particularly the BA 5590), as they had in the Vietnam and Gulf Wars.

One of the important Fort Monmouth systems used in Afghanistan was the phraselator. Developed in conjunction with DARPA, this system translated the English voice into Dari, Pashto, Arabic and other languages using fixed phrases from force protection and medical domains. This system was critical in OEF because there were not enough trained linguists on the ground. Fort Monmouth continued to assist DARPA in providing new domain vocabularies and developing a two-way phraselator capability.

Fort Monmouth developed a prototype demo unit for "down well" viewing in Afghanistan. The system was an immense success with the troops and was first deployed to Afghanistan in March 2003. Fort Monmouth engineers deployed to Afghanistan in October 2002 to support the Combat Service Support Automated Information System Interface (CAISI), a set of deployable wireless LAN equipment. Fort Monmouth civilians supported this system in a brigade support area near Kandahar. Eventually, the success of the equipment and the increase in Soldier morale led to it being installed in fifteen additional remote locations in Afghanistan. This system was named one of the top ten Army inventions of 2003.

Fort Monmouth sent a team to SWA to control crisis action planning, resolve financial and appropriations issues, establish a contracting office in a high threat environment, and provide administration of war contingency contracts. This team was responsible for successfully equipping joint forces in the region, standing up the Afghan Army logistics system and institutions, and accelerating the local production of supplies to help increase self reliance and build the local economy.

One of Fort Monmouth's tenant activities, the 754th Explosive Ordnance Disposal Detachment, has been deployed to Afghanistan twice in support of Operation Enduring Freedom. Their mission in Afghanistan was to dispose, render safe and advise about explosive hazards and ordnance. Based out of Kandahar, their main customers included Special Forces groups, the Air Force and teams from the 82nd Airborne Division. During their first deployment, the 754th disposed of 652,000 pounds of explosive ordnance and responded to twenty-six incidents.

Problems and lessons learned during the first year of OEF were similar to those to be encountered in Kuwait and Iraq the following year. Some of these problems were attributed to extreme weather conditions. Preparations for deployment turned arduous because engineers and administrative personnel were responsible for completing all documentation and orders for not only government personnel but also all the contractors sent to Afghanistan. Depending on military transport to get to Kandahar Airport was often a problem and caused lengthy delays. It regularly took personnel well over a week to get to the theater. Problems also arose in protecting Fort Monmouth equipment from excessive heat. This problem was addressed in part with Hex solar shades and Modular Ammunition solar shades (MASS). However, this equipment could only be requested through the Soldier and Biological Chemical Command (SBCOM) and approved by HQDA. Demand far exceeded sup-

754th Explosive Ordnance Detachment

ply.[235] Further support to OEF can be found in the on-going support to the Global War on Terror (GWOT) section.

TRANSFORMATIONS AND REALIGNMENTS IN 2002

CECOM again found itself in the midst of an Army reorganization effort in 2002. TIM (Transformation of Installation Management) looked at the way the Army managed its posts, camps and stations and sought to centralize management. For CECOM, this meant that the garrison commander would report to a regional office rather than to the Commanding General of CECOM. The Army believed that centralizing installation management into regions would provide for more streamlined funding, a tighter focus on similar installation issues within a defined geographic region, and establishment of better standards for installations. DA also intended to improve support to the Army Transformation and the readiness of Soldiers through TIM.

The U.S. Army Installation Management Agency (USAIMA) would direct overall Army installation operations. Regional offices would manage all Army installations and garrisons within a geographical area. There would be seven regional offices under TIM, with Fort Monmouth's Garrison belonging to the Northeast Regional Office at Fort Monroe, Virginia. The new installation management and realignment was implemented on 1 October 2002. Formal realignment of employees did not occur until October of 2003.

NETCOM (the Network Enterprise Technology Command), established in October of 2002, was another Army-wide realignment effort that affected CECOM. In the same way TIM centralized garrison management, NETCOM centralized the management of the Army's information technology and networks. NETCOM was aligned in several geographic regions, each with a Regional Chief Information Officer (RCIO). CECOM would receive its support from the Northeast Region, located at Fort Monroe, VA. The Directorate of Information Management (DOIM) came under the operational control of IMA while technical control resided with NETCOM.[236]

In 2002, the Research, Development and Engineering Command (RDECOM) was created. This new subordinate command of AMC was established under the direction of AMC Commander General Paul J. Kern and stood up, provisionally, on 1 October 2002. MG John Doesburg, former Commander of the Soldier and Biological Chemical Command (SBCCOM), assumed responsibility for the development and implementation of RDECOM. The mission of this new Command was to field technologies that sustained America's Army as the premier land forces in the world.

Operational control of the Major Subordinate Command (MSC) R&D activities transferred to RDECOM, effective 1 May 2003. This included the Tank-Automotive and Armaments Command Research, Development and Engineering Center (RDEC), the Edgewood and Natick RDEC, the Aviation and Missile RDEC and the CECOM RDEC (renamed the Communications Electronics RDEC or CERDEC). Portions of the CECOM LRC (sustainment engineering) and the CECOM Software Engineering Center (software engineering) were also

affected. The Combatant Commander Interoperability Program Office (CIPO) and specific Deputy Chief of Staff for Resource Management positions supporting the RDE financial program additionally transferred to RDECOM.

The CERDEC became the Army's information technologies and integrated systems center. Its Command and Control Directorate; Intelligence and Information Warfare Directorate; Night Vision and Electronic Sensors Directorate; and Space and Terrestrial Communications Directorate worked together to develop and integrate Command, Control, Communications, Computers, Intelligence, Surveillance, and Reconnaissance (C4ISR) technologies that enable information dominance and decisive lethality for the networked Warfighter.[237]

KNOWLEDGE MANAGEMENT

CECOM was truly a global organization by 2002, with only forty-seven percent of employees residing at Fort Monmouth. The CECOM Knowledge Center was unveiled in May 2002 to address this phenomenon. The "KC" was an internal knowledge-sharing portal intended to connect the global CECOM workforce.

The Knowledge Center stored information papers, trip papers, policies, and other documents in an easily searchable document library. It also offered collaborative workspaces and virtual meeting tools to facilitate project management.[238] With more than half of the workforce eligible to retire in five to ten years, knowledge management initiatives that targeted the preservation of tacit knowledge were enacted at both CECOM and at the Army level. On 4 October 2004, the CECOM/ CERDEC/PEO C3T/PEO IEW&S Knowledge Centers combined into a single portal now known as the Team C4ISR Knowledge Center.

ENTERPRISE SYSTEMS ENGINEERING

Enterprise Systems Engineering (ESE) was another initiative pursued in 2002. It quickly morphed into a concept and eventually became one of Fort Monmouth's top priorities. ESE was a single systems engineering effort that would tie all discrete integration initiatives together into an enterprise architecture solution. Fort Monmouth perceived this need and set up a Systems Engineering Team that compiled a "systems engineering handbook" and prototyped a system to address questions with enterprise implications.[239]

C4ISR ON-THE-MOVE

Another initiative pursued in 2002 was the C4ISR On-the-Move Demonstration. An integrated Command, Control, Communications, Computers, Intelligence, Surveillance, and Reconnaissance (C4ISR) system–of–systems to increase the lethality and survivability of the lighter platforms of the Future Combat System (FCS) would be pivotal to the success of Army Transformation.

The two-week demonstration took place at Fort Dix, NJ, conveniently located forty miles from Fort Monmouth. Fort Dix provided ample space to showcase the system's capabilities. A number of VIPs attended the demonstrations to include GEN Eric Shinseki CSA, and the exercises were considered an overall success by those who participated and observed.[240]

BATTERY MISSION

In the lead up to strikes against Iraq, the foresight of some senior Fort Monmouth leaders avoided the critical shortage of batteries experienced during Desert Storm. In anticipation of increased demands for lithium batteries, Fort Monmouth initiated actions to ensure a continued supply of batteries during deployment and actual operations. In November 2002, Fort Monmouth identified that the funding required in order to ramp up production prior to the OIF conflict would amount to $56.3 million. The Command received this funding in December 2003 and immediately put it on contract. As a result, overall BA-5590 production increased from 60,000 batteries per month to nearly 125,000 batteries per month by April 2003. Production was to continue increasing to 300,000 batteries per month to fill shortfalls while continuing to meet CENTCOM AOR requirements. In order to expedite the delivery of batteries from the factory to the foxhole, Fort Monmouth arranged for direct shipments to Kuwait from its major lithium producer, Saft, as well as its rechargeable producer, Bren-Tronics. Shipments went directly to Charleston AFB for airlift to Kuwait, bypassing the supply depot and saving several days of ship time. The use of rechargeable batteries was promoted for non-deployed forces and for selected missions in the U.S. Central Command Area of Responsibility (CENTCOM AOR). In order to facilitate the CENTCOM use of rechargeables, an existing battery charging van was shipped from Fort Benning along with battery chargers and batteries to allow centralized, high-volume charging in the AOR. This charging van remained in place to help reduce the consumption of non-rechargeable lithium batteries.

Battery Solutions for all Millitary C4ISR Needs

Despite these accomplishments, the Acting Principal Assistant for the Deputy Under Secretary of Defense for Logistics and Materiel Readiness, Bradley Berkson, issued a memorandum on 30 January 2004 that stripped Fort Monmouth of its lithium battery mission and transferred it to the Defense Logistics Agency (DLA). The decision was based upon the fact that stocks of the BA5X90 lithium battery ran extremely low during Operation Iraqi Freedom. Berkson believed the transfer would improve battery availability as the DLA Defense Supply Center Richmond (DSCR) was the integrated materiel manager for all batteries and the DLA funded critical consumables. The effective date for the transfer was no later than 30 September 2004. According to the MOA, the Army would continue to be the technical face to the user for technical issues and would be responsible for battery policy, standardization, and design

integrity and stability. DLA would take over inventory management and oversee the contracts. The transition would be handled by forming a joint team with representatives from both sides. The former Vice Chief of Staff, General John Keane, had opposed the move citing several years of funding shortfalls for the shortage. The Army also feared the move would pose a risk to troop readiness.

LIFE CYCLE MANAGEMENT COMMANDS AND C-E LCMC/CECOM LCMC

On 2 August 2004, Claude M. Bolton, Jr., Assistant Secretary of the Army for Acquisition, Logistics and Technology (AL&T), and General Paul J. Kern, Commanding General of AMC, signed a memorandum of agreement to formalize the Life Cycle Management Initiative. That initiative established life cycle management commands by aligning the AMC systems-oriented major subordinate commands such as CECOM with the Program Executive Offices (PEOs) with which they worked. The result of the initiative in the C4ISR arena was the formation of the Communications-Electronics Life Cycle Management Command (C-E LCMC; later, CECOM LCMC). This would link more closely than ever before CECOM, the PEO for Command, Control and Communications-Tactical, and the PEO for Intelligence, Electronic Warfare and Sensors. The Communications-Electronics Research, Development and Engineering Center (CERDEC) would maintain an operational link with the C-E LCMC to create a unified vision across the acquisition, research, development and sustainment communities. This unified vision would provide a single face to the Warfighter throughout the total life cycle of systems and equipment. The C-E LCMC stood up on 2 February 2005.

In the words of Major General Michael R. Mazzucchi, "A Life Cycle Management Command is completely and totally dedicated to providing Warfighters the best equipment and services in the shortest possible time, and provides the most sustainable equipment, effectively and efficiently, by being the best stewards of the resources the Nation has entrusted to us."[241]

When he assumed command in July 2007, Major General Dennis L. Via modified the command's name to CECOM Life Cycle Management Command in view of the years of name recognition CECOM had acquired across the Army.[242]

BRAC 2005

The National Defense Authorization Act for fiscal year 2002 authorized the Department of Defense (DoD) to pursue a Base Realignment and Closure (BRAC) round in 2005, a complex analysis and decision process that involved virtually all levels of DoD management, from installation through major command and component/agency headquarters to OSD. All bases, posts and installations were considered. On 13 May 2005, the Department of Defense recommended the closure of Fort Monmouth and the realignment of CECOM LCMC elements at Fort Monmouth to Aberdeen Proving Ground in Maryland. The recommendation affected 4,653 civilians and 620 military personnel at Fort Monmouth. Despite aggressive state and local lobbying, the Base Realignment and Closure (BRAC) Commission approved the DoD's recommendation on 24 August 2005. The BRAC recommendations to close Fort Monmouth and realign

CECOM LCMC elements at Fort Monmouth to Aberdeen Proving Ground, Maryland became law on 9 November 2005. The transition of the workforce to Maryland was scheduled to take place by 2011.

The BRAC recommendations affected not only CECOM LCMC elements but also other tenants at Fort Monmouth. The Commission approved relocating the U.S. Army Military Academy Preparatory School to West Point, NY; the Joint Network Management System Program Office to Fort Meade, MD; and the budget/funding, contracting, cataloging, requisition processing, customer services, item management, stock control, weapon system secondary item support, requirements determination, integrated materiel management technical support inventory control point functions for consumable items to the Defense Supply Center Columbus, OH. Further, it recommended relocating the procurement management and related support functions for depot level reparables to Aberdeen Proving Ground, MD, and designating them as Inventory Control Point functions, detachment of Defense Supply Center Columbus, OH, and relocating the remaining integrated materiel management, user, and related support functions to Aberdeen Proving Ground, MD. Information Systems, Sensors, Electronic Warfare, and Electronics Research and Development and Acquisition (RDA) were recommended for relocation to Aberdeen Proving Ground, MD. Elements of the Program Executive Office for Enterprise Information Systems would be consolidated with existing elements at

Fort Belvoir, VA.

The 13 May 2005 DoD recommendations resulted in a net gain of 275 positions at TYAD. TYAD would become the DoD Center of Industrial and Technical Excellence for communications and electronics equipment. New workload would be received from each of the Armed Services. The recommendations included: Lackland Air Force Base in Texas relocating work on computer maintenance, cryptographic equipment, electronic components and radios; the Naval Weapons Station, Seal Beach, California relocating work on the maintenance of electronic components, fire control systems and components, radar and radios; the Marine Corps Logistics Base in Barstow, California relocating work on the maintenance of electronics components, electro-optics/night vision/ forward looking infrared systems, fire control systems and components, generators, ground support equipment, radar and radios; and Red River Army Depot, Texas relocating the maintenance of tactical vehicles.

With regard to the Supply, Storage, and Distribution Management Reconfiguration, the DoD recommended realigning "Tobyhanna Army Depot, PA, by consolidating the supply, storage, and distribution functions and associated inventories of the Defense Distribution Depot Tobyhanna, PA, with all other supply, storage, and distribution functions and inventories that exist at Tobyhanna Army Depot to support depot operations, maintenance, and production. Retain the minimum necessary supply, storage, and distribution functions and inventories required to support Tobyhanna Army Depot, and to serve as a wholesale Forward Distribution Point. Relocate all other wholesale storage and distribution functions and associated inventories to the Susquehanna Strategic Distribution Platform." The BRAC Commission found the DoD recommendation for TYAD con-

sistent with the final selection criteria and the Force Structure Plan and approved the recommendation.

LEAN SIX SIGMA

In addition to providing critical C4ISR systems to the GWOT, command activities embraced Lean initiatives that supported Army Transformation goals. Not an acronym, Lean is instead a way of thinking that leads to continuous process improvement designed to banish corporate waste and maximize profit. Lean thinking evolved mainly from "Lean production," an approach pioneered by Toyota following World War II and later adopted by firms in a range of industries engaged in mass production. In 2003, impressed by Lean's high success rate in private industry, AMC Commander General Paul Kern mandated that Lean Thinking be extended to all processes throughout his command, particularly at the depots. In 2005, Lean Thinking was merged with another process improvement initiative, Six Sigma, to become Lean Six Sigma (LSS), and was adopted Army-wide under the Business Transformation umbrella. Lean Six Sigma principles stress the elimination of non value-added steps from a process and the reduction of waste caused by defect or variance. Over the years, the command launched dozens of Lean and Lean Six Sigma projects aimed at not only saving millions of dollars, but ultimately saving Soldier's lives through the rapid deployment of quality products and support to the field.[243]

SUPPORT TO THE GLOBAL WAR ON TERROR

Operations in Iraq began on 19 March 2003 with joint strikes by the U.S. and Great Britain designed to disarm Iraq of its weapons of mass destruction and remove the regime from power. CECOM logisticians began preparing for strikes against Iraq

BRAC Town Hall, Pruden Auditorium, November 2006

in early October 2002 by forming an Anticipatory Logistics Cell (ALC) to identify potential spare and repair part shortfalls. The ALC developed a list of C4ISR systems expected to be deployed by the Army, Special Operations Command and Marines. Supply supportability assessments were conducted, enabling logisticians to identify potential spare and repair part shortfalls. During the war, the ALC investigated high priority spare requests from Iraq and accelerated deliveries.

The organizations of the CECOM Life Cycle Management Command fielded and maintained a wide variety of equipment during the GWOT. In fact, the command managed half of the nationally stock numbered items in the Army inventory: 55,874.[244] These included items like frequency hopping tactical radios, satellite-linked computers inside vehicles, sophisticated sensors, and electronic jamming systems.[245]

In all, close to a million requisitions were processed between 11 September 2001 and mid 2007 against 13,439 different stock numbers. The majority of these requisitions were high priority shipments. Significant shipments included tactical satellite, Mobile Subscriber Equipment, SINCGARS, night vision, global positioning systems, Firefinder, aircraft navigation system, batteries, and aircraft survivability equipment.[246]

Hundreds of CECOM LCMC civilians, contractors, and military personnel deployed to the Southwest Asia Theater at any given time. This included deployments from among the command's population of over 200 world-wide Logistics Assistance Representatives (LAR). Many of these LAR deployed more than four times, with several LAR deploying as many as seven times.[247] LAR provided technical and logistical assistance for C4ISR Systems. As the number of command deployments significantly increased over the course of the GWOT, an automated accountability system, Roll Call, was developed and implemented for deployed personnel. The system was used to track and record command deployments and to ensure total command personnel accountability on a daily basis.

The CECOM LCMC continued operating its Emergency Op-

Mike Anthony, SCR, LSE commander, and CECOM LARs, Camp Speicher, Iraq. Note the RAID aerostat in the air, top right. August 2007

erations Center (the name was subsequently shortened to Operations Center) twenty-four hours a day, seven days a week with three shift rotations.

Two Electronic Sustainment Support Centers (ESSC) were deployed to Kuwait. The two ESSC consisted of sixty-five logistics and maintenance personnel. Both centers became operational 1 March 2003 at Camp Arifjan, Kuwait. ESSC provided a robust embedded and regional logistics and maintenance support capability for C4ISR systems. A forward repair activity was also established in Qatar. The forward repair activity was designed to halve turn around time for the repair of STAMIS/TIER III (computer hardware) used in SWA. The activity moved to Kuwait prior to the start of OIF. As of 2007, two to three managers from the CONUS and OCONUS ESSC, together with assigned government and contractor Field Ser-

An electronics mechanic at Tobyhanna Army Depot tests the AN/PRC-112D pilot survival radio

vice Representatives (FSR), are deployed to one of the eleven ESSC and service provider operating locations in Afghanistan, Iraq, and Kuwait.[248]

The CECOM LCMC depot Forward Repair Activities (FRA) expanded on-site repair of C4ISR systems as the GWOT continued, bringing the total to twenty-five permanent and eleven long-term temporary FRA providing support worldwide, with eight in Southwest Asia. In addition to providing on-site repair, the FRA also provided warranty processing, spares management, and upgrade services. All FRA had reach-back capability to Tobyhanna Army Depot (TYAD). TYAD backed its customers with the ability to deploy with the Logistics Support Element (LSE) thereby supporting them in the theater of operations.[249]

Contract awards and modifications were expedited to satisfy urgent war needs. In the first year of OIF, twenty-seven awards totaling over $63 million were made, including urgent requirements for Force XXI Battle Command Brigade and Below System, lithium batteries, antennas, transceivers, secure en-route communications packages, near term digital radios, laser detecting sets, shortstop electronic protection systems, single channel ground and airborne radio systems, joint tactical terminals. By FY07, the CECOM LCMC Acquisition Center obligated $14.5 billion on contracts and completed 25,400 contract actions. The command's security assistance experts initiated over $1 billion in new foreign military sales of C4ISR systems during the same period.

The command responded to 4,713 materiel release orders in

2003. These materiel release orders were mostly emergency call-ins to support Joint Chiefs of Staff special projects.

As the insurgency in Iraq intensified, Fort Monmouth's Public Affairs Office began to arrange casualty assistance services for northern and central New Jersey service members' families. These services were provided for families of Soldiers who were either killed in action or in non-combat accidents while deployed.

Over one hundred flash and immediate priority messages were responded to concerning compromise, emergency replacement and additional requirements for cryptographic key material in 2003. Fort Monmouth assisted NETCOM and the local DOIM with transmitting message traffic to the theater and expediting the shipment of Iridium secure telephones to units in Iraq.

One of the iconic features in the early days of the war was the televised daily briefing from the Central Command (CENTCOM) command center in Qatar, seen on news networks across the world. The Information Systems Engineering Command (ISEC) teamed with NETCOM to upgrade this command center and to make the infrastructure operational as well as developing, acquiring and performing quality control for this center.

Software changes were made to a variety of systems to include COMSEC equipment, ASAS, Guardrail, and the artillery fire control codes. In FY07, two hundred and fifty software releases were fielded, incorporating over 4,000 requirements. The command provided software support for the majority of the Army's deployed systems. The command's software experts also sup-

Soldier using Blue Force Tracking

ported Secretary of State Colin Powell for his address to the United Nations on 5 February 2003, replicating 1,000 copies of a CD for his multi media presentation to the Security Council entitled "IRAQ-Failing to Disarm."

Friendly fire incidents were virtually eliminated in OIF through the use of Blue Force Tracking and the Force XXI Battle Command Brigade and Below Command Control System. Developed and fielded by PEO Command, Control, Communications Tactical and supported by the C4ISR team, these systems gave commanders unprecedented situational awareness on the bat-

tlefield and allowed them to synchronize their forces. Combat and thermal identification panels, Phoenix Infrared lights, and GLO tape infrared reflective material- all Team C4ISR fielded items- also reduced incidents of fratricide during OIF.

4th ID Soldiers are instructed on the installation of combat ID panels, April 2003

The AN/ALQ-144 Infrared Jammer made U.S. aviators, helicopters, and aircraft safer than ever during OIF. These jammers were mounted on the fuselage of helicopters. They emitted signals to decoy heat-seeking missiles and caused them to detonate in the air and miss their targets.

The PEO for Intelligence Electronic Warfare and Sensors provided critical support on the Guardrail intelligence system that allowed Warfighters to locate threats and keep coalition forces safe during OIF. Guardrail is an airborne intelligence collection system that provided support to early entry forces, forward deployed forces and military intelligence. Guardrail is called a "common sensor" because it could intercept both classes of signal: Communications Intelligence (COMINT), low frequency radio transmissions and cell phone calls; as well as Electronics Intelligence (ELINT), and radar transmissions. Command software engineers resolved a significant software problem during OIF that was causing the ELINT precision location subsystem of the Guardrail to crash and therefore not be able to locate electronic emitters. The software that supports the Guardrail system was designed to support a particular way of flying. In Iraq, the Guardrail could not fly the way it was designed to, due to various geopolitical boundaries and restrictions in the region. It consequently could not identify targets that were being reported by Special Forces on the ground. The C4ISR software team was able to reprogram the software in just forty-eight hours to enable the Guardrail to accurately identify, targets once again while operating within geopolitical restrictions.

Challenges encountered in the first year of OIF included complications with several C4ISR systems due to environment, equipment age, and lack of trained personnel. Personnel eventually deployed in theater to support systems in an attempt to rectify these problems. They fully inspected equipment before shipping, strictly followed technical manual guidelines for sand environments, and kept parts and risk kits in theater. A shortage of repair parts also occurred due to problems with transportation. Items took a long time to get to the CONUS de-

BG Boles discusses future plans for relief and deployment of AMC LAR

parture sites. They arrived palletized at the theater distribution center, which further increased the amount of time it took to get the part to the recipient. LAR reported significant shortages of communication and transportation equipment. A communication and transportation package was recommended for future LAR as part of an initial deployment package.[250]

In 2005, the LCMC chartered a group of leaders on the ground to ensure that the fielding of C4ISR systems to the various units was both successful and consistent with objectives and timelines in the Army Campaign Plan. These leaders were known as Trail Bosses. Today, the command has Senior Command Representatives (SCR) collocated with Army Field Support Brigades. They are responsible for resolving any issues with C4ISR equipment in their area of responsibility. Trail bosses perform the same function as SCR but are located at the active component (AC) division level. LAR are located in forward deployed areas with the units they are assisting.[251] Some LAR in Southwest Asia performed Trail Boss duties at the brigade level in addition to their LAR mission.[252]

As the GWOT went into its second year, the need to repair battle damaged C4ISR equipment became a high priority. The CECOM LCMC "Reset" mission entailed the repair, replacement or recapitalization of equipment from units returning from OEF/OIF in a very short timeframe to ensure the unit was ready for its next deployment. The requirement was to bring all committed force structure equipment back to a desired level of combat effectiveness. The Reset mission, begun in July 2003 and formalized in 2004, involved coordinating C4ISR requirements and providing overall management of the induction of systems requiring depot level repairs.[253] By mid 2007, the command had supported eighty-six weapons systems and Reset 82,072 COMSEC items and 45,191 other C4ISR items for a total of 127,263 items Reset. These Reset efforts supported over 600 battalion level units.[254]

Several C4ISR systems were considered so critical to Warfighters that they were intensively managed. These systems included the AN/TPQ-36 Firefinder Mortar Locating Radar, the AN/TPQ-37 Firefinder Artillery Locating Radar, the Lightweight Counter Mortar Radar, the Counter Remote Control Improvised Explosive Device Electronic Warfare, and Intelligence and Security Command focused systems. This "intense management" included tracking operational status by serial number, tracking deliveries and fielding, and reporting weekly at the Four-Star level.[255] Firefinder detected and located enemy mortar and artillery weapon firing positions. The Firefinder radar system proved instrumental in the first few days during the battle for Baghdad when the Iraqi Army, to avoid detection by Firefinder, held their mortar and artillery fire, rather than fire on allied troop positions.

PEO Intelligence, Electronic Warfare and Sensors (IEW&S) Counter-Remote-Controlled Improvised Explosive Device (IED) Electronic Warfare (CREW) Team met an urgent need for Electronic Counter Measure (ECM) devices to defeat the enemy's use of IEDs against coalition forces. The team implemented a "near real time" counter-IED program that could neutralize new IED threats as soon as they emerged. They successfully deployed ECM devices, established and staffed logistics support fielding offices in the theater of operations, executed hundreds of contractual actions valued in the hundreds of millions of dollars in response to numerous urgency statements, and conducted a formal source selection for the next generation ECM devices. One such vitally important device was the WARLOCK ECM test set. The sets protected Army convoys in Iraq, Afghanistan and other locales in Southwest Asia by detecting and detonating IEDs planted along roadsides. Tens of thousands of these systems were fielded.

PEO IEW&S designed, built, tested, shipped, installed, and integrated a situational awareness system known as the Persistent Surveillance and Dissemination System of Systems (PSDS2). This system linked many different sensors, such as infrared, radar, commercial security cameras, and unmanned aerial vehicles, and allowed control of many of these sensors from within division headquarters. It provided the combatant commander and staff unprecedented situational awareness and the ability not only to respond to hostile actions, but also to enact better-coordinated responses to hostile and suspicious activity through the coordination and control of many varied sensors.

This PEO also aggressively maintained the Prophet program in order to continue to field Prophet Systems per HQDA mandated timelines. They sustained and supported fielded and deployed systems, applied quick reaction technical insertion capabilities to address theater specific requirements, and continued the system design and development of the next generation Prophet systems. The PEO continued to field Prophet Block I systems to every unit deploying in support of Operations Enduring and Iraqi Freedom, beating Army mandated modularity transformation timelines and enabling the Army to transform while simultaneously prosecuting the Global War on Terrorism. The Prophet provided the division, brigade combat team, Stryker Brigade Combat Team, and Armored Cavalry Regiment commanders with near-real-time force protection, situational awareness, and electronic attack capabilities to support the Army vision, current, and future force requirements. It was mounted on the heavy High Mobility Multi-purpose Wheeled Vehicle. It could also operate in a dismounted mode (e.g. uses a SIGINT man-pack) for airborne insertion and early entry operations. The Prophet's primary mission is force protection, by performing electronic sensing, and using direction finding to provide emitter lines-of-bearing.

The Forward-Looking Infrared (FLIR) programs were among the most technically and programmatically complex in the Army, providing state-of-the-art, second generation FLIR sensor, night vision capability to the Warfighter. In brief, FLIR referred to "an airborne, electro-optical thermal imaging device that detected far-infrared energy, converted the energy into an electronic signal, and provided a visible image for day or night viewing." These sensors were hailed as life saving, allowing Warfighters to see clearly at long ranges during varied atmospheric and battlefield conditions and provide battlefield dominance to Abrams, Bradley and Stryker platforms. During the period 2003-2005, PEO IEW&S fielded hundreds of different FLIR devices to Army and Marine Corps units. All fieldings occurred on time, or ahead of schedule, including many short notice HQDA directed fieldings to units in Iraq and Afghanistan.

The Lightweight Counter Mortar Radar-Army (LCMR-A), also managed by PEO IEW&S, provided 360 degrees of azimuth coverage and was used to detect, locate, and report hostile locations of enemy indirect firing systems. The LCMR-A, a digitally connected, day/night mortar, cannon, and rocket locating system, could be broken down, installed in man-packable carry cases, and shipped worldwide without damage by ground, rail, water, and air.

The LCMR-A was a spiral enhancement to the existing LCMR, which was originally designed to operate as a stand-alone ca-

Lightweight Counter Mortar Radar

Famous Firts

2001: *The world's smallest infrared cameras, attached to the end of PVC pipe, were deployed to the World Trade Center and used by rescue workers to look through voids in the rubble.*

Twenty-four hours after the original improvised search cameras were deemed lacking, engineers delivered cameras capable of looking 360 degrees around.[257]

2001: *Fort Monmouth adapted the Laser Doppler Vibrometer (a technology being explored for landmine detection) to monitor shifts in building structures at the World Trade Center site.*

The PDV-100 was the first laser Doppler vibrometer with digital signal processing.[258]

2002: *The CERDEC's Agile Commander Advanced Technology Demonstration (ATD) was named one of AMC's top ten greatest inventions of 2002.*

The ATD provides a scaleable digital command and control system with battle planning and execution monitoring functions.[259]

2003: *The BA-8180/U Zinc-Air Battery was developed and had up to 55-amp hours of capacity at twelve volts.*

One six pound BA-8180/U connected to a reusable interface adapter could replace up to five BA-5590s for extended missions. The BA-8180/U Zinc-Air Battery was cited as one of the top ten greatest U.S. Army inventions of 2003.

2003: *The Portable Omni-Directional Well Camera System was developed.*

This system was a lightweight, waterproof, battery-powered camera system designed for safer, faster, and more efficient inspection of areas that are unfit or unsafe for human inspection. The system provided soldiers with a safe method for rapidly inspecting wells and other underground locations up to three hundred feet deep and at a radius of twenty feet. The system could look 360 degrees around. It was powered by a BA 5590 battery for up to ten hours. The Well Camera System was also cited as one of the top ten greatest U.S. Army inventions of 2003.[260]

2004: *The Army recognized the Lightweight Counter Mortar Radar as one of the top ten Army inventions of 2004.*

It provided 360 degrees of azimuth coverage and is used to detect, locate, and report hostile locations of enemy indirect firing systems.

2005: *The Countermeasure Protection System (CMPS) was named one of the Top Ten Army inventions of 2005.*

Eighteen Warlock-CMPS units were delivered to Fort Monmouth for fielding to the Rapid Equipping Force in May 2005. Development of the Warlock-CMPS system directly led to the Warlock Increment 2 system commonly known as CREW-2. The CMPS featured advanced electronic warfare subsystems to counter the two predominant classes of radio-controlled Improvised Explosive Device threats used in the Global War on Terror. An optimized architecture was used to provide a maximum protection radius while minimizing the overall system cost and prime power consumption requirements. The system design supported future enhancements with minimal hardware redesign.

2005: *The Dual Band Antenna was named one of the Top Ten Army inventions of 2005. Approximately 8,000 units were delivered by the end of 2005.*

The antenna covered an unprecedented wideband frequency span from a single antenna structure. This was the first antenna that could be designated as the Army Common Antenna capable of performing multiple communications and electronic warfare functions critical in the Global War on Terror. The antenna provided an omnidirectional pattern over the entire frequency band. The antenna had a two-port design to interface with multiple communications and electronic warfare systems. In addition to wideband coverage, the antenna was designed to radiate from the upper eighteen inches of the radome. Such a design allowed for maximum performance while reducing the degrading effects of vehicle obstruction.

2006: *The Remote Urban Monitoring System was named one of the top ten Army inventions of 2006.*

RUMS hardware combined emerging technologies in Wireless Local Area Network technology, night-vision cameras and unattended ground sensors to eliminate false alarms. Tripped sensors transmitted an alarm signal to the camera module and operator after video and audio from multiple camera modules confirmed the unattended ground sensor's alarm signal.

pability for Special Forces. Unlike the Firefinder systems, the LCMR did not have a separate search and track beam. Instead, it performed a 'track while scan' operation. The CERDEC originally developed it, and the Army recognized the original LCMR as one of the Army's "Top Ten Greatest Inventions" of 2004. The CECOM LCMC team is responsible for the sustainment of all deployed LCMR systems.

The Army Battle Command System (ABCS) represented one of PEO Command, Control, and Communications Tactical's (C3T) top priorities as it maintained support of the GWOT. Prior to 1995, several independent projects were worked to leverage the rapid growth in Internet-related technologies and to develop systems that improved command and control capabilities in several battlefield functional areas. Then, in 1995, PEO C3T began working with the Training and Doctrine Command and elements of the 4th Infantry Division at Fort Hood, TX, to develop the ABCS. The ABCS provided commanders with the battle command architecture necessary to gain and maintain the initiative and successfully execute missions assigned by the National Command Authority. It joined eleven communications subsystems together onto one platform, making all interoperable. Some of these systems included the Force XXI Battle Command Brigade and Below, Global Command and Control System-Army, Maneuver Control System and All Source Analysis System. The biggest difference between previous ABCS versions and the ABCS 6.4 was that ABCS 6.4 improved and automated data sharing and horizontal interoperability among the systems. Soldiers, the requirements community, material developers, product managers, industry, software programmers, engineers, technicians, the test community, trainers, and combat systems all participated in the ABCS 6.4 testing.

The Joint Network Node (JNN) represented another PEO C3T priority. A highly transportable and mobile communications system, JNN supported the new transformational force structure. The JNN and associated user access cases provided enhanced video, voice, and data capabilities. The JNN connectivity was comprised of the Joint Network Node, Unit HUB Node and the Battalion Command Post Node. JNN afforded the Warfighter a communications network down to the Battalion level by allowing the Soldier to mimic connection capabilities used in an office and to make direct use of internet based applications. The Army started JNN in 2004 as a way to disseminate tactical communications down to the battalion level for troops in Iraq. In the summer of 2007, the Army announced that JNN was now the first of four increments of the Warfighter Information Network-Tactical (WIN-T). Former C-E LCMC, CG MG Michael Mazzucchi said, "The newly restructured WIN-T program will move the Army toward its goal of providing Soldiers down to the company level secure data, voice and imagery while on-the-move over great distances and varied terrain."[256]

HURRICANE KATRINA

C4ISR systems were as indispensable in 2005 for natural disaster recovery as they were in the Global War on Terror. The C-E LCMC responded immediately with C4ISR systems and support following Hurricanes Katrina and Rita in addition to its on-going support of the Global War on Terror.

Hurricane Katrina was a category five hurricane that hit the Gulf coast region of the United States, primarily affecting citizens in Louisiana, Alabama and Mississippi. Katrina made landfall on the Gulf Coast on 29 August 2005. Over 1,600 people died as Katrina made its way across land as a category three storm. To assist in recovery efforts, C-E LCMC provided generators ranging from 10KW to 840 KW prime power units and communications systems. When the storm disrupted the Defense Information System Agency hub in New Orleans, C-E

Celebrity Notes:

Bruce Springsteen

rehearsed for several days at Fort Monmouth's Expo Theater in July 2002. He stopped to take photos and sign autographs for Fort personnel at the theater.

Bon Jovi

entertained a full house at the Expo Theater in October 2005. The band members, Jon Bon Jovi, Richie Sambora, Tico Torres and David Bryan met with MG Michael Mazzucchi and garrison officials and visited with Cadet Candidates from the United States Military Academy Preparatory School.

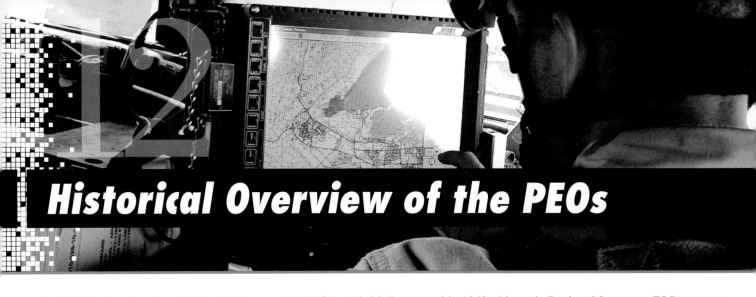

Historical Overview of the PEOs

When the Army Materiel Command (AMC) was initially created in 1962, thirty-six Project Managers (PM) were established to manage the development of major weapons systems and equipment. Prior to 1962, Project Managers were mostly used by industry. Throughout the 1970s most of AMC's commanding generals recognized the need to use PM to manage major materiel development programs. However, some commanders reduced the number of PM after recognizing that PM authority ran counter to traditional Army rank structure. There were several periods when commanders sought to reduce PM autonomy and subordinate them to headquarters and command staff.[262]

The 1 May 1987 implementation of the Goldwater-Nichols Department of Defense Reorganization Act of 1986 removed Project Managers from AMC control and placed them under Program Executive Officers, who reported directly to the Army Acquisition Executive (the Assistant Secretary of the Army for Research, Development, and Acquisition). The commodity commands continued to provide functional services to the PEO and their PM under the matrix support concept. The Communications-Electronics Command (CECOM) at Fort Monmouth supported three PEO: Command and Control Systems (CCS), Communications Systems (COMM), and Intelligence/Electronic Warfare (IEW).

FY88 represented the first year of full scale PEO operation as they prepared to become Department of the Army Field Activities, completing the break from CECOM. The new PEO structure improved the management of most PM programs. However, in creating the PEO concept, it removed the PEO from the development and acquisition support provided by AMC for their systems.[263] The fielding of Mobile Subscriber Equipment (MSE) and the award of the SINCGARS Second Source were among some of the most significant accomplishments that year because of their cost, their ability to affect the future of the Army on the battlefield, and the timely and cost effective ways in which they were handled.[264]

As of August 2004 the PEO were once again part of an Army reorganization initiative and became part of Life Cycle Management Commands. In the C4ISR arena, this initiative brought together CECOM, PEO C3T and the PEO IEWS under the new Communications-Electronics Life Cycle Management Command (C-E LCMC). While each PEO and CECOM remained part of their respective reporting chains, the Army intended the initiative to integrate significant elements of the Army Acquisition, Logistics and Technology (ALT) communities in order to provide products to the Soldier faster, make good products even better, and minimize life cycle cost.[265] The mission of all organizations remained the same: to develop, acquire, test, field

and sustain effective, suitable and survivable C4ISR capabilities from the soldier in the combat zone all the way back to the national leadership.

PROGRAM EXECUTIVE OFFICE FOR COMMAND, CONTROL, AND COMMUNICATIONS-TACTICAL (PEO C3T)

The history of PEO C3T can be traced through various organizational changes and Army restructuring initiatives. Formerly known as PEO Command, Control and Communication Systems which was formed from the merger of PEO Communications (COMM) and PEO Command and Control Systems (CCS), today, the organization plays a key role in fielding digital battlefield systems and sustaining them with in-theater support. The mission of the PEO is to rapidly develop, field, and support leading edge, survivable, secure and interoperable tactical, theater and strategic command and control and communications systems through an iterative, spiral development process that results in the right systems, at the right time and at the best value to the Warfighter.

Headquartered at Fort Monmouth, N.J., PEO C3T is responsible for seven Project Management (PM)/Product Management (PdM) offices: PM Battle Command (BC), Product Director Counter Rocket, Artillery and Mortar (C-RAM), PM Mobile Electric Power (MEP), PdM Network Operations-Current Force (Netops-CF), PM Tactical Radio Communications Systems (TRCS), PM Warfighter Information Network-Tactical (WIN-T), and PM Force XXI Battle Command Brigade and Below (FBCB2). The Special Projects Office (SPO)/Northeast Regional Response Center (NRRC) is also assigned to PEO

C3T. PEO C3T's workforce of over 2,300 employees is comprised of military, civilian, CECOM matrix, and support contractors. The PEO's total annual budget exceeds $2.8 billion.

In the late 1980s, PEO COMM was staffed by 250 military and civilian employees, managed over one hundred programs, and had a budget of $2.9 billion. It was responsible for PM Global Positioning System (GPS), PM Multi Service Communications Systems (MSCS), PM Mobile Subscriber Equipment (MSE), PM Position Location Reporting System/Tactical Information Distribution System (PLRS/TIDS), PM Regency Net (RN), PM Satellite Communications (SATCOM), PM Single Channel Ground and Airborne Radio System (SINCGARS) and PM Single Channel Objective Tactical Terminal (SCOTT).[266]

PEO CCS was staffed by 364 military and civilian employees and had an annual budget of $1.9 million. It was responsible for six PMs: PM Air Defense and Control Systems (ADCCS), PM All Source Analysis System (ASAS), PM Combat Service Support Control System (CSSCS), PM Common Hardware/Software (CHS), PM Field Artillery Tactical Data Systems (FATDS), and PM Operations Tactical Data Systems (OPTADS). Only three of these PMs were located at Fort Monmouth. ADCCS was located at Redstone Arsenal, AL; ASAS at Fort McLean, VA; and CSSCS at Fort Belvoir, VA. In FY90, the Army Acquisition Executive disestablished the PEO for Strategic Information Systems and directed PEO CCS to assume its responsibilities.[267]

PEO C3S supported Operations Desert Shield and Desert Storm from 1990-1991. The Advanced Field Artillery Tactical Data System (AFATDS) was used in Desert Storm and today commanders continue to use the system to plan and execute fires during each phase of action.

Under the Army's 1995 Digitization Master Plan all PEO and PM were responsible for: providing periodic digitization reviews, developing a plan to migrate to the DoD technical architecture, providing modeling and simulation support as appropriate for digitization, supporting the experimentation process and accomplishing specified installation kit responsibilities.

PEO CCS was directed to: manage the acquisition of hardware, software, and systems engineering support for integrated command and control systems, with support from AMC; manage the Army Digitization Office (ADO) system integration effort, with support from AMC; prepare in coordination with the ADO, an applique Experimentation Master Plan (EXMP); develop the COE documentation; provide modeling and simulation support as appropriate for digitization; and develop and enforce the standards of COE and prepare, in coordination with the ADO, a capstone EXMP which integrated the test programs of the following PEOs: COMM, CCS, Armored Systems Modernization, Aviation, Missile Defense and IEW.

PEO COMM was directed to provide the communications infrastructure needed to support reliable, horizontal and vertical seamless connections. This included the following networks: Enhanced Position Location Reporting System (EPLRS), SINCGARS, Joint Tactical Data System (JTIDS) and Mobile Subscriber Equipment (MSE); as well as the Marine Corps and Air Force communications equipment and commercial communication equipment; manage the development of the tactical internet in accordance with Army's Technical Architecture, with support from AMC; define the communications protocols and standards for COE; and Provide modeling and simulation support as appropriate for digitization.[268]

PEO Command, Control and Communication Systems (C3S) was formed from the merger of PEO COMM and PEO CCS in 1995. This merger sought to integrate the acquisition management of C3I Systems for the digitized battlefield, Force XXI, and Warfighters from the laboratory to the foxhole. The mission of PEO C3S was to rapidly develop, field, and support leading edge, survivable, secure and interoperable tactical, the-

Aerial photo of the Central Technical Support Facility (CTSF), Fort Hood, Texas

ater and strategic command and control and communications systems.

Task Force XXI, which digitized the Army for the first time, began in 1994. The 4th Infantry Division (4ID) was the experimental force, which used the digital systems at Fort Hood, Texas. System-related issues arose when the division first began to train. The products were sent to a facility at Fort Hood, which became the Central Technical Support Facility (CTSF). The challenges were resolved in the CTSF labs, rather than in the field.

About six months after the inception of Task Force XXI, PEO C3S briefed the Armed Services Committee which earmarked $2 million into the PEO's budget, so it could digitize the Army. The 4ID became the Army's first digitized division. The systems they received included: the Maneuver Control Systems (MCS) (displayed current situation reports, intelligence and contact reports); FBCB2-BFT (provided a graphical representation of friendly vehicles and aircraft on a topographical map or satellite image of the ground); All Source Analysis System (ASAS) (automated the processing and analysis of intelligence data from all sources); Integrated Meteorological System (IMETS) (an automated weather system used to receive, process and disseminate weather observations and forecasts); and the Combat Service Support Control System (CSSCS) (now known as the Battle Command Sustainment and Support System (BCS3), which provided commanders with a visual layout of battlefield logistics). Many of these systems still exist in the suite of systems known as Army Battle Command Systems (ABCS 6.4).

By September 2006, the Program Executive Office for Command, Control and Communications Tactical (PEO C3T) had procured over half a million Single Channel Ground and Airborne Radio Systems (SINCGARS).

ABCS existed prior to the creation of Task Force XXI. However, the use of ABCS in platforms was limited. Many platforms had contained voice only systems such as SINCGARS. The effort to digitize the Army added FBCB2 to platforms such as M1 tanks, Bradley fighting vehicles and HMMWVs.

PEO C3S took the lead in Army Knowledge Management initiatives in 2000. The goal of the initiative was to use informa-

FBCB2

tion technology to leverage Army wide innovation in services, processes and knowledge creation. A geographically dispersed workforce as well as rapidly changing technology necessitated knowledge management collaborative efforts. PEO C3S' Knowledge Center was activated as one of the Army's first three original pilot programs.[269]

The Warfighter Information Network Tactical (WIN-T) in use at the National Training Center, Fort Irwin, CA, in August 2007

The story behind the Warfighter Information Network Tactical (WIN-T) Increment One (formerly the Joint Network Node-Network) began with the launch of Operation Enduring Freedom in Afghanistan in 2001. The system was developed as an immediate response to the need for a communications pipe that allowed Warfighters to communicate further than they could see. The necessity for this capability was emphasized when General William S. Wallace led the Army in the invasion of Baghdad, Iraq during Operation Iraqi Freedom (OIF), which began in 2003. Wallace recognized that the pace of the war outran the coalition forces ability to communicate, which revealed a gaping hole in the way that they fought. The WIN-T Increment replaced the twenty-year-old Mobile Subscriber Equipment (MSE) network. It has since provided Battalion-level and above Warfighters with the ability to connect to the Army's digitized systems, voice, data and video via a satellite Internet connection. Future increments of WIN-T will provide communications on-the-move. WIN-T Increment One officially became a program of record in 2007.

In early 2003, the Army rapidly cobbled together a suite of digital Battle Command systems to support the initial fight. PM

BC was told to develop and field a "good enough" capability to the entire Army, focus future money on capabilities for the joint military force and add other capabilities to that base over time. This marked the initiation of ABCS 6.4. Since then, technologies have evolved and the military expects to one day field a singular suite of digital capabilities across all U. S. military forces. ABCS 6.4 is fielded in most modular Brigade Combat Teams in the active force.

PEO C3S was renamed in 2004 and became the PEO for Command, Control and Communications- Tactical (PEO C3T).

By September 2006, the PEO had procured over half a million SINCGARS radios, 8,458 Enhanced Position Location and Reporting Systems (EPLRS) and 1,400 High Capacity Line-of-Sight (HCLOS) radios in support of the GWOT.

C3T systems proved indispensable during Hurricane Katrina relief efforts. In September 2005 PEO technical experts drove a mobile, satellite-based communications system from Fort Monmouth to Camp Shelby, MS to provide support during the aftermath of the hurricane. The team used the system to restore communications for first responders after flooding in New Orleans, LA had shut down all terrestrial communications. The system allowed communications to go from a commercial domain to a military domain, which allowed first responders from separate communications networks to communicate with one another. PM Mobile Electric Power also provided power generation to first responders during their recovery efforts.

Today, PEO C3T is involved in critical work supporting Global War on Terror efforts through fielding situational awareness and mobile communication systems. In FY06, the Counter Rocket, Artillery, and Mortar system achieved the first combat intercept of a hostile mortar. This was the first instance of the Phalanx weapon system defeating a hostile round. By 2007, PM Mobile Electric Power had fielded over 45,000 Tactical Quiet Generators to Army units. C3T has integrated its approach towards fielding with its CECOM LCMC counterparts using the five-phased Unit Set Fielding process. Field service representatives and digital systems engineers provide direct system support to the Warfighter both by phone and on-site. The PEO continued to support the Army's progression as a modular force by providing the systems and communications pipes that allow Warfighters to see, hear and conduct the battle.

A full fleet of Tactical Quiet Generators provided by Project Manager, Mobile Electric Power (PM MEP)

Company delivers 75,000th SINCGARS

submitted by PEO Communications

ITT Aerospace/Communications Division, Fort Wayne, Ind., marked the continued success of the Single Channel Ground and Airborne Radio System (SINCGARS) by delivering to the U.S. Army its 75,000th ground radio system last week. More than 110,000 ITT SINCGARS have been ordered or delivered worldwide.

Accepting the milestone ITT radio, Army Acquisition Executive Gilbert F. Decker told ITT employees, "I was impressed today to see your dedication to quality. That's really important to our soldiers around the world, because your radios are critical. They are backbones and they have to be reliable."

"Your radios continue to perform with reliability at least six times what we asked for," agreed Brig. Gen. David R. Gust, program executive officer for Communications Systems.

"The great support we get from you is magnified by the thanks we get from the soldiers about the value of this radio," Gust said, "especially the digitization improvements to make it more data friendly. That will help our Chief of Staff's vision to digitize the battlefield."

SINCGARS is the standard combat net radio for the Army and Marine Corps. It has been purchased by the U.S. Navy and Air Force, as well as national defense forces in Europe, Asia and the Middle East. ✪

Monmouth Message, 18 November 1994

PROGRAM EXECUTIVE OFFICE FOR INTELLIGENCE, ELECTRONIC WARFARE AND SENSORS (PEO IEW&S)

PEO IEW&S is headquartered at Fort Monmouth, NJ. The mission of the PEO is to provide a persistent and integrated surveillance and reconnaissance capability which enables actionable intelligence at the point of decision, empowering all to understand and act.

PEO IEW&S originally began with five Project Managers: PM Electronic Warfare/Reconnaissance, Surveillance and Target Acquisition (EW/RSTA), PM Joint Surveillance Target Attack Radar System (JSTARS), PM Night Vision and Electro-Optics (NVEO), PM RADAR and PM Signals Warfare (SW) (initially located at VHFS and relocated to Fort Monmouth in October 1997). The PEO would later add a Project Manager for Control and Analysis Center (CA), located at VHFS. In the early nineties the PEO had a staffing strength of about twenty military and 200 civilians, supported by about 160 contractors. Components of PEO IEW&S were located at Fort Monmouth, NJ, Vint Hill Farm Station, VA, Redstone Arsenal, AL, Hanscom Air Force Base, MA, Fort Belvoir, VA, Pentagon, Washington DC, Wright Patterson Air Force Base, OH, Motorola, Scottsdale, Arizona, and in Seoul, Korea.

The majority of PEO activity in FY88 was directed toward system fielding. The Remotely Monitored Battlefield Sensor System (REMBASS) had a full release approved in October 1987 and was fielded to Korea and other areas. A depot contract for software support for the system was awarded in September 1988. The Meteorological Data System (MDS) Independent Evaluation Report was completed in February 1988, followed by initial system training at Fort Sill, OK. Full release was approved in May 1988 followed by fieldings to different areas. PM RADAR achieved a major milestone when the first prototype Block II product-improved AN/TPQ-36 underwent a totally successful Live Fire Test at Yuma Proving Ground in the fourth quarter of FY88. The Block II system repackaged the existing components of the AN/TPQ-36 into a single vehicle configuration. During FY88, DA approved a refined concept for incremental improvements to the Firefinder equipment. Over twenty contractual actions were completed with the Hughes Aircraft Company, the prime contractor for PM Firefinder. In

AN/TPQ-36 Firefinder Radar

addition, several actions were processed with other contractors to provide functional support to the Firefinder program.[270]

PEO COMM, CCS, and IEW&S acquired and fielded some 200 systems in FY89. The PEO during the early 1990s continued to achieve the functional and technical integration of the systems for which they held responsibility. They marked progress as well in synchronizing the acquisition and fielding of assigned systems, thus reducing the "dead time" that often delayed the completion of one program pending the availability of products from another. This allowed for the acceleration of fielding and modification of equipment needed by troops in Operation Desert Shield/Storm.

PEO IEW&S made significant contributions in FY90 to drug traffic interdiction efforts, Operation Just Cause in Panama and made deployments to Saudi Arabia as part of Operation Desert Shield. The DoD called upon PEO IEW&S to supply much of the equipment the government required to detect and monitor illicit drug traffic. Most of the sensor acquisition programs identified in the Counternarcotics Program Objective Memorandum were under IEW&S auspices.

PEO IEW&S Headquarters Staff

IEW&S accelerated fielding and modified equipment to meet the needs of commanders who were in or about to deploy to Saudi Arabia. IEW&S systems fielded to the Persian Gulf included Quickfix, Trailblazer, Teammate, Trafficjam, Tomcat, the Technical Control and Analysis Center (TCAC), Tactical Intelligence Gathering and Relay (TIGER) system, Guardrail, Firefinder radars and night vision aids. In addition to fielding these systems to Army units, Tomcat was also issued to the Marine Corps. Night vision technology proved critical in Operation Desert Storm. The 24th ID Commander, MG Barry McCaffrey, commented, "our night vision technology provided us the most dramatic mismatch of the war." A March 1991 *Newsweek* article summarized the Firefinder radars' usefulness, "When an Iraqi Battery fired a round, a U.S. Army Q-37 radar would sight it and feed the battery's coordinates to computers that directed the American guns. It took less than a minute to drop a counterround on the Iraqis. Many of them soon stopped firing. To pull the lanyard was to invite death."

Thirty-eight GPS receivers were provided to Military Intelligence (MI) Battalions in Southwest Asia as well as ten Friartuck sets. Friartuck was a laptop computer with a special card to enhance the MI mission capability. The JSTARS also proved its worth in Desert Storm, providing detailed information on Iraqi troop movements from 15 January 1991 until the cease fire on 27 February 1991. JSTARS detected sixty vehicle convoys, the Iraqi withdrawal from Kuwait and located a downed F-16 pilot as well as mobile SCUD units. The 3D Army Commander, BG John Stewart, remarked, "JSTARS was the single most valuable intelligence and target collection system in Desert Storm."[271] The overall readiness of IEW&S systems fielded in SWA during Operation Desert Storm/Shield was maintained at approximately eighty percent.[272]

AN/PVS-7 Night Vision Goggle

PM Joint Surveillance Target Attack RADAR System

The success of the Army's nighttime invasion of Panama in Operation Just Cause (1989) was due, in large measure, to the use of the AN/PVS-7 Night Vision Goggle. With this device, soldiers were able to enter Panama undetected, in the early hours of the morning, under the cover of darkness. Following the operation, PEO IEW&S personnel and unit commanders were called upon to testify before Congress on the use and performance on the Night Vision goggle during the initial hours

of the invasion. Firefinder radars also worked extremely well during Just Cause.[273]

In FY91 the production contract for the Shortstop program was issued. This system provided mobile, electronic countermeasures designed to protect personnel and high value targets from the most predominant of indirect fire threats without operator intervention. Electronic counter measure protection systems that would later become critical to protecting the lives of Warfighters in the Global War on Terror, were based on Shortstop technology.[274]

From the late 90s until today, PEO IEW&S has provided key support to operations in Kosovo by providing Warfighters with critical systems such as the Airborne Reconnaissance Low (ARL), Guardrail/ Common Sensor (GR/CS), Joint Service Work Stations (JSWS), TESAR Imaging Radar, Medium Ground Station Module, Night Vision Goggles and the Hunter UAV system. These systems have enabled critical events to be examined in near real time. An Aviation commander in Kosovo, BG Dick Cody, Task Force Hawk, remarked on the importance of Night Vision Goggles, "had it not been for the ANVIS-6 goggles, I am convinced we would have sustained several wire strikes and possibly one or two mid-air collisions."

As the twenty-first century dawned, IEW&S provided key support to the Army's new Brigade Combat Team (BCT) structure, a brigade size force deployable worldwide within ninety-six hours. The following systems were fielded to the initial BCT in Fort Lewis, WA: Night Vision Goggles, Monocular Night Vision Devices, Sniper Night Sights, the JSTARS Common Ground Station (CGS), the Hunter UAV System, and Prophet systems. Future fieldings would also include the Long Range Advanced Scout Surveillance System (LRAS), Driver's Vision enhancer (DVE), Thermal Weapon Sight (TWS), Lightweight Laser Designator Rangefinder (LLDR), Lightweight Video Reconnaissance System (LVRS) and the Shadow 2000 Tactical Unmanned Aerial Vehicle (TUAV).[275]

Prophet next generation communications intelligence and electronic warfare system

Today PEO IEW&S is involved in critical work supporting Global War on Terror efforts through fielding sophisticated intelligence, electronic warfare and sensor systems to the joint Warfighter. Some of these systems are highlighted below:

The Common Missile Warning System (CMWS) is managed by the Project Director, Aircraft Survivability Equipment (ASE). The system functions as a stand-alone system with the capability to detect missiles and provide audible and visual warnings to the pilot(s). When installed with the IR Laser Jammer and Improved Countermeasure Dispenser (ICMD), it activates expendables to decoy/defeat IR-guided missiles. An Apache pilot wrote - "I wanted you to know that your product saved my life today. I'm an Apache Longbow pilot deployed to Iraq and while on a mission today I was fired upon. The on-board CMWS deployed and defeated the missile saving myself and my copilot." A total of 1,491 CMWS were fielded and deployed in 2007.

The Product Manager Counter RCIED Electronic Warfare (CREW) developed and fielded ground-based electronic countermeasure (ECM) devices that neutralize the pervasive Improvised Explosive Device (IED) threat encountered in OEF and OIF. PM CREW fielded 11,717 CREW systems to OEF/OIF in FY07. The CREW ILS team from the Program Office won the 2007 Army Acquisition Excellence Award for Equipping and Sustaining Soldiers Systems.

The Persistent Threat Detection System (PTDS) is managed by PM Robotics and Unmanned Sensors (RUS). PTDS is a Quick Reaction Capability (QRC) aerostat-based surveillance, detection, and communication relay system that integrates existing battlefield sensors to provide an automatic "slew to cue" capability. The PTDS allows quick reaction forces to find, fix, track, and engage direct/indirect fire threats. The first PTDS was deployed to theater in October 2004 in support of OIF. As a result of its success, PTDS was designated as one of the Army's "Top Ten" greatest inventions for 2005. Additionally, the PM office received funding in August 2006 to build, deliver, field, operate and sustain seven additional PTDS systems in support of OEF/ OIF. All seven systems have been fielded in FY07 with PTDS being the first aerostat capability to operate in the harsh operating environment and altitude of Afghanistan.

The Persistent Surveillance Dissemination System of Systems (PSDS2) is managed by PM Robotics and Unmanned Sensors (RUS). The PSDS2 is a video exploitation, dissemination and cueing sensor fusion system deployed in OEF/OIF as a Quick Reaction Capability (QRC). It networks existing sensors into a Network Centric architecture enabled by elements of the Army's emerging Distributed Common Ground System (DCGS) enterprise architecture to support persistent surveillance and rapid dissemination of actionable information for the Warfighter. The system provides visualization tools to present live streaming video in context for a more intuitive situational awareness display and common operating picture. PM RUS fielded its 2nd PSDS2 system to OIF in 2007 and the sensor feeds for this system increased fifty-six percent from thirty-four to fifty-three sensors. The user accounts for this system in 2007 increased from 220 to over 1,000; a 400% increase in

Airborne Reconnaissance Low (ARL)

the number of Warfighters that have access to real-time video for situational awareness, collaboration, and direct action. Additionally, a PSDS2 capability was established in OEF for the first time in 2007 with over 500 users having access to multiple imagery feeds. In January 2007, the Institute for Defense and Government Advancement presented PM RUS and the PSDS2 Team with the 2007 Network Centric Warfare 1st Place Award as the Best U.S. Government program.

The Rapid Aerostat Initial Deployment (RAID) is managed as a QRC under PM RAID. The RAID system employs a variety of platforms (aerostat, tower, and mast) and sensor suites to provide unprecedented Elevated Persistent Surveillance in support of Intelligence, Surveillance and Reconnaissance (ISR) needs. RAID was designed and developed in response to a Vice Chief of Staff of the Army initiative to provide base security cells 360-degrees, hi-resolution, day/night surveillance capability. RAID provides enhanced target recognition and situational understanding. The RAID System supports a variety of missions ranging from force protection to border surveillance. RAID has been instrumental in the successful identification and tracking of individuals engaged in attacks against U.S. and coalition sites. As a result, commanders are requesting additional RAID assets as fast as they can be produced. Over 130 RAID systems have been deployed to Iraq and Afghanistan.

The Lightweight Counter Mortar Radar (LCMR) is managed by PM Radars. The LCMR, with its automatic detection, tracking, and weapon location capability is a critical force protection system that can both compute a mortar's point of origin (weapon location), and predict mortar impact, while providing both in-flight mortar location, as well as continuous 360 degree protective coverage. The LCMR was named one of the top ten Army inventions of 2004. Over 270 LCMRs were fielded since 2006.

Long Range Advanced Scout Surveillance System 3

The Long Range Advanced Scout Surveillance System 3 (LRAS3) is managed by PM FLIR. The LRAS3 provides long range target acquisition and far target location capabilities to Armor and Infantry Scouts enabling them to conduct reconnaissance and surveillance missions while remaining outside of threat acquisition and engagement ranges. LRAS3 exports targeting information to battlefield command and control via FBCB2. A total of 257 LRAS3 systems were fielded in 2007.

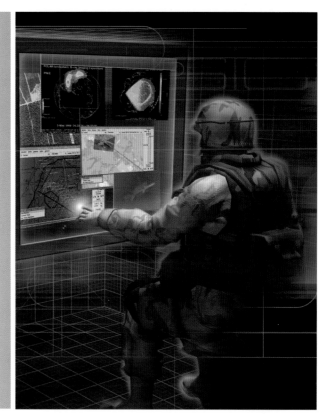

Distributed Common Ground System - Army (DCGS-A)

The Distributed Common Ground System - Army (DCGS-A) is managed by PM DCGS-A. DCGS-A is the Army's premier future force intelligence system. DCGS-A is the net-centric intelligence, surveillance, reconnaissance (ISR) component of the Army's future force and the primary system for ISR tasking, posting, processing, and using information about the threat, weather, and terrain at the joint task force and below. DCGS-A will contribute to visualization and situational awareness and understanding and is the ISR gateway to joint, interagency, allied, coalition, and national data, information, intelligence, and collaboration. It will provide access to theater and national intelligence collection, analysis, early warning and targeting capabilities and emphasizes the use of reach and split-based operations to improve accessibility to data and reduce the forward footprint. DCGS-A will consolidate capabilities found in the current force systems and select Quick Reaction Capability (QRC) products fielded in support of the GWOT. Early operational capability was established in the five DCGS-A fixed sites and DCGS-A V2/V3 systems were fielded and operational in support of OEF/OIF.

Task Force Observe, Detect, Identify and Neutralize (ODIN) is managed by PM Aerial Common Sensor (ACS). This effort in support of the GWOT provides for the integration, deployment and oversight of complex manned and unmanned aerial sensor systems and coordinates their execution. Eight original systems were successfully integrated into the Task Force ODIN architecture (four distinct airborne sensor systems with four distinct ground systems) to provide viewing, analysis, and dissemination of full motion video in support of Warfighter Reconnaissance, Surveillance and Target Acquisition missions in OIF.

Landmarks and Place Names

INTRODUCTION

This chapter describes Fort Monmouth's major landmarks and memorials. The post celebrates its rich history not only with monuments and plaques, but also with its gates, streets and buildings.

Scientists, engineers, program managers, logisticians and support staff here have delivered technological breakthroughs and provided superior support to our Soldiers, Sailors, Airmen, Marines, and Coast Guardsmen for almost a century. The post commemorates its historic signal/C4ISR mission at almost every turn. This mission initially began with signal training and today encompasses the full life cycle of C4ISR system research, development, engineering, logistics, contracting, fielding, acquisition and sustainment support.

THE MAIN POST

Nestled in Monmouth County, New Jersey, Fort Monmouth lies fifty miles south of Manhattan, sixty miles northeast of Philadelphia and about five miles from the Atlantic Ocean. Fort Monmouth's 1125.44 acres include hundreds of housing units, dozens of administrative buildings, several laboratory buildings, and an eighteen-hole golf course. Over half of that land, 636.1 acres, comprise the area known as the "Main Post".[1]

PARKS AND MEMORIALS

The following parks and memorials can be found on the Main Post.

Hemphill Parade Ground

The Hemphill Parade Ground is located just inside the West Gate (Johnston Gate) on the North Side of the Avenue of Memories. Dedicated on June 8, 1954 by General Orders Number 70, it honors COL John E. Hemphill, Commanding Officer of Camp Alfred Vail from December 1920- September 1, 1925. COL Hemphill laid out the master plan for permanent post construction, which occurred after his departure.[6]

Dean Field

The baseball field on the North Side of the Avenue of Memories became "Dean Field" on May 19, 1959 per General Orders Number 58. The designation honors SGT William H. Dean, Jr., a member of Headquarters Company 3rd Battalion, 330 Infantry. He died in combat on December 7, 1944 in Grobhau, Germany.[2]

Soldiers Park

Soldiers Park is located at the intersection of the Avenue of Memories and Wilson Avenue, opposite the Bowling Center. It is dedicated to "the Fort Monmouth Soldiers and civilians who deployed and fought worldwide and to the families who kept the home fires burning bright."

Avenue of Memories

The monuments and trees lining the Avenue of Memories pay homage to the Signal Corps Soldiers who gave their lives during WWII. The Army dedicated the Avenue on April 6, 1949 when the first marker was placed in memory of MAJ Edmund P. Karr. Originally designated as "Memorial Drive," new markers and trees have been added over the years. On Memorial Day, 1999, the Fort Monmouth community re-affirmed its commitment to those Soldiers honored on the Avenue of Memories.[xiv]

Husky Brook Pond

Husky Brook Pond is located near the Nicodemus Gate. It underwent a massive cleanup from 1966 through the early 1970s that sought to convert a ten-acre wasteland into a handsome, useful recreation area. Today, the area boasts several picnic sites and tranquil views.[7]

Dunwoody Park

Dunwoody Park is located at the intersection of Brewer and Malterer Avenues with Sherrill Avenue. Per General Orders Number 82, dated September 22, 1950, this park memorializes BG Henry H.C. Dunwoody (1842-1933). BG Dunwoody served as the Chief Signal Officer in Cuba from December 22, 1898- May 24, 1901. Under his leadership, the U.S. Army reconstructed, extended, and modernized the entire Cuban Telegraph System. Dunwoody Park features the Spanish-American War Memorial.[3]

Voris Park

Voris Park is located between Russel and Allen Avenues immediately west of Jagger Park in the Officer Family Housing Area to the north of the East Gate entrance. Dedicated on September 27, 1957 by General Order Number 76, the park honors COL Alvin C. Voris (1876-1952). Voris commanded the post from April 30, 1937- August 1, 1938.[12]

Jagger Park

Jagger Park is located in the Officer Family Housing Area to the North of the East Gate entrance. Designated sometime prior to 1935, it honors 1LT H.R. Jagger, 304th Field Signal Battalion, 79th Division, who died in action in France, 1918.[8]

Cowan Park

Cowan Park is located inside the East Gate near Russel Hall. Dedicated on June 24, 1961 by General Orders Number 48, Cowan Park memorializes COL Arthur S. Cowan, who commanded Camp Alfred Vail from September 16, 1917- June 28, 1918 and Fort Monmouth from September 2, 1929 - April 30, 1937. COL Cowan consolidated the Signal Corps Laboratories and is remembered for his astute leadership during the formative years of the Signal School.[1]

Augenstine Memorial

The Augenstine Memorial is located just south of Barker Circle. Dedicated in 1951, the bench and its plaque originally had a background of three dogwood trees. The bench memorializes Chief Warrant Officer Edwin Daniel Augenstine, a native of West Long Branch, NJ with over 17 years of Army service. Augenstine served at Fort Monmouth from 1936 to 1942, and then deployed to Europe and Manila. He died of an unknown tropical disease in November 1945.[13]

Greely Field

Located between Sherrill and Saltzman Avenues (immediately west of Russel Hall), Greely Field honors MG Adolphus W. Greely, Chief Signal Officer from 1887-1906. MG Greely led the Signal Corps expedition to the Arctic in 1881.[4] The field, dedicated by General Orders Number 22 on April 6, 1949, comprises part of the Historic District and features the World War II Memorial. The memorial is located on the infield of the old Monmouth Park Racetrack.[5]

Breslin War Memorial

The Breslin War Memorial resides just inside the Community Center's fence. Dedicated at Fort Monmouth on April 25, 1961, the stone monument memorializes those service members "who did not return" from war. Two private citizens, Pat and Sandy Breslin, originally erected the memorial on private property. After losing their lease, the memorial sat in storage until coming to its resting place, here.[16]

Van Kirk Park

Designated by General Orders Number 24, dated June 21, 1943, Van Kirk Park is located between Brewer and Malterer Avenues (opposite the post field house). First Lieutenant John Stewart Van Kirk died in combat on November 30, 1942 in Djedeida, Tunisia. He had previously attended Officer Candidate School at Fort Monmouth.

A memorial granite bench was erected at Van Kirk Park without ceremony in the early 1950s pursuant to the wishes of the father, the donor. The park also features the Purple Heart Memorial.[11]

Purple Heart-Memorial

The Purple Heart Memorial honors the recipients of the nation's oldest military decoration. It resides in Van Kirk Park and was dedicated c. 2004.[25]

MAIN POST PARKS AND MEMORIALS (Continued)

Wright Memorial

The 10th Field Signal Battalion Association and the 7th Division Association of WWI veterans dedicated a beech tree and plaque to their founder, E. Frederic Wright (1899-1974), on May 21, 1977.

The 10th Field Signal Battalion organized at Fort Monmouth, then called Camp Alfred Vail, on July 10, 1917. The Battalion was assigned to the 7th Division on December 6, 1917 and departed from this installation on August 17, 1918. It saw front line action in France from October 8 to November 11, and had its colors decorated by GEN John J. Pershing.

WWII Memorial

The WWII Memorial is located at the northern border of Greely Field. It honors those members of the Signal Corps who gave their lives during WWII, and was dedicated at the celebration of the 35th Anniversary of Fort Monmouth on October 4, 1952.[31] A large number of donors contributed to the building of the monument, including relatives and friends of the honored dead, and members of Fort Monmouth and other Signal Corps installations.

Howitzers

There are two historic Howitzer guns on the grounds of the U.S. Military Academy Preparatory School (USMAPS; off of Abbey Road).[21]

Kain Memorial

The Kain Memorial is located near Building 1104, behind Soldiers Park. It memorializes Wesley L. Kain, who died in action on December 16, 1944. Originally located in Brooklyn, NY, the deceased's family requested that it be moved here in September 1994.[22]

USMAPS Memorial

A large memorial on the grounds of the U.S. Military Academy Preparatory School (off of Abbey Road) memorializes those cadets lost in times of war with a list of those killed in each war.[29]

Battle of the Bulge Monument

This monument honors the men and women of the United States Armed Forces who participated in the Battle of the Bulge. Dedicated on May 6, 2001, it is located at the intersection of Wilson Avenue and the Avenue of Memories (diagonal to the Bowling Center).[15]

Centennial Time Capsule

The U.S. Army Signal Corps Centennial Time Capsule installation ceremony occurred in front of Russel Hall on September 16, 1960, to commemorate the first centennial of the Corps. The Capsule is to be opened June 21, 2060. It contains items depicting the status of military communications in 1960, as well as historical material showing the origins of the Corps and progress made during its first hundred years.[17]

Mudd Memorial

The Fort Monmouth community dedicated a maple tree and plaque in front of Russel Hall (Garrison Headquarters) on August 31, 2007 in memory of former Garrison Chief of Staff George W. Mudd "for his faithful service to Fort Monmouth- 1975-2003."[23]

Rodman Guns

The two Rodman guns located in Cowan Park are the only two 8" Rodman guns in the U.S. Army Museum System worldwide.[26]

They came to Fort Monmouth from Fort Hancock on nearby Sandy Hook in the spring of 1950, as Fort Hancock closed for the first time.

At Fort Hancock, the guns were located west of and across the road in front of Officers Row Quarters No. 12, the Commanding Officer's Quarters. There, they served as lawn ornaments to mark the Commanding Officer's residence for many years.[27]

Spanish American War Memorial

The granite Spanish American War Memorial is located in Dunwoody Park. The U.S. Veteran Signal Corps Association, Spanish War Division presented it to Fort Monmouth at their 50th Annual Reunion on September 22, 1950. The monument memorializes the officers and men of the Regular and Volunteer Signal Corps, U.S. Army, who established and maintained communications throughout the Spanish-American War, the Philippine Insurrection, and the China Relief Expedition.[28]

Vietnam Memorials

The post features two Vietnam veterans' memorials. The entrance to the nature walk across the street from Building 977 and behind Gosselin Family Housing features a plaque reading, "Dedicated to the men and women who served during the Vietnam War, 1959-1975." The nature walk was dedicated in the Spring of 2001.[30]

The Defense of Freedom Memorial is located in front of Building 1207 along the Avenue of Memories. It lists the names of Soldiers killed in Vietnam, and was dedicated sometime during that War.

Pigeon Memorial

The birdbath Pigeon Memorial was located on the east side of Malterer Ave. near buildings 550 and 551. It stood in what was then a wooded area in commemoration of the winged couriers who "got the message through" during WWI, WWII, and Korea. The Signal Corps Pigeon Breeding and Training Section was located here from 1919-1957 to prepare those birds for war. Fort Monmouth dedicated the memorial on July 14, 1960, as a part of the post's celebration of the U.S. Army Signal Corps Centennial.[24] The Memorial has subsequently been removed.

Holocaust Memorial

The Holocaust Memorial Garden is located to the right of the Main Post chapel when you are facing that building. Jewish War Veterans donated and planted a tree there in 1992. The next year, they added a plaque at the base of the tree.[19]

A sculpture for the Memorial Garden, created by Brian Hanlon and donated by the Jewish Federation of Greater Monmouth County, was unveiled on the day of the post's 2005 Holocaust Remembrance Program.[20]

D-Day Memorial

The D-Day Memorial commemorates the 40th Anniversary of D-Day and is located in front of the Van Deusen Library. This dedication occurred on June 6, 1984.[18]

MAIN POST ENTRANCES AND BUILDINGS*

Scriven Hall (Building 270)

Scriven Hall, completed in 1929, is located on Allen Avenue near Officer Family Housing. Today, it serves as Bachelor Officers' Quarters, but it once housed the Fort Monmouth Officers Club and NCO Open Mess. Per General Orders Number 46, dated June 15, 1950, Scriven Hall memorializes Brigadier General George P. Scriven, Chief Signal Officer from 1913-1917. BG Scriven pioneered the development of the Army Air Service.[48]

National Historic Reports show that Building 270 was built "of concrete block and faced with red brick. It has a hipped roof and hipped dormers, and a one-story porch that spans three-quarters of the facade and is ornamented with wood balustrades."[49]

Blair Hall (Building 259)

Blair Hall, dedicated March 4, 1969, memorializes Fort Monmouth scientist and inventor COL William R. Blair.[45] COL Blair posthumously entered the New Jersey Inventor's Hall of Fame in 2004. He is considered the father of American radar.[46] Scouting groups now use the building as a meeting facility.

Gardner Hall (Building 271)

These Bachelors Officers'/ Visitors' Quarters were constructed in October 1931 and dedicated for BG John Henry Gardner on June 5, 1945. Gardner's Army career included service as Director of the Aircraft Signal Service and as Assistant Chief, Procurement and Distribution Service, Office of the Chief Signal Officer.[50]

National Historic Reports show that, like Building 270, Building 271 was built "of concrete block and faced with red brick. It has a hipped roof and hipped dormers, and a one-story porch that spans three-quarters of the facade and is ornamented with wood balustrades."[51]

Commander's Residence (Building 230)

Building 230 was constructed for the Commanding Officer in 1936 at the western end of Voris Park. National Historic Reports show that this Georgian revival style building "is seven bays wide and has a one-story garage wing. A pedimented gable roof with lunette and a porch with double columns topped by a balustrade mark the entrance. The building also features paired end chimneys and a dentil cornice."[43]

West Gate

The West Gate, or Johnston Gate, is located at the intersection of Routes 35 and 537 in Eatontown. It marks the most utilized entrance to the Main Post. Upon driving through the gate on the Avenue of Memories, buildings 1200- 1214 are to the left. To the right is an information area for visitors where one can find maps, phone numbers, and telephones.

Originally dedicated c. 1961 and re-dedicated on April 18, 1986, the Johnston Gate memorializes COL Gordon Johnston. COL Johnston first enlisted in the Army during the Spanish American War. He joined the Signal Corps in 1903 and received the Congressional Medal of Honor for gallantry in battle.[33]

Officer Housing

Officer housing is located north of Greely Field, "lining Russel and Allen Avenues and surrounding a landscaped open space known as Voris Park. Officer Housing includes single family dwellings constructed for field officers and duplex family dwellings constructed for company officers." National Historic Reports show that single family officer houses (Buildings 215, 216, 221, 224, 229) were designed in the colonial revival style and "feature an ornamental, generally pedimented, entry, a dentil cornice, and gabled dormers. In addition, these units include an attached garage wing. Duplex officer housing units (Buildings 211-214, 218-220, 222-223, 225-228) are symmetrical, eight-bay buildings with paired central main entries, each with a gabled pedimented surround."[xliii]

Abramowitz Hall (Building 114)

Abramowitz Hall, also known as the Athletic Center or Field House, serves as Fort Monmouth's Physical Fitness Center. Dedicated on June 7, 1985, the building honors Lieutenant Colonel Reuben Abramowitz, who served at Fort Monmouth as a radio instructor. It is said that LTC Abramowitz's innovative methods of training radio operators influenced every man trained during and after World War II.[34]

Non-Commissioned Officer Housing

Housing for non-commissioned officers (NCO) is located south of the parade ground (Buildings 233-258). National Historic Reports show that "this residential area comprises 25 duplex units. The standardized Quartermaster design employed for NCO housing is simple and symmetrical. All the structures are two stories and four bays wide. The buildings are brick and have either a gabled or hipped roof; all examples include two end chimneys. Some examples have one-story brick or wood-frame sun porches attached to either ends of the structure. Other buildings include enclosed entry vestibules. Construction of the group was completed in four periods: Building 233 was completed in 1929; Buildings 234-239 were completed in 1931; Buildings 240-246, in 1932; and Buildings 247-258, in 1934."

"Thirty-three detached, one-story garages are located in the main cantonment area. In the officer housing area, the garages are two-car garages constructed of brick with slate hipped roofs. Multiple bay garages were constructed for bachelor officers and non-commissioned officers. In the NCO housing area, the garages are wood-frame buildings with asphalt shingled hipped roofs."[44]

*Current OPSEC conditions prevent some building locations from being identified

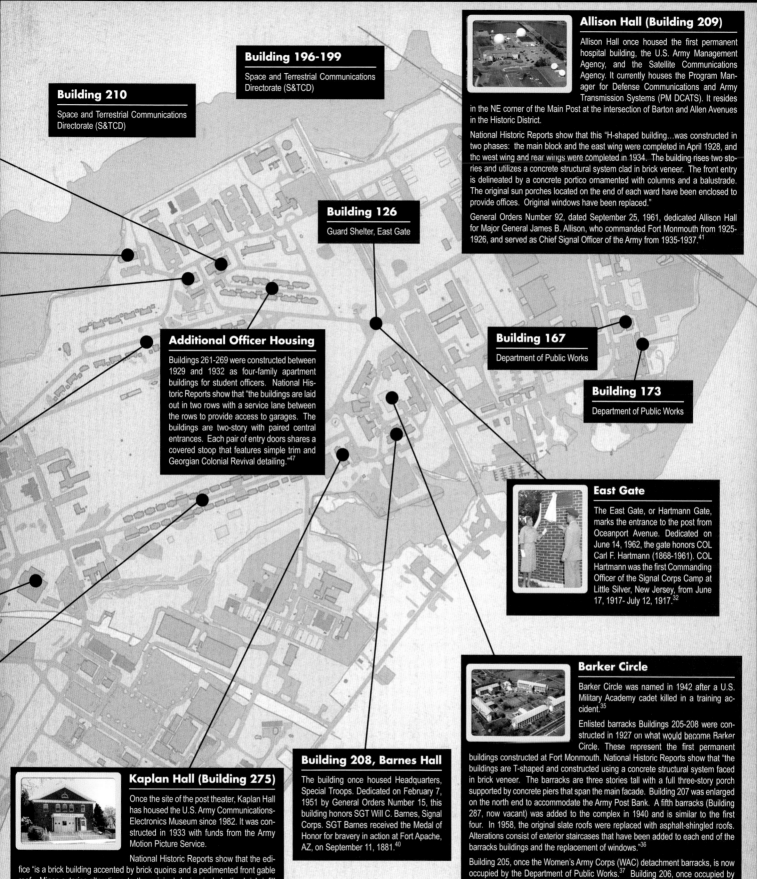

Building 210

Space and Terrestrial Communications Directorate (S&TCD)

Building 196-199

Space and Terrestrial Communications Directorate (S&TCD)

Allison Hall (Building 209)

Allison Hall once housed the first permanent hospital building, the U.S. Army Management Agency, and the Satellite Communications Agency. It currently houses the Program Manager for Defense Communications and Army Transmission Systems (PM DCATS). It resides in the NE corner of the Main Post at the intersection of Barton and Allen Avenues in the Historic District.

National Historic Reports show that this "H-shaped building...was constructed in two phases: the main block and the east wing were completed in April 1928, and the west wing and rear wings were completed in 1934. The building rises two stories and utilizes a concrete structural system clad in brick veneer. The front entry is delineated by a concrete portico ornamented with columns and a balustrade. The original sun porches located on the end of each ward have been enclosed to provide offices. Original windows have been replaced."

General Orders Number 92, dated September 25, 1961, dedicated Allison Hall for Major General James B. Allison, who commanded Fort Monmouth from 1925-1926, and served as Chief Signal Officer of the Army from 1935-1937.[41]

Building 126

Guard Shelter, East Gate

Additional Officer Housing

Buildings 261-269 were constructed between 1929 and 1932 as four-family apartment buildings for student officers. National Historic Reports show that "the buildings are laid out in two rows with a service lane between the rows to provide access to garages. The buildings are two-story with paired central entrances. Each pair of entry doors shares a covered stoop that features simple trim and Georgian Colonial Revival detailing."[47]

Building 167

Department of Public Works

Building 173

Department of Public Works

East Gate

The East Gate, or Hartmann Gate, marks the entrance to the post from Oceanport Avenue. Dedicated on June 14, 1962, the gate honors COL Carl F. Hartmann (1868-1961). COL Hartmann was the first Commanding Officer of the Signal Corps Camp at Little Silver, New Jersey, from June 17, 1917- July 12, 1917.[32]

Barker Circle

Barker Circle was named in 1942 after a U.S. Military Academy cadet killed in a training accident.[35]

Enlisted barracks Buildings 205-208 were constructed in 1927 on what would become Barker Circle. These represent the first permanent buildings constructed at Fort Monmouth. National Historic Reports show that "the buildings are T-shaped and constructed using a concrete structural system faced in brick veneer. The barracks are three stories tall with a full three-story porch supported by concrete piers that span the main facade. Building 207 was enlarged on the north end to accommodate the Army Post Bank. A fifth barracks (Building 287, now vacant) was added to the complex in 1940 and is similar to the first four. In 1958, the original slate roofs were replaced with asphalt-shingled roofs. Alterations consist of exterior staircases that have been added to each end of the barracks buildings and the replacement of windows."[36]

Building 205, once the Women's Army Corps (WAC) detachment barracks, is now occupied by the Department of Public Works.[37] Building 206, once occupied by the Military Police Detachment, currently houses the Directorate for Resource Management.[38] Building 207, once occupied by the 389th Army Band, currently houses U.S. Military Academy (USMA) Cadet Candidate Barracks.[39]

Kaplan Hall (Building 275)

Once the site of the post theater, Kaplan Hall has housed the U.S. Army Communications-Electronics Museum since 1982. It was constructed in 1933 with funds from the Army Motion Picture Service.

National Historic Reports show that the edifice "is a brick building accented by brick quoins and a pedimented front gable roof. Minor exterior alterations to the original design include the brick infill of a front facade window and the removal of the ticket kiosk from the front entrance."

Building 275 was dedicated for MAJ Benjamin Kaplan (1902-1952) on December 21, 1953 by General Orders Number 221. As post engineer, Major Kaplan oversaw construction of what is now known as the Historic District.[52]

A plaque reading "A Memorial to Homing Pigeons In Combat: Courage • Loyalty • Endurance" is affixed to the building.

Building 208, Barnes Hall

The building once housed Headquarters, Special Troops. Dedicated on February 7, 1951 by General Orders Number 15, this building honors SGT Will C. Barnes, Signal Corps. SGT Barnes received the Medal of Honor for bravery in action at Fort Apache, AZ, on September 11, 1881.[40]

MAIN POST BUILDINGS

Squier Hall (Building 283)

Squier Hall also comprises part of the Historic District and was completed in March 1935. Located on Sherrill Drive behind the McAfee Center, the building now houses the Program Executive Office for Enterprise Information Systems (PEO EIS) and the Defense Information Systems Agency (DISA). The building previously housed the headquarters of the Signal Corps Engineering Laboratory (from 1945-1955) and the Academic Building, Officers' Department, the Signal School.[54]

New York architects Rodgers and Poor designed Squier Hall. National Historic Reports show that the L-shaped building "consists of two functional parts: an administration and laboratory section and a rear shop section. The administration section of the building is two stories and has a steel structural frame faced in brick veneer. The front facade is characterized by the entrance portico that is composed of two-story concrete piers faced in brick. The windows are bands of hopper windows. The shop section is located in the rear of the building and is constructed with a steel frame, brick veneer, and industrial sawtooth roof. The laboratory was enlarged in 1947 and, again, in 1958."[55]

Initially known simply as the "Fort Monmouth Signal Laboratory," War Department General Orders Number 49 designated Building 283 as the "Squier Signal Laboratory" on June 28, 1945. Fort Monmouth General Orders Number 79, dated July 19, 1955, later redesignated the building "Squier Hall." Both designations honor Major General George Owen Squier, Chief Signal Officer, 1917-1923. MG Squier is credited with inventing the "wired wireless" basis for the modern carrier system.[58]

Building 288

Building 288, located to the east of Squier Laboratory, lies within the historic district. National Historic Reports show that it does not contribute to the district, however, "due to its later construction date." Fortunately, professionals have concluded that "the building does not constitute a serious intrusion because it relates in function to the Squier Laboratory and is not a strong element in the landscape."[59] Project Manager Signals Warfare currently occupies the building.

Building 295

Logistics and Readiness Center

Building 293

LRC Power Sources Branch Battery Test Facility

Smarr Hall (Building 291)

Smarr Hall currently houses the Readiness Directorate of the Logistics and Readiness Center (LRC). Dedicated on December 15, 1972, the building memorializes COL Albert W. Smarr, a Signal School instructor who died in a helicopter crash near Danang, Vietnam, in 1972.[60] Fort Monmouth re dedicated the building on June 19, 1998, at which point COL Smarr's family received the POW Medal, posthumously.[61]

Building 292

902nd Military Intelligence Group; 956th Transportation Company; Storage

Building 296

Project Manager Signals Warfare

Building 552

Community Center (the original recreation hall)

Building 555

Readiness Directorate, LRC (once a post theater)

Van Deusen Library (Building 502)

The Van Deusen Library commemorates the 37-year Army career of MG George Lane Van Deusen. MG Van Deusen graduated from the U.S. Military Academy in 1909 as president of his class. He served in various capacities at Fort Monmouth, his last being that of Commanding General of the Eastern Signal Corps Training Center (1942-1945). The building was dedicated on June 21, 1977, the 117th anniversary of the United States Signal Corps.[63]

Building 566

Recruitment Command

Building 563

PM Combat ID

Armstrong Hall (Building 551)

Armstrong Hall, once home to the Signal Corps museum, now serves as the Education Center of Fort Monmouth.[66] The Deputy Chief of Staff for Personnel (DCSPER), Human Resources, Army Continuing Education, and the Learning Resource Center reside there. Per General Orders Number 51, dated May 31, 1955, the building memorializes Major Edwin Howard Armstrong, a scientist and Soldier of the U.S. Army Signal Corps. Much of his work concentrated on the development of FM radio technology.[67] Major Armstrong, who held 42 patents, allowed the free use of them by the Army during World War II.

Bijur Hall (Building 361)

Bijur Hall functions as senior enlisted quarters. It was built and dedicated to CPT Arthur H. Bijur (1919-1945) per General Orders Number 19, dated July 12, 1965. CPT Bijur died in action on January 14, 1945 on Luzon Island, Philippine Islands, after he climbed from a foxhole to warn his men of impending enemy fire. Assigned to the 43rd Signal Company at the time of his death, he had previously trained at Fort Monmouth. Bijur Hall is located inside the East Gate on the north side of Allen Avenue near the Officer Family Housing.[63]

Black Hall (Building 360)

These distinguished Visitors Quarters were completed in 1956 and dedicated by General Order 45 on June 28, 1957. They memorialize BG Garland C. Black, who landed at Normandy as General Bradley's 12th Army Group Signal Officer and received numerous commendations. The building has also served as Bachelors Quarters and a guesthouse. It is located inside the East Gate on the south side of Allen Avenue near the Officer Family Housing.[lxii]

Building 363-365

Guest housing

Building 359

PM DCATS

Building 500

Chapel (Malterer Avenue), built in 1962

Building 501

Administrative space, formerly the site of Social Work and Mental Health Services (moved to Building 1075)

Building 550

Project Manager Distributed Common Ground System, Army (PM DCGS; once the post library)

Quartermaster Support Area

National Historic Reports show that "in the original design of the permanent facility, the quartermaster support area was separated from the main cantonment area... The buildings encompassed in this area were constructed to house the provisioning and maintenance functions for the installation."

The area included a quartermaster warehouse/commissary (Building 277, today the Supply Services Division of the Directorate of Logistics), a bakery (Building 276, also currently housing the Supply Services Division of the Directorate of Logistics), a quartermaster garage (Building 279, today, a repair shop), a utility shop [Building 280, now Tecom-Vinnell Services (TVS)], and a blacksmith shop (Building 281, now TVS administrative space). "The buildings were utilitarian in design, and reflected simple versions of the Georgian Colonial Revival Style. With the exception of the quartermaster warehouse and commissary, the buildings are one-story in scale. The quartermaster warehouse/commissary represents a standard warehouse design that can be traced directly to the late nineteenth century. In this plan, the first floor was devoted to the post commissary and the second floor was devoted to clothing stores. The quartermaster garage and utility shop are very similar in appearance. Both buildings have steel structural systems that support single-span steel roof trusses. Both have industrial steel sash windows. These buildings are in good condition, and only minor modifications have been undertaken to adapt these buildings to their current functions."

"When the main cantonment of Fort Monmouth was completed during the 1930s, the quartermaster support area was distinctly separate from the main cantonment. The land between the two areas remained an open space. During World War II, the area between the main cantonment and the quartermaster support area became the location for World War II mobilization buildings."

Fire Station (Building 282)

National Historic Reports show that the fire station was constructed in 1935 "according to a Quartermaster- standardized plan using Georgian Colonial Revival motifs. It is a two-story, T-shaped building with flanking one-story wings. The building facade features a pedimented gable roof with a lunette window. Bays for four fire trucks and an office are located on the first floor; a firemen's dormitory is housed on the second floor; and a guardhouse occupies the rear wing. Four large arched openings in the main facade allow egress for firefighting vehicles."[53]

Russel Hall (Building 286)

Russel Hall was built in the 1930s as a Headquarters building for the post.

Philadelphia architect Harry Sternfield designed the building, which was completed in 1936 in collaboration with the Office of the Constructing Quartermaster. National Historic Reports show that "It is an Art Deco building composed of a four-story central pavilion and three-and-a-half story flanking wings. The central portion is faced with Indiana limestone. Sculptured reliefs that depict the Signal Corps in the Civil War and World War I are located on either side of the central entrance doors. Above the entrance is a limestone relief of the Seal of the United States. The flanking wings, like the central portion of the building, are constructed with a concrete frame, but are faced with brick and have a limestone veneer on the first half story. The wings feature recessed, vertically articulated windows separated by decorative brick spandrels. Decorative bands of brick define the cornice. Alterations consist of exterior staircases that have been recently added to each end of the building."[lvii]

The building originally provided space for the post library, chaplain's office, telephone switchboard rooms, a court martial room, classrooms, and a large map and war game room. Today, the building acts as the Garrison Headquarters. It is located at the east end of Greely Field in the Historic District. Russel Hall memorializes MG Edgar Russel, Chief Signal Officer, American Expeditionary Force (AEF), World War I, May 1917- July 1919.[lviii]

Moorman Hall (Building 362)

Building 362 serves as Bachelor Officers' Quarters. It was built in 1965 and named for WWI era Signal Corps officer Frank Moorman in 1967. In addition to his distinguished military career, Frank Moorman fathered MG Frank W. Moorman, who later commanded Fort Monmouth from 1963- 1965.[64] Moorman Hall is located inside the East Gate on the north side of Allen Avenue near the Officer Family Housing tennis courts.

MAIN POST BUILDINGS (Continued)

Vail Hall (1150)

Vail Hall houses the Directorate of Information Management (DOIM) and the Information Technology Services Directorate (ITSD). The building was constructed in 1952 and per General Orders Number 19, dated June 21, 1956, memorializes Alfred Vail (1807-1859), the distinguished New Jersey inventor whose great mechanical and financial contributions to wire communications substantially accelerated the first experiments in telegraphy. Vail was a close associate of Samuel Morse.[76]

Building 1209

Logistics and Engineering Operations, LRC; Software Engineering Center; Directorate for Corporate Information

1200 Area

The 1200 Area features the barracks and classroom buildings originally constructed for the Signal School in 1953. "Backfill" following the departure of the Signal School included the Military Academy Preparatory School (1975), the Chaplain Center and School (1979-1996), the 513th Military Intelligence Brigade, (1982-1993), and the Federal Bureau of Investigation's regional computer support center. The location was rehabilitated in 1996 to house elements of the U.S. Army Communications Electronics Command (CECOM).

Building 686

Thrift Shop

Building 601-4

I2WD Laboratory, Warehouse, and Storage

Building 1208 (Allensworth Hall)

Building 1208 currently houses the Acquisition Center and Logistics and Readiness Center HQ. Fort Monmouth dedicated the building to LTC Allen Allensworth in 1989. LTC Allensworth was a former slave who went on to reach the highest military rank bestowed upon an African American to that date.

Building 675-8

I2WD Administrative Space

Building 1212

United States Military Academy Preparatory School

Building 1210

Software Engineering Center and Kathy's Catering

Building 682

U.S. Army Military Affiliate Radio System Station

Building 1203

Federal Bureau of Investigation (FBI)

Building 1213-14

Software Engineering Center (SEC)

Building 671

Fort Dix Criminal Investigation Division

Building 1204-05

United States Military Academy Preparatory School (USMAPS)Space

Building 689

Bowling Center

Building 1202

Communications Directorate, LRC

Building 1200

Command and Control Systems- Avionics, LRC

Building 1215

Expo theater

Building 750

Transportation Motor Pool

Building 1201

Intelligence, Electronic Warfare and Sensors, LRC

Building 761

Equal Employment Opportunity (EEO) Administrative Space

Pruden Auditorium (Building 1206) and Mallette Hall (Building 1207)

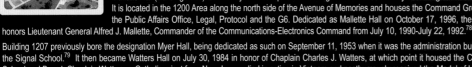

Pruden Auditorium was dedicated in 1987 to LTC Aldred A. Pruden, an Army Chaplain who fought for the right to wear his rank insignia (which was against Army policy). Pruden also planned and organized the Army's first training school for chaplains.[77]

Mallette Hall (Building 1207) is currently the Headquarters building of the CECOM Life Cycle Management Command. It is located in the 1200 Area along the north side of the Avenue of Memories and houses the Command Group, G3, the Public Affairs Office, Legal, Protocol and the G6. Dedicated as Mallette Hall on October 17, 1996, the building honors Lieutenant General Alfred J. Mallette, Commander of the Communications-Electronics Command from July 10, 1990-July 22, 1992.[78]

Building 1207 previously bore the designation Myer Hall, being dedicated as such on September 11, 1953 when it was the administration building for the Signal School.[79] It then became Watters Hall on July 30, 1984 in honor of Chaplain Charles J. Watters, at which point it housed the Chaplains School and Board. Chaplain Watters, a Catholic priest from New Jersey, died in action in Vietnam and posthumously received the Medal of Honor.[80]

Building 788

DCSPER Classroom

Building 789

Garrison Base Realignment and Closure (BRAC) Office

Building 1107
Union Meeting Space

Building 1108
Fort Dix Criminal Investigation Division

Building 1123
Military Pay and Identification

Building 1124
Car Wash

Building 1102
Intelligence, Electronic Warfare and Sensors Laboratory/Storage

Building 1103
Union Hall/ Administrative Space

Building 1104
Project Manager Signals Warfare Administrative Space

Building 1105
Legal Services/Inspector General Administrative Space

McAfee Center (Building 600)

The McAfee Center, Building 600, houses the Intelligence and Information Warfare Directorate (I2WD) of the Communications-Electronics Research, Development, and Engineering Center (CERDEC). It is located on Sherrill Avenue. Dedicated on July 28, 1997, the Center honors renowned physicist Dr. Walter McAfee. Dr. McAfee held numerous supervisory positions during his 42 years at Fort Monmouth, and participated in the now-famous Diana Project that resulted in man's first contact with the moon.[68]

Lane Hall (Building 702)

Lane Hall functions as the Community Activity Center. Fort Monmouth dedicated the building in 1983, after the old Enlisted Mess Hall that had originally memorialized Lane (dedicated in 1950) was razed. Lane Hall memorializes Private Second Class Morgan D. Lane, the first member of the Signal Corps to receive the Medal of Honor. PVT Lane fought with the Union Army during the Civil War and was rewarded for capturing the Confederate Flag from the Gunboat Nansemond near Jetersville, Va.[69]

Building 1000
Post Exchange

Building 1001
Shopping Center, Four Seasons

Building 1002
Postman Plus/Specialty Shop/Concession Shop

Building 1003
Optical Shop/Flower Shop

Building 1006
First Atlantic Federal Credit Union

Building 1007
Commissary

Building 1010
Morale, Welfare, and Recreation Tickets and Tours

Building 800
PM Aerial Common Sensor (ACS) Administrative Space

Building 909-18
PM WIN-T

Building 975
Warehouse; once occupied by the Quartermaster's Warehouse[71]

Building 906
S&TCD, PM Warfighter Information Network Tactical (WIN-T)

Building 977
Department of Public Safety; once the Guard House[73]

Building 616
Garrison Administrative Space

Building 901
Deputy Chief of Staff for Personnel

Building 699
Auto Repairs, gas station

Building 976
Storage; once occupied by the Cold Storage Plant[72]

Building 1005
Post Office

Building 801
Gear to Go (outdoor recreation center)/PM ACS Administrative Space

Building 822
Burger King

Building 826
Internal Review Office Administrative Space

Building 814-21
USMAPS Athletic Facilities

Building 1077-8
Housing

Building 812
Army Community Services, dedicated November 18, 1986[70]

Building 810
Veterinarian Treatment Facility

Patterson Army Health Clinic (Building 1075)

Medical care at Patterson Army Health Clinic (PAHC) has been available in one form or another since 1958, and continues today as an integral part of the operation of Fort Monmouth. PAHC provides ambulatory and preventive health care services to approximately 10,000 eligible beneficiaries. Located in Building 1075, PAHC sees approximately 120 patients per working day. The primary focus of PAHC is to increase the health and wellness of the population through preventive health services. PAHC also supports two outlying health clinics: Ainsworth U.S. Army Health Clinic, Fort Hamilton, NY and the Mills Troop Medical Clinic, Fort Dix, NJ.

The present facility was opened and dedicated in 1958. General Order 40, dated April 24, 1958, named it in honor of Major General Robert Urie Patterson, United States Army Medical Corps (1877-1950). After completing medical studies at McGill University, MG Patterson graduated with honors from the Army Medical School in 1902. He received two Silver Stars for conspicuous gallantry in action in the Philippines. During World War I, the British, Italian, Czechoslovakian and Serbian governments all decorated MG Patterson. He also received the Distinguished Service Medal. During the postwar years, his assignments included Instructor, U.S. Army War College; General Staff, War Department; Medical Director, U.S. Veteran's Bureau; Executive Officer, Office of the Surgeon General and Commander of the Army and Navy Hospital, Hot Springs, Arkansas. He served a four-year term as Surgeon General of the Army and retired in 1935. Major General Patterson died on December 6, 1950.[74]

The new Fort Monmouth Community-Based Outpatient Clinic (CBOC) opened on July 21, 2004. The CBOC encompasses 6,200 square feet on the third-floor wing of the Fort Monmouth Patterson Army Health Clinic. It is expected to accommodate about 10,000 visits per year.[75]

A plaque next to the Primary Care Clinic entrance to the building is dedicated to Medical Department Activity Soldiers who served in Desert Storm.

MAIN POST STREETS

Saltzman Ave

MG Charles M. Saltzman, Chief Signal Officer from 1924-1928, served under President Hoover as chair of the Federal Radio Commission and was instrumental in laying the groundwork for the Federal Communications Commission. General Order 8, March 22, 1943, designated Saltzman Avenue.[99]

Telegraph Ave

Telegraph Avenue commemorates the first field telegraph wagon given to front line troops of the Union Army Signal Department under BG Myer. From that date through World War I, the telegraph served the Army as its primary means of long-distance communication.[101]

Crystal Ave

Crystal Avenue commemorates the 1948 production, here at Fort Monmouth, of the first synthetically produced large quartz crystals. The crystals were used in the manufacturing of electronic components, and made the U.S. largely independent of foreign imports for this critical mineral.

Nealis Ave

SGT John J. Nealis' lengthy and distinguished military career included service in Europe in the 102nd Field Signal Battalion from May 1918 to March 1919. He received the British Military Medal, the Distinguished Service Cross, and the Purple Heart and Oak-leaf Cluster. General Orders Number 6, dated May 7, 1956, designated Nealis Avenue.[96]

North Dr

SGT Ludlow F. North died in the Philippines in 1900, ambushed while repairing telegraph lines cut by insurgent forces.

Echo Ave

Echo Avenue is located between Tindall and Cockayne Avenues and runs parallel to Todd Avenue.

Vanguard Road

In 1958, solar cells developed by scientists at Fort Monmouth powered the Vanguard I Satellite for more than five years.

Whitesell Ave

On February 25, 1944, Tech 4 Joseph L. Whitesell received the Asiatic-Pacific Medal and the Bronze Service Star for service in the Central Pacific Area. As a member of the 295th Joint Assault Signal Company, he died in Action on Saipan, July 10, 1944, and posthumously received the Silver Star and the Purple Heart. General Orders Number 9, dated May 7, 1956 dedicated the Avenue.[103]

Avenue of Memories

General Order 22, April 9, 1949, and General Order 31, December 7, 1949, designated the Avenue of Memories in honor of the officers and men of the Signal Corps who gave their lives during World War II in the service of their country. The Avenue of Memories turns into Tinton Avenue immediately outside the main gate.[84]

Nicodemus Ave

LTC William J. L. Nicodemus helped BG Myer inaugurate the "wig-wag" method of visual signaling in the Army during the 1860 expedition against the Navaho nation. He helped Myer organize the Signal Corps, served as Commandant of the Signal School in Georgetown, DC, and was Acting Chief Signal Officer from 1863-1864.

Irwin Ave

COL Jack Irwin enlisted in the Ohio National Guard on June 19, 1916. Having served in the Signal Corps from July 1931 to April 1936, Irwin died in action during World War II. General Order 42, November 30, 1945, designated Irwin Avenue.

Malterer Ave

CPT John A. Malterer was noted for his contributions to the development of Signal Corps radio procedure and practice. He was serving as chief of the Signal School's Radio Division when he died in 1927.[94] General Orders Number 11, dated April 13, 1943, designated Malterer Avenue.

Carty Ave

BG John J. Carty, Signal Reserve, served during World War I as Director of Wire Communications for the Allied Expeditionary Forces. He received the Distinguished Service Medal.[87]

Allen Ave

As Chief Signal Officer, 1906-1913, BG James Allen established the Army Air Service. He purchased the first military plane, built by the Wright Brothers.[83] During the Spanish-American War, Allen established the first telegraph link between the U.S. and its forces in Cuba.

Sanger Ave

COL Donald B. Sanger (1889-1947) became Chaplain of the Signal Corps on November 5, 1917.

Wallington Ave

COL Merton G. Wallington acted as Assistant Commandant of the Enlisted Men's Department, Eastern Signal Corps School, during World War II. General Orders Number 28, dated July 3, 1942 designated Wallington Avenue.[102]

Barker Circle

Cadet Ernest S. Barker, United States Military Academy, died in 1942 during a training flight accident at West Point. General Order 56, December 2, 1942, designated Barker Circle.

Gosselin Ave

First Sergeant Alexander Gosselin, 2nd Field Signal Battalion, received the Distinguished Service Cross[90] on December 4, 1918 for heroism during the Meuse-Argonne offensive.

Murphy Dr

LTC William H. Murphy, a pioneer in aircraft radio, died in action during World War II in the Western Pacific area. Camp Murphy, Florida, site of the Signal Corps Aircraft Warning School, also memorialized him.[95] General Order 28, July 3, 1942, designated Murphy Drive.

Moonshot Dr

Moonshot Drive runs from Malterer Avenue to Murphy Drive.

Lane Ave

PVT Morgan D. Lane was the Signal Corps' first Medal of Honor winner, having merited the award for capturing a Confederate Flag from the Gunboat Nansemond, near Jetersville, Virginia.

Rasor Ave

COL Winchell I. Rasor commanded the 51st Telegraph Battalion in World War I. He died at MacDill Field, Florida, in 1942. General Order 56, December 2, 1942, designated Rasor Avenue.

Heliograph St

Heliograph Street is named for the heliograph, an apparatus for telegraphing by means of the sun's rays flashed from a mirror.

Stephenson Ave

In World War I, SFC Claud Stephenson of the 2nd Field Signal Battalion, "with unusual coolness and bravery, went forward with the first wave, constructing and maintaining his lines of communications under heavy ... fire and constantly encouraging his men until he was killed."[100]

Alexander Ave

In 1859, LT E. P. Alexander helped General Myer demonstrate his system of visual signaling between Fort Hamilton, Sandy Hook, and Twin Lights. At the outbreak of war, Alexander joined the Confederate Army and became its first Chief Signal Officer (as a Brigadier General).[82]

Cockayne Ave

SGT Albert H. Cockayne died in 1900 during the Philippine Insurrection when he and an assistant attempted, without escort, to repair a telegraph line cut by insurgents.[88]

MAIN POST STREETS (Continued)

Sherrill Ave

BG Stephen H. Sherrill served as Commandant of the Eastern Signal School and Commanding General of the Eastern Signal Corps Training Center and Fort Monmouth, October 1944 through December 1945.

Battery Ave

This avenue honors Fort Monmouth's long standing mission to support military batteries.

Messenger Ave

Messenger Avenue was dedicated to all the messengers of the Signal Corps who sacrificed life and limb to "get the message through."

Rittko Ave

PFC Theodore E. Rittko entered the service in April 1952. He distinguished himself in action against an armed enemy on June 29, 1953, near Kumhwa, North Korea, and posthumously received the Silver Star, the Purple Heart, and the Korean Service Medal with two Bronze Service Stars. General Order Number 8, dated May 7, 1956, designated Rittko Avenue.[98]

Semaphore Ave

Semaphore Street memorializes the semaphore, an apparatus for visual signaling or a system of visual signaling by two flags held, one in each hand.

De Rum Ave

CPL Howard P. De Rum, 102nd Field Signal Battalion, died in action near Ronssoy, France on September 29, 1918. He posthumously received the Distinguished Service Cross for extraordinary heroism in action. General Orders Number 7, dated May 7, 1956, designated De Rum Avenue.[89]

Wilson Ave

Native Bolomen attacked and killed CPL Wilson in 1900, while he was repairing cut lines on Bohol Island, Philippines.

Diana Lane

"Diana" was the name of the 1946 project that resulted in the first radio bounce off the moon.

TIROS Ave

TIROS sent first televised weather photographs of the earth's cloud cover and weather patterns to the giant sixty foot "Space Sentry" antenna at Camp Evans.

Abbey Ave

Technician Fifth Grade Claude W. **Abbey**, 90th Signal Company, died in France on November 10, 1944 from wounds received the previous day in combat. He posthumously received the Silver Star for gallantry in action. General Order 10, dated May 7, 1956, designated the Avenue.[81]

Radio Ave

Radio Avenue commemorates the Signal Corps' adoption of the radio as a means of communication soon after Marconi demonstrated its practicability. The Signal Corps installed the first fixed-site radio station in America in 1899 and established its Radio Laboratory at Fort Monmouth (Camp Vail) in 1918.

Lockwood Ave

LT James B. Lockwood, until his death, was second in command of Greely's 1881-1884 expedition to the Arctic. Lockwood Island, which he discovered in 1882, was then the Nation's northernmost outpost.[93]

Signal Ave

The avenue honors Fort Monmouth's roots as a Signal Corps training camp.

Housing Ave

Fort Monmouth's housing office and guest housing accommodation is located here.

Barton Ave

LTC David B. Barton (1901-1944) was the Assistant Director of Training Literature at Fort Monmouth in 1943. He died in action in Italy in early 1944.[85]

Brewer Ave

COL John H. Brewer, a 1924 graduate of the United States Military Academy and a 1933 graduate of the Signal School Officers Course, died on or about May 12, 1943 during an airplane flight over New Guinea. General Orders Number 24, dated June 21, 1943, designated Brewer Avenue.[86]

Russel Ave

MG Edgar Russel served as the Chief Signal Officer, American Expeditionary Force (AEF), World War I, May 1917- July 1919.

Harmon Ave

Air Corps General Harmon died in combat in World War II.

Hildreth Ave

CPL Hildreth, 4th Field Signal Battalion, was wounded during World War I in the Battle of the Marne. Having received first aid, he returned to the field only to die in hand-to-hand combat.[92]

First Ave

This avenue is located in the Barker Circle Barracks area.

Communication Ave

This avenue recognizes Fort Monmouth's historic association with communications technology.

Courier Ave

The Courier Satellite proved high-volume communications, up to 100,000 words per minute, could be relayed through space.

Army Lane

This lane intersects with Cockayne and Alexander Avenues.

Tindall Ave

MAJ Richard G. Tindall, Jr., Signal Officer of the 92nd Infantry Division during World War II, died in action during the Italian Campaign. General Order 33, September 6, 1945, designated Tindall Avenue.

Guardrail Ave

The Guardrail is a signals intelligence location/collection system.

Todd Ave

Insurgents killed SFC Robert J. Todd during the Philippine Insurrection while he was in charge of a construction crew repairing lines at Amuling, Luzon.

400 AREA BUILDINGS AND STREETS

Construction of four hangars and two airfields began in the 400 Area in December 1917. Squadrons of the United States Army Air Service arrived here in 1918. Experiments with aircraft radios, radio direction finding, and aerial photography required ninety to ninety-five flights a week. Flying activities transferred to Hazelhurst Field, Long Island, by the end of the year, but enormous headway had been made at Camp Vail in adapting radio to aircraft.

Following the departure of the flying activity in 1918 and until the completion of the 1200 Area in 1953, the hangars served as classroom buildings for the Signal School.[104] Before and after its service as an airfield, the site functioned as a polo field.[105]

The last of the hangars was razed in 1983. A Hangar One Memorial sign along Oceanport Avenue marks the spot to this day.

The 400 Area also featured the U.S. Army Signal Corps Pigeon Breeding and Training Section lofts. The lofts were located across Riverside Ave. from the marina, where there is now a playground. Homing pigeons served the Army as couriers in WWI, WWII, and Korea.

400 Area with Alfred Vail Era Planes

Pigeon Breeding and Training Center

Hazen Ave

MG William B. Hazen served as Chief Signal Officer from 1880 until his death on January 16, 1887.[109]

Burns Ave

In World War I, SGT Kenneth K. Burns of the 2nd Field Signal Battalion, "with unusual coolness and bravery, went forward with the first wave, constructing and maintaining his lines of communications under heavy ... fire and constantly encouraging his men until he was killed."[107]

Evans Ave

LTC Paul W. Evans commanded the 101st Field Signal Battalion in World War I during the Champagne-Marne, Aisne-Marne, and St. Mihiel operations. He died in the Canal Zone on April 10, 1936 while on duty as Department Signal Officer.

Fisher Ave

COL Benjamin F. Fisher helped BG Albert J. Myer organize the Signal Corps during the Civil War and served as Chief Signal Officer, 1864-1866, during the final months of war and the demobilization. General Order 19, June 5, 1945, designated Fisher Avenue.[108]

Fraser Ave

2LT N. W. Fraser died in action during World War II. General Order 28, July 3, 1942, designated Fraser Avenue.

Selfridge Ave

LT Thomas E. Selfridge became the first Army officer to die in an air accident when, in 1908, he crashed while flying with Orville Wright.[109] This street was, appropriately, the site of the airfield built in 1918 for the Radio Lab's 122nd Aero Squadron.

Wade Ave

1LT LaVerne L. Wade died in action while fighting the Japanese in the Philippines on December 30, 1941.[111] General Order Number 28, dated July 3, 1942 designated Wade Avenue.

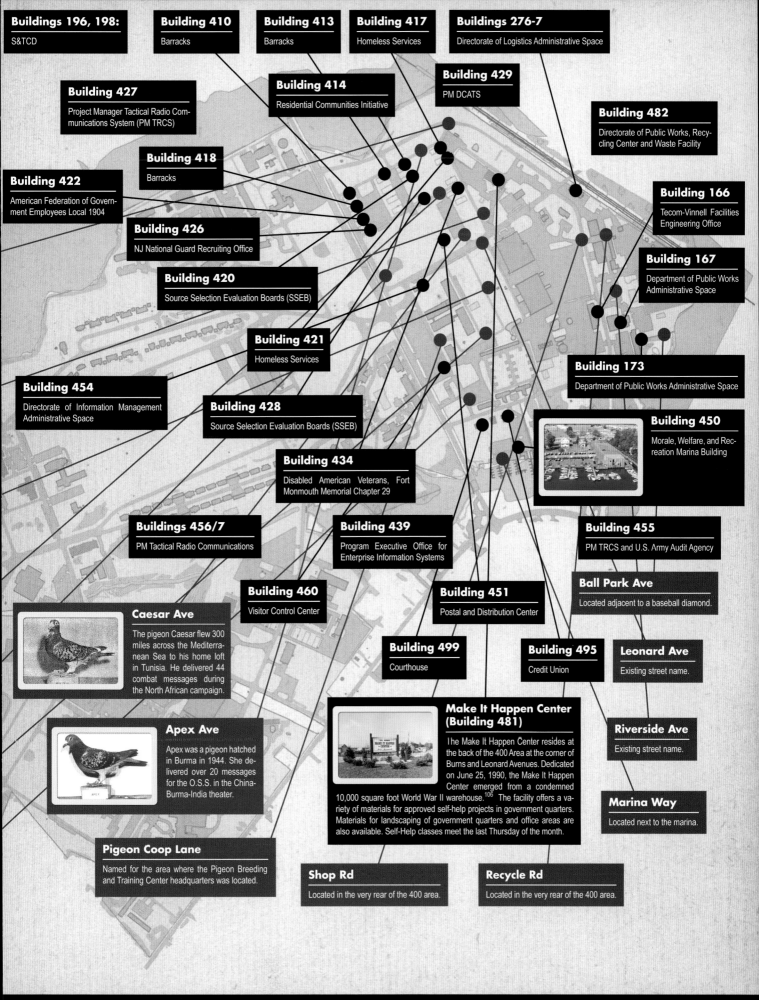

Buildings 196, 198:
S&TCD

Building 410
Barracks

Building 413
Barracks

Building 417
Homeless Services

Buildings 276-7
Directorate of Logistics Administrative Space

Building 427
Project Manager Tactical Radio Communications System (PM TRCS)

Building 414
Residential Communities Initiative

Building 429
PM DCATS

Building 482
Directorate of Public Works, Recycling Center and Waste Facility

Building 418
Barracks

Building 422
American Federation of Government Employees Local 1904

Building 166
Tecom-Vinnell Facilities Engineering Office

Building 426
NJ National Guard Recruiting Office

Building 167
Department of Public Works Administrative Space

Building 420
Source Selection Evaluation Boards (SSEB)

Building 421
Homeless Services

Building 173
Department of Public Works Administrative Space

Building 454
Directorate of Information Management Administrative Space

Building 428
Source Selection Evaluation Boards (SSEB)

Building 450
Morale, Welfare, and Recreation Marina Building

Building 434
Disabled American Veterans, Fort Monmouth Memorial Chapter 29

Buildings 456/7
PM Tactical Radio Communications

Building 439
Program Executive Office for Enterprise Information Systems

Building 455
PM TRCS and U.S. Army Audit Agency

Ball Park Ave
Located adjacent to a baseball diamond.

Building 460
Visitor Control Center

Building 451
Postal and Distribution Center

Caesar Ave
The pigeon Caesar flew 300 miles across the Mediterranean Sea to his home loft in Tunisia. He delivered 44 combat messages during the North African campaign.

Building 499
Courthouse

Building 495
Credit Union

Leonard Ave
Existing street name.

Apex Ave
Apex was a pigeon hatched in Burma in 1944. She delivered over 20 messages for the O.S.S. in the China-Burma-India theater.

Make It Happen Center (Building 481)
The Make It Happen Center resides at the back of the 400 Area at the corner of Burns and Leonard Avenues. Dedicated on June 25, 1990, the Make It Happen Center emerged from a condemned 10,000 square foot World War II warehouse.[106] The facility offers a variety of materials for approved self-help projects in government quarters. Materials for landscaping of government quarters and office areas are also available. Self-Help classes meet the last Thursday of the month.

Riverside Ave
Existing street name.

Marina Way
Located next to the marina.

Pigeon Coop Lane
Named for the area where the Pigeon Breeding and Training Center headquarters was located.

Shop Rd
Located in the very rear of the 400 area.

Recycle Rd
Located in the very rear of the 400 area.

THE CHARLES WOOD AREA BUILDINGS AND STREETS

The Charles Wood area consists of 489.34 acres acquired by the Army in 1941 to accommodate the wartime expansion of the Signal Corps Replacement Training Center. Personnel constructed a cantonment area for 7,000 troops within ninety days of the purchase. That cantonment area included sixty barracks, eight mess halls, nineteen school buildings, ten office buildings, a recreation hall, a Post Exchange, an infirmary, and a chapel. The Army dedicated the area to the memory of LTC Charles W. Wood, Assistant Executive Officer at Fort Monmouth. LTC Wood died suddenly on June 1, 1942 while on temporary duty in Washington. The property, portions of which formerly comprised the Monmouth County Country Club (originally, Suneagles Country Club), included, in addition to the Olmstead Gardens area, the areas now occupied by the Myer Center and the golf course.[112]

Colin Kelly Field

CPT Colin P. Kelly, Jr., won the first Distinguished Service Cross of World War II on December 9, 1941 when, on his return from a bombing mission, two enemy fighters jumped his plane and shot it down. All the crew, except Kelly, bailed out. General Order 33, July 15, 1942, designated Kelly Field. Joe DiMaggio made a guest appearance during the re-dedication in 1959 and demonstrated his famed swing for an enthusiastic crowd of Junior League ball players.

Myer Center (Building 2700)

The Myer Center, also known as "the Hexagon," was constructed in 1954 to house the Signal Corps Labs. The building lacks two of its six sides, purportedly because some of the funds were misappropriated to construct facilities for the Atmospheric Sciences Laboratory at Fort Huachuca, Arizona.

The building has recently undergone extensive renovation.[115] It houses the HQ of the Communications-Electronics Research, Development, and Engineering Center (CERDEC), the Program Executive Office for Command, Control and Communications – Tactical (PEO C3T), and the Program Executive Office for Intelligence, Electronic Warfare and Sensors (PEO IEWS); Chenega Technology Services Corporation (audio visual services); Department of Public Works, S&TCD, Command and Control Directorate (C2D), and PM Joint Computer-Aided Acquisition & Logistics Support (JCALS) administrative space; a Federal Emergency Management Agency Office; Directorate for Corporate Information personnel; and a cafeteria. Dedicated in the late 1980s, it memorializes the founder of the U.S. Army Signal Corps.

While assigned as an assistant surgeon in the Regular Army of the United States, Dr. Albert J. Myer devised a military visual signaling system that the Army adopted in 1860. Dr. Myer became Signal Officer of the subsequently created Signal Division, and received the rank of Major. On March 3, 1863, President Abraham Lincoln signed legislation making the Signal Corps a separate military branch. Major Myer became first Chief Signal Officer with the rank of Colonel. He would continue to lead the Signal Corps for two and a half decades.[116]

The Myer Center roof features a Dymaxion Deployment Unit (DDU). Richard Buckminster Fuller received a patent for these corrugated metal structures in 1944. The Signal Corps bought 200 of them, and "seem to have been Fuller's only significant customer." Dozens of the units were used at Fort Monmouth and nearby Camp Evans for radar experimentation and storage. The DDU at the Myer Center is the only "igloo" remaining on Fort Monmouth property, and is by all accounts extremely rare.

Olmstead Gardens

The Olmstead Gardens were constructed in the late 1950s/early 1960s to accommodate enlisted Soldiers and their families. They honor the memory of Major General Dawson Olmstead, who commanded the Signal Center and Fort Monmouth from 1938 to July 1941. This area was located on the south side of Tinton Ave and has been demolished.[118]

Building 2241

Family Child Care

Building 2900

Visitor Center

Chapel (Building 2275)

The Chapel was built in 1942. It is of the standard 700 series Quartermaster Corps design. The chapel was decommissioned in 2008.[114]

Building 2560

Fire Station

Heliport Drive

Located off of Corregidor Rd.

Building 2290

Child Development Center (CDC)

Building 2525

Project Manager Net Ops; Project Manager Force XXI Battle Command Battalion/Brigade and Below (FBCB2)

Building 2539-40

Safety Administrative Space

Building 2704

Command and Control Directorate (C2D), CERDEC

Building 2566

Youth and School Age Services

Watson Road

Colonel Paul Edwin Watson, Signal Corps died at Fort Monmouth on 18 September 1943 while on active duty. A laboratory in Eatontown previously honored him.

Building 2535

Test Facility

Quartz Ave

In 1948 the first synthetic large quartz crystals were produced at Fort Monmouth. The crystals were used in the manufacturing of electronic components, and made the U.S. largely independent of foreign imports for this critical mineral.

Building 2705

PM Future Combat Systems (PM FCS)

Building 2719-21

754th Explosive Ordnance Detachment

Building 2507

Motor Pool

Building 2628-9

Fire Training Center

Building 2627

Pistol Range

Building 2506

Paint Shop

Building 2707-8

PEO C3T Administrative/Storage Space

Building 2504

Command and Control Fabrication and Integration Facilities

Laboratory Road

Laboratory Road reflects the research and development work done in laboratories at Fort Monmouth.

Building 2503

Machine Shop

Pearl Harbor Ave

Japanese zeros attacked the naval base at Pearl Harbor on the "day of infamy," December 7, 1941.

Building 2501

Fire Academy

Building 2502

Sheet Metal Shop

Building 2508

Department of Public Works

Wake Road

The U.S. Marine detachment on Wake Island mounted a heroic resistance to the Japanese during the attack of December 8, 1941, but surrendered on December 23, when the Japanese landed on the island in force.

Hope Road

Hope Road runs North and South and essentially divides the Suneagles Golf Course from the bulk of the Charles Wood Area.

Mariveles Road

Mariveles, on the southern tip of Bataan, was site of a U.S. Navy Base, hastily built in 1941. The Japanese took the base in April 1942.

Bataan Ave

Bataan Peninsula on the Island of Luzon (West of Manila Bay) was America's last stronghold in the Philippines during the Japanese invasion in WWII.

Hemphill Road

COL John E. Hemphill, Commanding Officer of Camp Alfred Vail (December 1920 - September 1925), was instrumental in getting the camp designated a permanent post. Hemphill died August 17, 1948.

Corregidor Road

Corregidor, site of Fort Mills, was the largest of four fortified islands in Manila Bay. On May 7, 1942, from this location, General Wainwright broadcast surrender instructions for the Philippine Islands over station KZRH.

Corput Plaza Dr

During his 35-year military career, MG Rex Van Den Corput, Jr.'s assignments included Director of the Signal Corps Laboratories (1941-1944), Chief Signal Officer of the European Command (1952-1953) and Chairman of the Joint Communications-Electronic Committee. He died March 12, 1960.[119]

Building 2567

Shoppette

Guam Ln

The Island of Guam, attacked on December 8, 1941 by Japanese naval and air forces, was the first American possession to fall into enemy hands.

Building 3001

Outreach Center (Howard Commons Area)

Radio Way

The first radio laboratories were built at Fort Monmouth in 1917. Fort Monmouth developed various backpack and vehicle FM radios that contributed to the allied victory in WWII.

Satellite Road

This road honors Fort Monmouth's long standing work supporting communications and electronics for satellites.

Academy Drive

This drive marks the home of Building 2501, the Fire Academy.

Lowther Dr

LTC Ralph L. Lowther died in action in Belgium, January 14, 1945, while serving as Signal Officer of the 75th Infantry Division. He posthumously received the Bronze Star. General Order 27, May 27, 1947, designated Lowther Drive.

Maxwell Place

An existing street name.

Gibbs Hall (Building 2000)

Gibbs Hall (the Fort Monmouth Officer's Club) began as a private country club known as "Suneagles," built by Max Phillips in the 1920s. The country club consisted of a clubhouse (which is still largely intact as part of Gibbs Hall), an eighteen-hole golf course, a polo field, and an airfield. The Army acquired the site, along with the rest of the Charles Wood Area, in 1941. Officer housing beyond the clubhouse on Megill Drive arose between 1949 and 1955. First designated in 1947 and dedicated in 1950, Gibbs Hall memorializes Major General George S. Gibbs, Chief Signal Officer from 1928 to 1931.[113]

Building 2018

Mulligans Restaurant

Megill Dr

COL Sebring C. Megill (b. 1875) served during World War I as Signal Officer with the 78th, 79th, and 81st Divisions, for which he received the Victory Medal with battle clasps for St. Mihiel, Meuse-Argonne, and the Defensive Sector.

Wig Wag Road

This road commemorates the military flag signaling system devised by Albert J. Myer, founder of the U.S. Army Signal Corps.

Building 2568/9

Charles Wood pool/Bath House

Pinebrook Road

Pinebrook Road runs along the southernmost edge of the Charles Wood Area.

Mitchell Dr

MAJ (retired, COL) George E. Mitchell commanded the Signal Corps Camp, Little Silver, from July 12 to September 15, 1917, when he departed for overseas service with the American Expeditionary Force. General Orders Number 44, dated March 25, 1954, designated Mitchell Drive.[121]

Howard Commons

The Howard Commons, initially known as Eatontown Gardens, consists of 600 family housing units in fifty-two buildings constructed from 1953-1954 by Wherry Housing. The complex was renamed for Congressman James J. Howard in recognition of his long-time support for Fort Monmouth and his contributions in Congress to the welfare of Soldiers and the Army. The 1995 Base Closure and Realignment Commission mandated the closure of Howard Commons.[117] Today, the area stands vacant.

Helms Dr

COL George W. Helms commanded Camp Alfred Vail from June 28, 1918 to December 15, 1920. He died November 30, 1946. General Orders Number 42, dated March 25, 1954, dedicated this Drive.[120]

Off Post

CAMP COLES/COLES AREA

World War II brought significant change to the Signal Corps Laboratories at Fort Monmouth. On June 30, 1940, the organization had a staff of just eight officers, fifteen enlisted men, and 234 civilians. By June 30, 1941, the civilian strength had grown to 1,227. Fifteen months later, in December 1942, the Labs had 14,518 military and civilian employees on board.

Clearly, growth of the magnitude experienced by the Labs between June 1940 and December 1942 could not be achieved without a corresponding expansion of facilities. To this end, in 1941, the government acquired several remote sites. For Field Laboratory #1, later designated "Camp Coles Signal Laboratory," the government acquired rights to 46.22 acres of land west of Red Bank, at Newman Springs and Half Mile Roads. The Office of the Chief Signal Officer earmarked more than $700,000 for construction on the site, which the government purchased in June 1942 for $18,400. This laboratory was primarily responsible for ground communications technologies (radio and wire).

Camp Coles was dedicated on October 1, 1942, in honor of Colonel Roy Howard Coles, Executive Officer for the Chief Signal Officer of the American Expeditionary Forces in World War I. War Department General Orders Number 24, dated April 6, 1945, redesignated the site as "Coles Signal Laboratory." R&D activities at Coles Signal Laboratory ended about 1956, when the missions and personnel of the organization were moved to the newly constructed "Hexagon" in the Charles Wood Area. The site was occupied next by the U. S. Army Signal Equipment Support Agency, at which time (December 18, 1956) it was formally renamed "The Coles Area." Subsequently, until completion of the "Command Office Building" in Tinton Falls, the Coles Area served as the home of the U. S. Army Electronics Command's Procurement Directorate. The government disposed of the property in the mid-1970s.

COMMAND OFFICE BUILDING (CECOM BUILDING)

The Command Office Building was a privately owned, government-leased facility located at the end of Tinton Avenue. Before this building officially opened in the mid-1970s, offices of the Electronics Command (ECOM) were scattered throughout the post, mostly in World-War II vintage "temporaries," with the Procurement Directorate (about 1,000 employees) in

Former CECOM Office Building

the Coles area, four miles North, and the National Inventory Control Point (about 2,000 employees) in Philadelphia. With the opening of the Command Office Building, ECOM closed the Coles Area and moved the Philadelphia operations to Fort Monmouth. Activities vacated the building by December 1998 as directed by the 1993 Base Realignment and Closure Commission.[122]

DEAL TEST SITE

The property known as the Deal Test Site is a 208 acre parcel of land in Ocean Township, Monmouth County, NJ. It is two miles inland from Deal and bound by Deal Road on the south, Whalepond Road on the east, Dow Avenue on the north, and private property to the west.

U.S. Government activity at the Deal Test Site began in the mid-1950s with a continual series of leases ranging from one to three years. The U.S. Army Engineers, District of New York was the original lessee; however the using agency was the U.S. Army Electronics Command, headquartered at Fort Monmouth. Approximately eight miles separated the Main Post of Fort Monmouth from this test site.

The Deal Test Area was often in the news in the late 1950s and 1960s because of its excellent facilities and performance in monitoring satellites. It was, for a period, one of the prime tracking stations of the North Atlantic Missile Range. When Sputnik I was launched in October 1957, the Deal area was the first government installation in the United States to pick up and record the Russian signals. An elaborate monitoring facility was set up in time to monitor Sputnik II. Once again, Deal was the first American station to receive the signals.

Space achievements followed rapidly. All satellites, both American and Russian, were monitored and logged continuously by Deal personnel, as were all missiles launched from Cape Kennedy. The Deal Area was the communications center for COURIER, the first large capacity active communication satellite, which had been developed at Fort Monmouth. It was also instrumental in the TIROS I and II weather satellites. Its space availabilities dropped off gradually as NASA and the Air Force set up their own monitoring and tracking facilities.

In compliance with Army and DoD directives to abandon excess leased real estate, ECOM terminated their lease with the site owners effective 30 June 1973. The Deal facilities and personnel were moved to the government-owned Evans Area.[123]

DIANA SITE

A large, dish-type antenna just south of the Marconi Hotel in the Evans Area marks the site of the Diana Project. Here, in January 1946, Signal Corps scientists successfully bounced a signal off the surface of the moon and proved the feasibility of satellite communications. The Diana antenna resembled a pair of bedsprings on a tall mast, reminiscent of early RADAR antenna design.

The dish-type antenna now on the site, which is known as the "Space Sentry," is not the Diana antenna. However, it has a history of its own. The Signal Corps erected it in the 1950s as America was preparing to launch its first communications satellites, which were also being developed at Fort Monmouth. When the Soviet Union launched SPUTNIK I in 1957, Signal Corps engineers used this antenna to track the Soviet satellite and monitor its signals.

EVANS AREA

The Evans Area, in what is today Wall Township, N.J., consisted of some 253 acres about 15 minutes from Fort Monmouth. Prior to WWI, the Marconi Wireless Telegraph Company of America established a transatlantic radio receiving station there. In addition to a number of tall antenna masts, which have long since disappeared, the Marconi Company, in 1914, constructed a 45-room brick hotel for unmarried employees and two brick houses for company officials. Their red tile roofs make these buildings easily identifiable by air. The Army purchased the site in 1941 for Field Laboratory #3, the radio position finding section of the Signal Corps Labs, which at the time was stationed at Fort Hancock (on Sandy Hook). This laboratory handled the Army's top-secret radar projects. Apart from the Marconi Hotel, houses, an operations building, and a small redbrick laboratory building, most of the buildings in the Evans Area, including a collection of radar antenna shelters, date from the World War II era.

The 1993 Base Realignment and Closure (BRAC) mandate that closed the Evans Area for the Department of the Army had a scheduled completion date of 1998. The transfer process, however, continues.

Today, the Evans Area is listed on the National Register of Historic Places. The National Park Service approved 37 acres of land and historic buildings for use by Infoage, a group of cooperating non-profit organizations "dedicated to the preservation and education of information age technologies." This

Evans Area

group seeks "to develop an interactive learning center focused on the information age technologies at historic Camp Evans." They aim to inspire people of all ages to learn from the past and improve the future.

Brookdale Community College also claims land in the Evans Area, utilizing several renovated buildings as classrooms for the New Jersey Coastal "Communiversity."

The Evans Area memorializes LTC Paul W. Evans, who commanded the 101st Field Signal Battalion in World War I during the Champagne-Marne, Aisne-Marne, and St. Mihiel operations. LTC Evans died in the Canal Zone on April 10, 1936 while on duty as Department Signal Officer.[124]

LTC P. W. Evans
(1889-1936)

Diana Site, 1946

Commanding Officers at Fort Monmouth

COMMANDING OFFICER'S QUARTERS

Building 230 was designed especially for the Commanding Officer of Fort Monmouth in the late 1930s.

Public Law 67, 73rd Congress, 16 June 1933 mandated that construction on military posts continue despite the Great Depression. The second phase of permanent construction thus occurred on the post here from 1934 through May 1936. Russel Hall, then Post Headquarters, was among the buildings completed during this period.

Commander's Quarters, c. 1920

Also completed was Building 230. A description of the building from Signal Corps Bulletin Number 94, January – February 1937 follows:

"The quarters are 61 by 30 feet in area with a double garage, 19 by 24 feet in area, on the south and a porch two stories in height, 10 by 41 feet in area, on the west side. On the first floor there is a reception hall in the center and a large dining room and living room opening to the right and left. The living room, reception hall, and dining room extend the full length of the building on the west. A kitchen and pass pantry are located on the front of the building to the left of the hall and a library to the right. On the second floor there is a master's bedroom, dressing room, three bedrooms, a maid's room, and a total of four bathrooms, one of which is the maid's bath. The basement provides space for a coal storage room, boiler room, and laundry. A separate entrance to the boiler room for the removal of ashes, supplying of coal, access to laundry, etc., is provided. There is a large attic which is unfinished except for flooring. There is also an attic over the garage."

War Department records show that the building, completed on 7 October 1935, cost $14,500. COL Arthur S. Cowan, as the eighth Commanding Officer of Fort Monmouth (September 1929-April 1937), first occupied the new Commanding Officer's Quarters.

Commander's Quarters Today

39

★★

Major General Dennis L. Via

Major General Dennis L. Via serves as the Commanding General, CECOM Life Cycle Management Command (CECOM LCMC). As Commander, General Via leads a world-wide organization of over 10,000 military and civilian personnel responsible for coordinating, integrating and synchronizing the entire life-cycle management of the C4ISR systems for all of the Army's battlefield mission areas - maneuver control, fire support, air defense, intelligence, combat services support, tactical radios, satellite communications, and the warfighter information network.

Prior to assuming command, General Via served as Commanding General, 5th Signal Command, and United States Army, Europe and Seventh Army (USAREUR) Chief Information Officer/Assistant Chief of Staff, G6 (CIO/G6).

General Via is a native of Martinsville, Virginia. He attended Virginia State University in Petersburg, Virginia, where he graduated in May 1980 as a Distinguished Military Graduate, and received his commission as a Second Lieutenant in the Signal Corps. He holds a Master's Degree from Boston University. General Via is a graduate of the United States Army Command and General Staff College, and the United States Army War College.

General Via began his career with the 35th Signal Brigade, XVIII Airborne Corps, Fort Bragg, North Carolina. Key assignments included Commander, 82nd Signal Battalion, 82nd Airborne Division; Commander, 3rd Signal Brigade and III Corps Assistant Chief of Staff, G6; Division Chief, Joint Requirements Oversight Council (JROC) Division, Office of the Deputy Chief of Staff, Army G8, Headquarters, Department of the Army; Director, Global Operations, Defense Information Systems Agency (DISA); and Deputy Commander, Joint Task Force-Global Network Operations (USSTRATCOM).

General Via's military awards and decorations include the Distinguished Service Medal, the Defense Superior Service Medal, two awards of the Legion of Merit, two awards of the Defense Meritorious Service Medal and five awards of the Meritorious Service Medal. The General is authorized to wear the Army Staff Identification Badge and the Master Parachutist Badge.

He and his wife, the former Linda A. Brown, have two sons, Brian and Bradley.

At the groundbreaking ceremony for the command's new facilities at APG, MG Via cited the Army's once-in-a-generation investment in the command's mission to provide superior C4ISR systems for Warfighters.

38

★★

Major General Michael Mazzucchi

Major General (MG) Michael R. Mazzucchi served as the Commanding General, Communications-Electronics Life Cycle Management Command (C-E LCMC) and Program Executive Officer for Command, Control, Communications-Tactical (PEO C3T) from June 2004 until his retirement in July 2007.

In this position he directed the development, acquisition, integration and sustainment- in short, the life cycle management- of the systems that tied together all of the Army's battlefield mission areas – maneuver control, fire support, air defense, intelligence, combat services support, tactical radios, satellite communications and the Warfighter information network. This included a major role in introducing new digital battle command technologies fielded to forces engaged in the global war on terrorism.

General Mazzucchi previously served as the CECOM Deputy for Systems Acquisition and Director, CECOM Systems Management Center and as Assistant Program Executive Officer for Integration and Director of the Central Technical Support Facility at Fort Hood, Texas. Other prior assignments included Project Manager, MILSATCOM and Product Manager, Tactical Satellite Terminals, both here.

A graduate of Purdue University with a degree in Electrical Engineering, he also holds a Masters of Science in Electrical Engineering from the Air Force Institute of Technology. His military education includes the Command and General Staff College; the Defense Systems Program Management Course; the Army War College; and the National Security Leadership Course.

Major General Mazzucchi has received the Legion of Merit medal (with Oak Leaf Cluster), the Defense Meritorious Service Medal (with Oak Leaf Cluster); the Meritorious Service Medal (with two Oak Leaf Clusters); and the Army Commendation and Achievement Medals. He is authorized to wear the Office of the Secretary of Defense Identification Badge and the Air Force Senior Space Operations Badge.

He and his wife, the former Linda L. Crenshaw, have a son, Steve, a daughter, Rachel.

MG Mazzucchi was the first Commander of the LCMC at Fort Monmouth

37

★★

Major General William H. Russ
JULY 2001 - JUNE 2004

Major General William H. Russ assumed his duties as Commanding General of the United States Army Communications- Electronics Command and Fort Monmouth, New Jersey on July 20, 2001.

Previously he was the Commanding General of the United States Army Signal Command of Fort Huachuca, Arizona and the Director for Programs and Architecture, Office of the Director of Information Systems for Command, Control, Communications and Computers in Washington, D.C.

Major General Russ is a graduate of Florida A & M University with a Bachelor of Science degree in Electronics. He holds a Master of Science degree in Public Administration from Shippensburg University in Pennsylvania.

His military education includes the Signal Basic and Advanced Courses, the Armed Forces Staff College and the United States Army War College.

After completion of Airborne Ranger School and the Signal Officer Basic Course, Major General Russ served as Communications Officer for the 1st Battalion, 32nd Armor, 3rd Armored Division, United States Army Europe and Seventh Army. After serving four years in Germany, Major General Russ attended the Signal Officer Advanced Course, Fort Gordon, Georgia, and, upon completion, was assigned as Assistant S-3, 67th Signal Battalion (Combat), Fort Gordon. He later served as Commander, Communications and Electronics, United States Army Joint Support Group - Joint Support Area, United States Forces, Korea; Instructor/Branch Chief, Officer Advanced Division, Officer Training Directorate, United States Army Signal Center and School, Fort Gordon; Personnel Assignments Officer, Signal Branch, Total Army Personnel Center, Alexandria, Virginia; Associate Director (Information Mission Area Steering Group), Executive Officer for the Deputy Chief of Staff for Information Management; Commander, 43rd Signal Battalion, 5th Signal Command, United States Army Europe and Seventh Army, Germany; Staff Officer, Office of the Director for Information Systems, Command, Control, Communications and Computers, United States Army, Washington, D.C.; Commander, 1st Signal Brigade, United States Forces, Korea; Executive Assistant, J-6, Joint Staff, Washington, D.C.; Secretariat, Military Communications-Electronics Board, Joint Staff, Washington, D.C.; and Deputy Director, Chief Information Office, Forces Command, Fort McPherson, Georgia.

His military awards and decorations include the Defense Superior Service Medal, Legion of Merit, Meritorious Service Medal (with four Oak Leaf Clusters), Army Commendation Medal (with Oak Leaf Cluster), Parachutist Badge and Ranger Tab.

MG Russ retired on 25 June 2004 in a ceremony held at the Pruden Amphitheater, Fort Monmouth.

> *Major General Russ led CECOM through the tragic events of September 11, 2001*

36

★★

Major General Robert L. Nabors
SEPTEMBER 1998 - JULY 2001

MG Nabors was born in Boston, Massachusetts, and grew up in Lackawanna, New York. He holds a Bachelor of Science degree in Systems Engineering from the University of Arizona; a Master of Science degree in Systems Management from the University of Southern California; and served as a Senior Fellow in the National Security Affairs Program at Harvard University. He is also a graduate of the Senior Officials in National Security Program at Harvard University. His military schooling includes the Signal Officer Candidate School, the Signal Officer Basic and Advanced Courses and the Armed Forces Staff College.

MG Nabors' initial duty assignment was with the 67th Signal Battalion at Fort Riley, Kansas. After a tour in Vietnam, MG Nabors served at Fort Dix, New Jersey; Aberdeen Proving Ground, Maryland; and Worms, Germany. In November 1979, MG Nabors was selected as Aide de Camp for the Commanding General, VII Corps. MG Nabors was attached to the J-6 Staff of the Combined Forces Command/United States Forces, Korea and subsequently served as the S2/3 of the 41st Signal Battalion. He was assigned to the Office of the Director of Plans, Programs, and Policy at the United States Readiness Command in December 1983, and was then selected to command the 509th Signal Battalion in Italy.

MG Nabors served as Special Assistant to the U.S. Army's Director of Information Systems for Command, Control, Communications, and Computers (DISC4J). He was also Chief, Integration Division, Architecture Directorate. Prior to assuming command of the 2nd Signal Brigade in December 1990, MG Nabors served as Deputy Commander, White House Communications Agency. MG Nabors served as the Executive Officer for the DISC4 before his assignment as Director, Single Agency Manager for Pentagon Information Technology Services. MG Nabors assumed command of the 5th Signal Command on 22 November 1995.

MG Nabors' awards and decorations include the Defense Superior Service Medal; Legion of Merit with four Oak Leaf Clusters; the Bronze Star Medal; Meritorious Service Medal with four Oak Leaf Clusters; the Joint Service Commendation Medal; the Army Commendation Medal with four Oak Leaf Clusters; the Department of the Army Staff Identification Badge; the Joint Meritorious Unit Award; and the Presidential Support Badge. MG Nabors is a member of the American Mensa Society.

MG Nabors retired on 20 July 2001.

MG Nabors and his wife, Valerie, have three adult children: Robert, Richard and Jonathan.

> *MG Nabors was the first African American Commanding Officer of Fort Monmouth*

35

⭐⭐

Major General Gerard P. Brohm

JANUARY 1995 - SEPTEMBER 1998

Major General Gerard P. Brohm assumed command of the Communications-Electronics Command and Fort Monmouth on 10 January 1995.

General Brohm was born in New York, New York. He enlisted in the U.S. Army in 1966 and took basic and advanced training as an infantryman. He entered Officer Candidate School in 1967 and was commissioned a Second Lieutenant in July of that year. He received a Bachelor of Arts degree in Literature from Seton Hall University. General Brohm also holds a Master of Science degree in Telecommunications from the University of Colorado. His military education includes the Signal Officer's Advanced Course, Command and General Staff College and the U.S. Army War College.

Previous troop assignments include platoon leader in Vietnam; two company commands at Fort Bragg; S-3 and later Executive Officer for the 41st Signal Battalion in Korea; Battalion Commander, 143rd Signal Battalion, 3d Armored Division; Brigade Commander, 93rd Signal Brigade, VII (US) Corps and Deputy Commanding General, U.S. Army Signal Center and Fort Gordon, Georgia.

Other previous staff assignments include Executive Officer for the Deputy Chief of Staff of Operations and Plans, U.S. Army Communications Command, Fort Huachuca, Arizona; Director of Combat Developments at Fort Gordon, Georgia and Chief, Communications Systems Section, Supreme Headquarters Allied Powers Europe, Belgium. Prior to his current assignment as Commander, U.S. Army Communications-Electronics Command and Fort Monmouth, he served as Director for Command, Control and Communications Systems, United States Pacific Command, Camp H.M. Smith, Hawaii, from July 1993 to January 1995.

General Brohm's awards and decorations include the Defense Superior Service Medal, Legion of Merit with Oak Leaf Cluster, Bronze Star Medal, Defense Meritorious Service Medal, Meritorious Service Medal with three Oak Leaf Clusters, the Army Commendation Medal with Oak Leaf Cluster and the Army Achievement Medal.

MG Brohm retired on 1 September 1998. He was at that time presented the Distinguished Service Medal, the Army's highest peacetime award, by GEN Johnnie E. Wilson, AMC Commander.

Major General Brohm and his wife, Ines, have four children: Maria Elena, Kathy, Jerry and Michael.

In his farewell address, MG Brohm said that he had three great loves in his life: His wife, his children, and the Army.

34

⭐⭐

Major General Otto J. Guenther

JULY 1992 - JANUARY 1995

(Then) Brigadier General (BG) Otto J. Guenther assumed command of CECOM and Fort Monmouth in July 1992. He received his second star that October.

He was born in Long Branch and raised in Red Bank, New Jersey. He completed the Reserve Officers Training Corps curriculum at Western Maryland College in 1963 and was commissioned a Second Lieutenant in the Signal Corps. General Guenther holds a Bachelor of Arts degree in Economics and a Master of Science degree in Procurement/Contract Management from Florida Institute of Technology. His military education includes the Signal Officer Basic Course, the Infantry Advanced Course, the Command and General Staff College, the Army War College, and the Defense Systems Management College. General Guenther has had extensive user and field experience, holding key command and staff assignments to include Platoon Leader, Company Commander, Battalion Executive Officer and Battalion Commander. Prior to his current assignment as Commander, U.S. Army Communications-Electronics Command and Fort Monmouth, he served as the Program Executive Officer for Communications Systems, Fort Monmouth

General Guenther's acquisition experience includes assignments as Deputy Director of Contracts, Defense Contract Administration Services Region, San Francisco and as Commander of the Defense Contracts Administration Services Office overseeing a major Defense contract for production of tracked vehicles including the Bradley Fighting Vehicle and armored personnel carriers. He was also the Deputy for R&D Procurement and Base Operations, Fort Monmouth, in 1983. That same year, he assumed Command of the Defense Contract Administration Services Region, New York. In 1985 he became Director of the Defense Acquisition Regulation System and Director of the (DAR) Council. General Guenther was reassigned to Fort Monmouth in December 1987 and became the Project Manager for Position Location Reporting System/Tactical Information Distribution System.

He has received the Defense Superior Service Medal with Oak Leaf Cluster, the Legion of Merit, the Bronze Star Medal with Oak Leaf Cluster, the Meritorious Service Medal with four Oak Leaf Clusters, the Joint Service Commendation Medal, the Army Commendation Medal with Oak Leaf Cluster, and the Secretary of Defense Identification Badge.

Upon leaving Fort Monmouth in January 1995, MG Guenther received his third star and became the Director of Information Systems for Command, Control, Communications, and Computers (C4) in the Office of the Secretary of the Army in Washington.

General Guenther and his wife have two daughters, Tracy and Debra.

In his farewell address, LTG Guenther assured the Fort Monmouth community that, as Director of Information Systems for C4, he would make sure they had plenty of work!

33 ★★ Major General Alfred J. Mallette

JULY 1990 - JULY 1992

(Then) Brigadier General Alfred J. Mallette assumed command of CECOM and Fort Monmouth on 10 July 1990. He received his second star in April 1991.

Mallette was born in Green Bay, Wisconsin, on 21 November 1938. Upon completing the Reserve Officer Training Corps curriculum at St. Norbert in 1961, he was commissioned a Second Lieutenant and awarded Bachelor of Science degrees in Physics and Mathematics. He also holds a Master of Science degree in Operations and Research Analysis from Ohio State University. His military education includes completion of the Signal Officer Basic Course, the Infantry Advanced Course, the U.S. Army Command and General Staff College, and the Industrial College of the Armed Forces.

General Mallette has held a variety of important command and staff positions in addition to that of Commander, U.S. Army Communications-Electronics Command. Key assignments have included: Commanding General of the 5th Signal Command and Deputy Chief of Staff for Information Management in USAREUR; Deputy Director of the Plans, Programs, and Systems Directorate, DISC4, Office of the Secretary of the Army; Deputy Commanding General of the U.S. Army Signal School; Commander of the 93d Signal Brigade, USAREUR; and Commander of the 8th Signal Battalion, USAREUR.

General Mallette also served in a variety of important career-building assignments. He was Chief of the Program Section, Information Systems Division, Allied Forces Central Europe; he was S-3 and later Executive Officer of the 121st Signal Battalion, 1st Infantry Division; he served as Chief of the Plans and Policies Division, U.S. Army Support Command, and he served in Vietnam in the Office of the Deputy Chief of Staff for Logistics/Criminal Investigation Division of the U.S. Military Assistance Command.

General Mallette's awards and decorations include the Legion of Merit with two Oak Leaf Clusters, the Bronze Star Medal with Oak Leaf Cluster, and the Meritorious Service Medal with Oak Leaf Cluster. General Mallette is also authorized to wear the Senior Parachutist Badge.

On 22 July 1992, the Army promoted Mallette to the rank of Lieutenant General (LTG) and assigned him to serve as Deputy Director General of the NATO Communications and Information Systems Agency.

General Mallette and his wife, Nancy, have three children: Scott, Randy, and Nicole.

LTG Mallette died on 15 August 1994. Building 1207 was dedicated as Mallette Hall on 17 October 1996 to honor him.

> *"This is the greatest collection of talent and dedicated workforce I have ever run into."*
>
> *LTG Alfred J. Mallette to the CECOM workforce at his change of command ceremony*

32 ★★ Major General Billy M. Thomas

MAY 1987 - JULY 1990

(Then) Major General Billy M. Thomas assumed command of CECOM and Fort Monmouth in May 1987.

He was born in Crystal City, Texas on 14 August 1940 and graduated from Killeen High School, Killeen, Texas. Upon completion of the Reserve Officers Training Corps curriculum and the educational course of study at Texas Christian University in 1962, he was commissioned a Second Lieutenant and awarded a Bachelor of Science degree in Secondary Education. General Thomas also holds a Master of Science degree in Telecommunications Operations from George Washington University.

His military education includes completion of the Signal Officer Basic and Advanced Courses, the Army Command and General Staff College, and the Army War College.

Prior to commanding here, he held a wide variety of other command assignments, including Commander of two companies: one in Germany, and one in Vietnam; Commander, 5th Signal Battalion, 5th Infantry Division (M); and Commander, 93rd Signal Brigade, VII Corps, United States Army Europe.

General Thomas also served in a variety of important staff assignments. He served as an Airborne Battle Group and Brigade Signal Officer in the 161st Airborne Division and served two years as the S-3, 447th Signal Battalion in Germany; Chief, 31M Radio Relay Course, Communications System Department, United States Army Signal School, Signal Planning Advisor with the Logistics Directorate, United States Army Military Assistance Command in Vietnam; and served in Thailand as Chief of Plans and Requirements, Office of the Assistant Chief of Staff for Communications Electronics, United States Support Activity Group. Returning to the United States, General Thomas next served as Management Directorate, United States Army Military Personnel Center; Assistant Chief of Staff, G-5 (Civil Affairs), 5th Infantry Division; and Special Assistant to the Dean, National Defense University. General Thomas later served as the Deputy Commanding General, U.S. Army Signal Center and School.

Prior to assuming command of the U.S. Army Communications-Electronics Command and Fort Monmouth on 15 May 1987, he was assigned to the Army staff as Deputy Director, Combat Support Systems, Office of the Deputy Chief of Staff for Research, Development and Acquisition.

Thomas' awards and decorations include the Legion of Merit (with one Oak Leaf Cluster), the Bronze Star Medal (with three Oak Leaf Clusters), the Meritorious Service Medal (with three Oak Leaf Clusters), the Joint Service Commendation Medal and the Army Commendation Medal. He is also authorized to wear the Parachutist Badge.

MG Thomas left Fort Monmouth on 10 July 1990 to assume duties as the Deputy Commanding General for Research, Development, and Acquisition of the Army Material Command in Alexandria, Va. He received his third star upon leaving Fort Monmouth.

> *When General Thomas joined the Army in 1962, he wanted to be a band leader.*

31

★★

Major General Robert D. Morgan
JUNE 1984 - MAY 1987

(Then) Brigadier General Robert D. Morgan became the thirty-first Commanding Officer of Fort Monmouth and the third Commanding General of CECOM on 26 June 1984. He was promoted to Major General on 10 September of that year.

MG Morgan was born in Buffalo, New York in March 1934. He graduated from Canisius College in Buffalo in 1955 and was commissioned through the ROTC as a Second Lieutenant in the Signal Corps. He graduated the Signal Officers' Basic Course (1956) and Signal Officers' Advance Course (1963). It appears that General Morgan is the first Commanding Officer of Fort Monmouth to have been an Army Aviator, having graduated Army Aviation Primary Flight Training Program at Fort Rucker, Alabama (1957-1958) and having served as Command Pilot with the U.S. Army Military Assistance Command, Vietnam (USAMACV) (1965-1966). His second tour in Vietnam was as Battalion Commander, 40th Signal Battalion, 1st Signal Brigade (1971-1972). Other assignments included Germany; Fort Huachuca, Arizona; Fort Rucker, Alabama; and the Pentagon.

Morgan returned to Fort Monmouth in June 1976 and served as Project Manager for Position Location Reporting System (PLRS); Deputy Commanding General for Research & Development and concurrently, Commander of the CECOM Research and Development Center; and Deputy Commanding General for Procurement and Readiness, CECOM.

MG Morgan left Fort Monmouth on 15 May 1987 for service at the Pentagon.

General Morgan is the first Commanding Officer of Fort Monmouth to have been an Army Aviator.

30

★★

Major General Lawrence F. Skibbie
OCTOBER 1982 - JUNE 1984

(Then) Major General Lawrence F. Skibbie became the thirtieth Commanding Officer of Fort Monmouth and the second Commanding General of CECOM on 28 October 1982.

Skibbie was born in Bowling Green, Ohio, in February 1932. He graduated from the U.S. Military Academy in the Class of 1954 and was commissioned a Second Lieutenant, Artillery. He subsequently completed the Artillery School Basic Course and the Basic and Advanced Ordnance School. As a junior officer he served in AAA billets; as radar and Nike officer at Fort Bliss, Texas; and in Ordnance billets at Aberdeen, Maryland, White Sands, New Mexico, and the U.S. Army Europe. General Skibbie served in Vietnam, 1968-1969, in the G-4, U.S. Army Vietnam, and also commanded the 63d Maintenance Battalion at DaNang. Assignments just prior to service at Fort Monmouth included tours of duty in the Department of Army in Washington and at the Armament Research and Development Command at Dover, New Jersey. From 1975 to 1977 he commanded the Rock Island Arsenal, U.S. Army Armament Command (later United States Army Materiel Readiness Command), and from July 1977 to October 1978 was Deputy Commander for Ammunition Readiness at Rock Island, Illinois.

Immediately before coming to Fort Monmouth, he served in the Office of Deputy Chief of Staff for Research, Development and Acquisition, Washington, DC, from 1978 to 1982.

General Skibbie ended his tour of duty at Fort Monmouth on 26 June 1984, upon his promotion to Lieutenant General and assignment as Deputy Commanding General for Readiness, Army Materiel Command (AMC).

"Nowhere have I been more impressed with teamwork... than here at CECOM."
- MG Skibbie to the workforce

29

★★

Major General Donald M. Babers
JUNE 1980 - OCTOBER 1982

(Then) Major General Donald M. Babers first came to Fort Monmouth as the second Commanding General of CERCOM and the twenty-ninth commander of Fort Monmouth in June 1980. Reorganization and the transition of CERCOM and CORADCOM resulted in the establishment of the U.S. Army Communications-Electronics Command (CECOM) on 1 May 1981. General Babers thus became the first Commanding General of CECOM.

Babers was born in Newkirk, New Mexico, in May 1931. He graduated from Oklahoma A & M College and was commissioned a Second Lieutenant, Ordnance Corps, in May 1954.

He attended the Associate Officers Ordnance Course and served with the 881st Ordnance Company in Europe. Subsequent assignments included duty with the 25th Artillery, Fort Sill, Oklahoma, and as assistant project manager, Combat Vehicles, Detroit Arsenal. He served in Republic of Vietnam, 1964-1965, in the Military Assistance Program, J-4, Military Assistance Command, Vietnam.

General Babers was Project Manager for the M60 Tank program and later director of Procurement and Production at the Tank-Automotive Command (TACOM), Warren, Michigan, before becoming deputy Commanding General at TACOM in July 1976.

He ended his tour of duty at Fort Monmouth on 28 October 1982 when he was promoted to Lieutenant General and assigned as Deputy Commanding General for Readiness, DARCOM, (now AMC), Alexandria, VA. In June 1984 he was reassigned as Commander of the Defense Logistics Agency, Alexandria, VA

MG Babers was the first Commanding General of CECOM

28 ★★ Major General John K. Stoner
NOVEMBER 1976 - JUNE 1980

Major General John K. Stoner, Jr. came to Fort Monmouth as the seventh Commanding General of ECOM and the twenty-eighth commander of Fort Monmouth in November 1976. ECOM was disestablished and two new commands were established: the Communications-Electronics Materiel Readiness Command and the Communications Research and Development Command on 1 January 1978. Stoner then became the first Commanding General of CERCOM and served in that capacity until his retirement in June 1980.

Stoner was born in Woodbury, New Jersey, in January 1929. He graduated from Drexel Institute of Technology and received a regular Army commission through the Reserve Officers Training Corps in July 1951 as a Second Lieutenant, Artillery. He completed the Chemical Corps School Advanced Course and served in Korea.

Other tours of duty included a year in Vietnam with the Military Assistance Command, Vietnam; Office of the Assistant Secretary of the Army in Washington; and Commanding Officer of Pine Bluff Arsenal in Arkansas and Commanding Officer of Edgewood Arsenal, Maryland. Stoner commanded the U.S. Army Materiel Management Agency in Germany from 1973 to 1976, and was Commanding General, 2d Support Command (Corps) in Europe before coming to Fort Monmouth.

> *"...the first responsibility of a leader is to prove himself to his followers. You can't demand their respect, you have to earn it." -MG Stoner in an interview upon assuming command, here.*

27 ★★ Major General Albert B. Crawford, Jr
AUGUST 1975 - NOVEMBER 1976

Major General Albert B. Crawford, Jr. became the sixth Commanding General of ECOM and the twenty-seventh commander of Fort Monmouth in August 1975.

Crawford was born in Tucson, Arizona, in February 1928. He graduated from the U.S. Military Academy in the Class of 1950 and was commissioned a Second Lieutenant, Signal Corps. His early assignments included the 93d Signal Battalion and 301st Signal Group, 7th Army, Europe. He graduated from the Signal Corps Officers Advanced Course in 1954.

Subsequent assignments in Europe followed as mobile digital computer team chief with the 7th Army (1959-1962) and assistant division signal officer with the 4th Armored Division (1962-1963). He served in Vietnam from 1968-1969 with the 1st Signal Brigade.

Crawford returned to Fort Monmouth in 1971 and served as project manager of Army Tactical Data Systems until 1975. He then served as Commander here until his retirement from the U.S. Army on 30 November 1976.

> *Only 46 years old when he assumed command here, Crawford was one of the Army's youngest Major Generals.*

26 ★★ Major General Hugh F. Foster, Jr
MAY 1971 - AUGUST 1975

Major General Hugh F. Foster, Jr. served as the twenty-sixth Commanding Officer of Fort Monmouth and fifth Commanding General of ECOM from May 1971 to August 1975.

Foster was born in Brooklyn, New York, in March 1918. He graduated from the U.S. Military Academy in 1941 and was commissioned a Second Lieutenant, Signal Corps. His first duty was with the 4th Signal Company, 4th Infantry Division (Motorized), at Fort Benning, Georgia.

Duty overseas in World War II included the 560th Signal Aircraft Warning Battalion in North Africa and Sicily. Other assignments included Signal Section of Allied Headquarters in Algiers, Fifth Army in Italy, and 53d Signal Battalion.

Postwar Signal Corps duty assignments included Camp Crowder, Missouri, commander of the 63d Signal Battalion in Austria, and the Army Electronics Proving Ground, Fort Huachuca, Arizona.

General Foster returned to Fort Monmouth for three years between June 1962 and June 1965 as project manager for Universal Integrated Communications/Strategic Army Communications. He served as signal officer of the United Nations Command and Assistant Chief of Staff, Communications-Electronics, Eighth Army, Korea, from 1965-1967; and commanded the newly created U.S. Army Communications Systems Agency from July 1967 to August 1969.

General Foster served in Vietnam as commander of the 1st Signal Brigade before returning to Fort Monmouth as Commander.

He retired from the U.S. Army in September 1975.

> *MG Foster served in World War II, Korea, and Vietnam.*

25

★★

Major General Walter E. Lotz, Jr.
SEPTEMBER 1969 - MAY 1971

(Then) Major General Walter E. Lotz, Jr. became the twenty-fifth Commanding Officer of Fort Monmouth and the fourth Commanding General of ECOM in September 1969.

Lotz was born in Johnsonburg, Pennsylvania, in August 1916. He graduated from the U.S. Military Academy in 1938 and was commissioned a Second Lieutenant, Signal Corps. His first assignment was with the 51st Signal Battalion at Fort Monmouth. He attended the Signal School, graduating in February 1941, and joined the 1st Signal Operations Company (Aircraft Warning) in training, here. He left in June 1941 with that unit for duty in Iceland.

During World War II Lotz served with the Ninth Air Force, later redesignated the XII Tactical Air Command, in the European Theater of Operations as assistant director of communications.

Lotz returned to Fort Monmouth for three years (1947-1950) as chief of the Meteorological Department at Evans Laboratory, and later became deputy director of the Laboratory. He became the first deputy Commanding General of ECOM, serving from August 1962 to October 1963.

He served in Vietnam as Assistant Chief of Staff, Communications-Electronics, Military Assistance Command, from September 1966 to January 1968. General Lotz served as Commanding General at the U.S. Army Strategic Communications Command at Fort Huachuca, Arizona, from January 1968 to September 1969. He returned to Fort Monmouth as Commander from September 1969 to May 1971, when he was nominated for promotion to Lieutenant General and assigned to NATO Headquarters in Brussels.

General Lotz retired from the U.S. Army in August 1974.

General Lotz's wife, Shirley, was the daughter of Roger B. Colton, a retired Major General and former director of the Signal Corps Laboratories, here. Colton made the landmark decision to use frequency modulation (FM) as opposed to amplitude modulation (AM) in military radios and oversaw the completion of the SCR-268 practical radar system.

24

★★

Major General William B. Latta
OCTOBER 1965 - SEPTEMBER 1969

Major General William B. Latta became the twenty-fourth Commanding Officer of Fort Monmouth and the third Commanding General of ECOM in October 1965.

Latta was born in El Paso, Texas, in October 1914. He graduated from the U.S. Military Academy in 1938 and was commissioned a Second Lieutenant, Signal Corps. His first assignment was with the 51st Signal Battalion at Fort Monmouth. He left in September 1939 for duty at Fort Sam Houston, but returned to Fort Monmouth in October 1940 to attend the Signal School. After graduation in 1941 he remained for four months as acting director of the Officers Candidate School.

During World War II General Latta commanded the 1st Armored Signal Battalion in the landing at Casablanca and participated in the campaigns in Sicily, Normandy, and the Rhineland. In May 1945 he became signal officer of the XXI Corps in Germany.

Subsequent tours of duty included the War Department General Staff; Signal Corps Procurement Agency in Chicago; Office of the Chief Signal Officer; signal advisor to the Chief of the Military Assistance Advisory Group, Taiwan; and Seventh Army in Stuttgart, Germany.

General Latta came to Fort Monmouth from duty as Deputy Chief of Staff for Communications and Electronics, North American Air Defense Command, in Colorado. He commanded here from October 1965 until September 1969, leaving Fort Monmouth to become the Commanding General, U.S. Army Strategic Communications Command at Fort Huachuca, Arizona. General Latta retired from the U.S. Army in August 1972.

The 1st Armored Signal Battalion, under Latta, provided communications for the Roosevelt-Churchill conference at Casablanca.

23 ★★ Major General Frank W. Moorman

JULY 1963 - OCTOBER 1965

Major General Frank W. Moorman became the twenty-third Commanding Officer of Fort Monmouth and the second Commanding General of ECOM in August 1963.

Moorman was born in Camp Jossam, Philippine Islands, in February 1912. He graduated from the U.S. Military Academy in 1934 and was commissioned a Second Lieutenant, Infantry. His early Infantry assignments were at Fort Washington, Maryland, and Schofield Barracks, Hawaii. After graduating from the Infantry School in 1938 he entered the Signal Corps School, and graduated in June 1939. He was then assigned to the 4th (later 6th) Signal Company at Fort Monmouth and remained with it until war was declared.

During World War II he served as signal officer with the 82d Airborne Infantry Division in Italy and at Normandy. In 1944 he served with the XVIII Corps (Airborne) as Assistant Chief of Staff, G-4.

During the Korean War Moorman served briefly on the Eighth Army Staff in Korea and later as secretary of the general staff at Headquarters Far East Command. He was Commanding General of the U.S. Army Electronics Proving Ground at Fort Huachuca, Arizona, from 1958 to 1960.

General Moorman returned to command Fort Monmouth and ECOM in August 1963. He served until 30 September 1965, when he retired from the U.S. Army.

Moorman Hall, Building 362, is named for MG Moorman's father, Frank, also a Signal officer.

22 ★★ Major General Stuart S. Hoff

AUGUST 1962 - JULY 1963

Major General Stuart S. Hoff assumed command here in August 1962.

Hoff was born in Muskogee, Oklahoma, in November 1914. He graduated from Texas A & M College in 1929 and received an Army Reserve commission in the Infantry. It was not until September 1940 that he was ordered to extended active duty as a Captain, and initially served as assistant signal officer at Fort Sam Houston, Texas.

During World War II he served as assistant signal officer with the Sixth Army in the Southwest Pacific from 1943 to 1946. He was integrated into the regular Army in 1946 and officially transferred to the Signal Corps in 1951. He had two subsequent tours of duty in the Far East in Japan and Korea.

He came to Fort Monmouth as commandant, U.S. Army Signal School, serving from August 1956 to October 1957. After other tours of duty in the Office of the Chief Signal Officer, Korea; U.S. Army Pacific, Hawaii; and as chief of Research and Development, Office of the Chief Signal Officer, Hoff returned to Fort Monmouth in August 1962.

Under the reorganization of the Army during 1962, the U.S. Army Electronics Command (ECOM) was established on 1 August 1962. With the establishment of ECOM, the Commanding General here became the commander of ECOM and of Fort Monmouth. General Hoff thus became the first Commanding General of ECOM and the twenty-second commander of Fort Monmouth. General Hoff served until his retirement on 31 July 1963. He died in August 1978.

MG Stuart S. Hoff was the first Commanding General of ECOM.

21 ★★ Major General William D. Hamlin

MARCH 1960 - JULY 1962

Major General William D. Hamlin became the twenty-first Commanding Officer of Fort Monmouth in March 1960.

Hamlin was born in Clinton, New York, in April 1905. He graduated from the U.S. Military Academy in 1929 and was commissioned a Second Lieutenant, Signal Corps.

He first came to Fort Monmouth in September 1929 for duty and later attended the Signal School, graduating in June 1931. Other duties as a Signal Corps officer followed at New Jersey Bell Telephone Company; 51st Signal Battalion at Fort Monmouth; Fort Jay, New York; Fort Sam Houston, Texas; and Headquarters, Third Army, in San Antonio Texas.

During World War II he served in the Office of the Chief Signal Officer as executive officer from 1941 to 1943 and at SHAEF Headquarters in Europe from 1943 to 1945. After the war he again served at Fort Monmouth in the Enlisted Men's School and as director of the Officers Department, Signal School. He had two additional tours of duty with the Office of the Chief Signal Officer. He was advisor on communications to the Korean Military Government in 1947-1948 and then signal officer with the IX Corps in Japan.

He returned to Fort Monmouth as commandant, the Signal School, in July 1954 and served until July 1956.

He was Commanding General of the U.S. Army Signal Supply Agency in Philadelphia from 1956 to 1957 and then signal officer at Headquarters, U.S. Army Europe, Heidelberg, from 1957 to 1960.

General Hamlin returned to command Fort Monmouth in March 1960 and served until July 1962. He retired from the Army on 30 December 1962 and died in April 1975.

The first High Capacity Communication Satellite, Courier, was developed and built under the supervision of the Fort Monmouth Laboratories the same year MG Hamlin assumed command.

20
★★

Major General Albert F. Cassevant
SEPTEMBER 1958 - FEBRUARY 1960

Major General Albert F. Cassevant became the twentieth Commanding Officer of Fort Monmouth in September 1958.

Cassevant was born in Biddeford, Maine, in June 1908. He graduated from the U.S. Military Academy in 1931 and was commissioned a Second Lieutenant, Coast Artillery Corps. For the next six years he had duty in Coast Artillery billets.

From August 1937 until June 1941, Cassevant was assigned to the Signal Corps Laboratories at Fort Monmouth as the Anti-aircraft Liaison Officer and a radar project officer.

During World War II he served in two Coast Artillery billets and then was assigned to organize and become chief of the Electronics Branch, Army Ground Forces in Washington in February 1943. In November 1943 he was detailed to the Signal Corps and was named Director of the Evans Laboratory at Fort Monmouth. In July 1944 he became Chief, Engineering Division, and then Assistant Commanding Officer of the Signal Corps Ground Signal Agency, Bradley Beach, New Jersey. He left for a short time on temporary duty in the Asiatic-Pacific Theater from September 1945 to August 1946. He transferred from the Coast Artillery Corps to the Signal Corps in October 1947.

He returned to Fort Monmouth in June 1948 as Director of the Evans Signal Corps Laboratories, serving until June 1950. Other duties in a Signal Corps capacity followed in the Office of the Chief Signal Officer, U.S. Army Forces Far East in Tokyo, and again in the Office of the Chief Signal Officer.

In October 1957 General Cassevant returned to Fort Monmouth as commandant of the U.S. Army Signal Center and School. He then commanded the post from September 1958 until his retirement from the Army in February 1960. General Cassevant died in April 1971.

> *One early radar pioneer called Cassevant, "the man in the Army who sold the ideas."*

19
★★

Major General W. Preston Corderman
JUNE 1957 - AUGUST 1958

Major General W. Preston Corderman became the nineteenth Commanding Officer of Fort Monmouth in June 1957.

Corderman was born in Hagerstown, Maryland, in December 1904. He graduated from the U.S. Military Academy in 1926 and was commissioned a Second Lieutenant in the Signal Corps.

He first came to Fort Monmouth in 1927 for duty with the 51st Signal Battalion and then attended the Signal School, graduating in June 1929. For the next decade he served in Signal Corps assignments at Fort Warren, Wyoming; Field Artillery School, Fort Sill, Oklahoma; Office of the Chief Signal Officer, Washington, DC; and Fort William McKinley, Philippine Islands, finally becoming Signal Corps advisor to the Philippine Government.

During World War II he was Chief Postal Censor, Office of Censorship, for a year and in February 1943 assumed command of the Army Security Agency at Arlington Hall, Arlington, Virginia, remaining in the latter post until 1946.

General Corderman served for two years in the Alaska Command as Assistant Director of Communications, Director of Communications, and finally as Chief of Staff of the Command. In October 1951 he assumed command of the Signal Corps Procurement Agency in Philadelphia, and became Commanding Officer of the Signal Corps Supply Agency in January 1952. From 1953 to 1957 Corderman was assigned to the Office of the Chief Signal Officer as Chief, Research and Development Division, and later as Deputy Chief Signal Officer.

General Corderman returned to command Fort Monmouth in June 1957 and served until 31 August 1958, when he retired from the U.S. Army after more than thirty-two years of active service.

> *In 1994, the Central NJ Chapter of the Retired Officers Association approved the establishment of a "W. Preston Corderman Lifetime Achievement Award."*

18
★★

Major General Victor A. Conrad
SEPTEMBER 1954 - JUNE 1957

Conrad was born in Hammond, Wisconsin, in October 1900. He graduated from the U.S. Military Academy, Class of 1924, and was commissioned a Second Lieutenant in the Signal Corps. His first Army assignment was here, at Camp Vail. He graduated from the Signal School at Fort Monmouth in June 1926 and remained at the Signal School as an instructor until 1931.

Subsequent duties as a signal officer were at Bolling Field in the District of Columbia; Patterson Field, Ohio; Panama Canal Zone; and with the Radio Section, Office of the Chief Signal Officer in Washington.

During World War II he served as Chief of the Signal Section, Allied Forces Headquarters in Algiers for a year, and returned to the States to command the Signal Corps Ground Signal Agency at Bradley Beach, New Jersey. He also served a short time as the commanding officer of the Signal Corps Labs at Fort Monmouth. After the war he returned to the Pacific and served in the Philippines-Ryukyus Command in Manila.

He later served in the Office of the Chief Signal Officer and in the Office of the Joint Chiefs of Staff in Washington.

Conrad returned to Fort Monmouth in July 1954 as commandant of the Signal School, but only served a week. He was acting Commanding Officer during General Lawton's convalescent leave. Conrad assumed command of Fort Monmouth on 1 September 1954 and served in that capacity until June 1957.

General Conrad was placed on the Temporary Disability Retired List on 1 April 1960, and on 1 July 1963 he was placed on the Army Retired List in the grade of Major General. He died in December 1964.

> *MG Conrad commanded Fort Monmouth the year that the Army deactivated the pigeon service. Pigeons sold here for $5 a pair in 1957, with "hero" pigeons going to zoos.*

17
★★

Major General Kirke B. Lawton
DECEMBER 1951 - AUGUST 1954

Major General Kirke B. Lawton became the seventeenth Commanding Officer of Fort Monmouth in December 1951.

Lawton was born in November 1894 in Athol, Massachusetts. He graduated from Worcester Polytechnic Institute in 1917. He entered the U.S. Army in August 1917 after being commissioned a Second Lieutenant of Infantry, Regular Army.

He first came to the Signal Corps and Camp Vail in January 1920 for signal training. He remained five years at Camp Vail with duty in the 15th Service Company and completed the Signal School Course in 1924. After a tour of duty in the Canal Zone he returned to the Signal School, remaining until the summer of 1931.

During World War II General Lawton served as director of the Army Pictorial Service in the Office of the Chief Signal Officer, and later was assigned to the Public Relations Division of the Supreme Headquarters, Allied Expeditionary Forces in Europe.

Lawton returned to Fort Monmouth in December 1951 as Commanding Officer and served until August 1954, when he retired from the U.S. Army. General Lawton died in October 1979.

> *Over 8,000 troops participated in the garrison review honoring MG Lawton's retirement.*

16
★

Brigadier General Harry Reichelderfer
APRIL 1951 – DECEMBER 1951

Brigadier General Harry Reichelderfer became the sixteenth Commanding Officer of Fort Monmouth in April 1951.

Reichelderfer was born in Peoria, Illinois, in March 1896. He attended the University of Illinois for three years and subsequently graduated from Yale University. He received an MS degree from Yale in 1928. He entered the U.S. Army in May 1917 and was commissioned a Second Lieutenant of Infantry, Regular Army.

He served with the American Expeditionary Forces (27th Infantry Regiment) in Siberia 1919-1920. Reichelderfer transferred to the Signal Corps in March 1921 and completed the Signal School at Fort Monmouth in 1925. In the following years he saw duty with the Sound Labs at Fort H. G. Wright, New York; Signal Corps Repair Section of Rockwell Air Depot, California; and Aircraft Radio Lab at Wright Field, Ohio.

During WWII he was the signal officer of the Third Army in Texas and the Sixth Army in the Pacific from New Guinea to Japan.

General Reichelderfer returned to Fort Monmouth in April 1949 when he became Commanding General of the Signal Corps Engineering Lab. He assumed command here in April 1951 and served in that capacity until December 1951. General Reichelderfer retired from the Army in November 1956 and died in August 1973.

> *BG Harry Reichelderfer enthusiastically supported memorialization efforts during his brief tenure as commander, even participating in the planning for the WWII memorial.*

15 ★★

Major General Francis H. Lanahan
JUNE 1947 - APRIL 1951

Major General Francis H. Lanahan became the fifteenth Commanding Officer of Fort Monmouth in June 1947.

Lanahan was born in October 1897 in Trenton, New Jersey. He served a year in the New Jersey National Guard before entering the U.S. Military Academy. He graduated with the Class of 1920 as a Second Lieutenant in the Coast Artillery Corps, but shortly thereafter transferred to the Field Artillery. General Lanahan first came to Fort Monmouth as a First Lieutenant for duty with the 15th Signal Service Company in September 1926. A year later he entered the Signal Corps School at Fort Monmouth, and graduated in June 1928. He was named Acting Director of the Department of Training Literature and Assistant Signal Officer of the Signal School. He was transferred from the Field Artillery to the Signal Corps on 21 November 1929. He subsequently served as Chief Signal Officer at Fort Benning, Georgia, and Chief Signal Instructor at Fort Leavenworth, Kansas.

During World War II he was assigned as duty Chief Signal Officer at Supreme Allied Headquarters, and promoted to Chief Signal Officer in 1945. On the dissolution of the Combined Headquarters on 15 July 1945 he was appointed Chief Signal Officer of the United States Forces in the European Theater, a position he occupied until he returned to the States in May 1947. General Lanahan returned to command Fort Monmouth from June 1947 - April 1951. General Lanahan retired from the Regular Army in March 1955 and entered private industry. He died in December 1975 at his home in Hillsdale, New Jersey.

> *General Lanahan initiated a new round of construction on 23 December 1948 to alleviate the serious shortage of quarters for enlisted men and officers. This would include housing on Megill Drive, overlooking the golf course; on Hope Road, and the Wherry Housing area on the southern edge of Camp Charles Wood.*

14 ★

Brigadier General Jerry V. Matejka
APRIL 1946 - JUNE 1947

Brigadier General Jerry V. Matejka assumed command of Fort Monmouth on 15 April 1946.

Matejka was born in Texas in August 1894. He graduated from the University of Texas in 1916 and received a regular commission as a Second Lieutenant in the Coast Artillery Corps. He was detailed to the Signal Corps in 1920, served in the Panama Canal Department, and then graduated from the Signal School in 1930. In August 1940 General Matejka was assigned to General Headquarters of the United States Army, and in May 1941, as a member of the Special Observers Group, was transferred to the United Kingdom. He subsequently was named Chief Signal Officer of the European Theater of Operations. After returning to the States, in July 1943, Matejka had a tour of duty in the Office of the Chief Signal Officer.

General Matejka came to Fort Monmouth as Commanding General of the Eastern Signal Corps Training Center in December 1945. Upon deactivation of the Center, Matejka became the fourteenth Commanding Officer of this post and served in that capacity until June 1947.

General Matejka retired as a Major General on 31 October 1955 after more than thirty-eight active years of Army service. He died in May 1980.

> *MG Matejka was the last Commanding Officer of the Eastern Signal Corps Training Center.*

13

Colonel Leon E. Ryder
NOVEMBER 1944 - APRIL 1946

Colonel Leon E. Ryder became the thirteenth Commanding Officer of Fort Monmouth on 27 November 1944.

Ryder was born in Maine in July 1891. He graduated from Norwich University in Vermont in 1916 and was commissioned a Second Lieutenant of Cavalry, Regular Army. He had prior service in the Maine National Guard. During the early 1920s he had three tours of duty with the Signal Corps at the Presidio of San Francisco. He transferred from the Cavalry to the Signal Corps in 1928 and, in 1929, completed the Signal Corps School Course. He became Executive Officer of Fort Monmouth in 1931, and served until his 1934 transfer to Governor's Island, New York. Ryder returned to Fort Monmouth in May 1943 for duty at the Signal Corps School.

Upon deactivation of the Eastern Signal Corps Training Center on 1 April 1946, its commander, Brigadier General Jerry V. Matejka, assumed command of Fort Monmouth, succeeding Ryder. Ryder remained at Fort Monmouth until his retirement from the Army on 30 November 1946. He died in February 1959.

> *COL Ryder commanded Fort Monmouth when on 10 January 1946 our Camp Evans facility took the first electronic step into space and bounced a radar signal off the moon using a modified SCR-271. It took the Diana radar just 2-1/2 seconds to reach the moon and return.*

12
Colonel James B. Haskell
SEPTEMBER 1942 - NOVEMBER 1944

Colonel James B. Haskell became the twelfth Commanding Officer of Fort Monmouth in September 1942.

Haskell was born in July 1891 in Minnesota. He graduated from the United States Military Academy in 1914 and was commissioned a Second Lieutenant in the Coast Artillery Corps. Haskell served as a temporary Major in France during World War I on General Pershing's staff. He transferred to the Signal Corps in 1924 and completed the Signal Corps Company Officers' Course the same year.

Colonel Haskell came to Fort Monmouth in September 1940 as a Lieutenant Colonel from instructor duties with the Georgia Tech Reserve Officers Training Corps. When the Replacement Center was activated in February 1941, he was appointed Executive Officer under then Colonel Van Deusen. He was assigned to the Signal Board in September 1941, was promoted to Colonel in February 1942, and became Executive Officer of Fort Monmouth.

When Brigadier General Van Deusen was assigned as the Commanding General of the Eastern Signal Corps Training Center, Haskell assumed command of Fort Monmouth in September 1942. He served as Commanding Officer until November 1944. Colonel Haskell retired from the Army on 31 August 1946.

COL Haskell served as Executive Officer of the Signal Corps Training Center when it was activated in February 1941.

11
Brigadier General George L. Van Deusen
AUGUST 1941 - SEPTEMBER 1942

(Then) Brigadier General George L. Van Deusen became the eleventh Commanding Officer of Fort Monmouth in August 1941.

Van Deusen was born in Passaic, New Jersey, in February 1888. He graduated from the United States Military Academy in June 1909 and was appointed a Second Lieutenant of Infantry. While he later became a Coast Artillery Corps Officer, his first assignment with the Signal Corps came when he commanded the 105th Signal Battalion at Camp Sevier, South Carolina, in 1917. He served in France with the 30th Division in 1918. Subsequent short tours of duty with the Signal Corps in the Office of the Chief Signal Officer followed.

Van Deusen transferred from the Coast Artillery to the Signal Corps in July 1920 and came to Fort Monmouth, then Camp Vail, for the first time in June 1921 as Assistant Commandant of the Signal School. Van Deusen returned to Fort Monmouth for various tours of duty in 1925, 1937 and 1940. He was assigned to command the new Signal Corps Replacement Training Center in January 1941, as a Colonel.

He became a Brigadier General in April 1941, and assumed command of Fort Monmouth that August. He concurrently served as commandant of the Signal Corps School from October 1941 to September 1942. He was succeeded as Commanding Officer of the Post in September 1942 by Colonel James B. Haskell and as commandant of the Signal Corps School by Colonel W.O. Reeder.

When the Eastern Signal Corps Training Center (ESCTC) was established at Fort Monmouth, Van Deusen became the first Commanding General of the Center in October 1942. He was promoted to the rank of Major General in May 1944 and served as Commanding General of ESCTC until December of that year. Major General Van Deusen became the Chief of Engineering and Technical Service in the Office of the Chief Signal Officer in Washington in January 1945. He retired from the Army on 31 August 1946 and died 12 January 1977.

Fort Monmouth dedicated the Post Library the "Van Deusen Library" in honor of General Van Deusen on 21 June 1977.

10
Brigadier General Dawson Olmstead
AUGUST 1938 - JULY 1941

(Then) COL Dawson Olmstead became the tenth Commanding Officer of Fort Monmouth in August 1938.

Olmstead was born in Corry, Pennsylvania, in May 1884. He graduated from the U.S. Military Academy in the class of 1906 and was commissioned a Second Lieutenant of Cavalry.

He attended the Signal School at Fort Leavenworth, Kansas, and graduated in July 1909. He served in the Office of the Chief Signal Officer in Washington from 1909 to 1912. Olmstead served for a short time in 1917 as Signal Officer, 83d Division, Camp Sherman, Ohio, and served from June to September 1918 in the Office of the Inspector General, GHQ, Chaumont, France.

Olmstead had five tours of duty with the Signal Corps during the 1920s and 1930s and two additional tours in the Office of the Chief Signal Officer. He came to Fort Monmouth as a Colonel in June 1938 for duty as commandant, Signal Corps School. In August 1938 he assumed command of Fort Monmouth. He was promoted to Brigadier General in October 1940, thus becoming the first general officer to serve as commander. General Olmstead served at Fort Monmouth until July 1941. He subsequently received his second star and became the Chief Signal Officer of the Army.

General Olmstead retired from the Army in January 1944 and died in September 1965.

The Olmstead Gardens, which were located on the south side of Tinton Ave, were constructed in the late 1950s/early 1960s to accommodate enlisted Soldiers and their families.

BG Dawson Olmstead was the first general officer to command Fort Monmouth.

Colonel Alvin C. Voris

APRIL 1937 - JULY 1938

Colonel Alvin C. Voris became the ninth Commanding Officer of Fort Monmouth on 30 April 1937

Voris was born in Illinois in January 1876. He entered the United States Army as a private in the 4th Regiment, Illinois Volunteers, in May 1898. He was commissioned a Second Lieutenant in the 4th Illinois Infantry on 23 December 1898. He served with the Signal Corps as early as 1903 and graduated from the Army Signal School in 1912. Voris served in France and Germany in World War I successively as Chief Signal Officer, I Corps, and Chief Signal Officer, Third Army, American Expeditionary Forces.

Colonel Voris assumed command of Fort Monmouth while concurrently serving as commandant of the Signal School until his retirement on 31 July 1938. He died in November 1952.

Voris Park, between Russel and Allen Avenues, was named in COL Voris' honor on 27 September 1957.

Colonel Arthur S. Cowan

SEPTEMBER 1929 - APRIL 1937

Colonel Arthur S. Cowan first commanded Camp Alfred Vail from September 1917 - June 1918. He was the third commander of the post at that time. Cowan, a graduate of the Class of 1899, U.S. Military Academy, later returned to Fort Monmouth to become commandant of the Army Signal School in June 1929. He then became the eighth Commanding Officer of Fort Monmouth that September while concurrently serving as commandant of the School.

Colonel Cowan's tenure as the eighth Commanding Officer of Fort Monmouth spanned the worst years of the Great Depression. In spite of this, or because of this, the second phase of permanent construction began. Field officers quarters; company officers quarters; a theater (now Kaplan Hall, Communications-Electronics Museum); an additional wing to the hospital (now Allison Hall); a Headquarters Building (Russel Hall); the Signal Corps Laboratory (now Squier Hall); a fire station; an incinerator, etc., were all completed for the Signal Corps by the Construction Division of the Quartermaster Corps and the Works Projects Administration (WPA) by the end of May 1936.

Colonel Cowan completed his service at Fort Monmouth in April 1937 and was transferred to Governor's Island, New York. He retired from the Army in April 1939, but was recalled for World War II. He again retired in 1942. Colonel Cowan died in June 1957. The area east of Russel Hall, containing the flagpole, became Cowan Park in June 1961.

Cowan Park, located inside the East Gate near Russel Hall, was dedicated on 24 June 1961 by General Orders Number 48.

COL Cowan served as the 3rd and 8th commander of Fort Monmouth.

Colonel George E. Kumpe

AUGUST 1926 - SEPTEMBER 1929

Colonel George E. Kumpe served as the seventh Commanding Officer of Fort Monmouth from August 1926 to August 1929.

Kumpe was born in May 1876 in Leighton, Alabama, and entered the United States Army as a Sergeant, 1st Montana Infantry, on 1 May 1898. He was honorably discharged on 23 March 1899 to accept a commission as Second Lieutenant in the 1st Regiment, Montana Infantry.

Kumpe served in the Philippines during the Spanish-American War and in France during World War I. He reported to Camp Alfred Vail for three months, September to December 1917, before leaving for France. He served in the Signal Office of the III Corps in Paris. He returned to Fort Monmouth as a Lieutenant Colonel in July 1926. In addition to serving as Commanding Officer from August 1926 to August 1929, he concurrently served as commandant of the Signal School.

Colonel Kumpe retired in August 1938 and died in November 1961.

During Kumpe's three year tour, the first phase of permanent construction began at Fort Monmouth. A number of field officers, company officers and NCO quarters; one BOQ; four family apartment houses; four barracks buildings and two wings of the Post Hospital had been completed before the end of 1929.

6

Colonel James B. Allison
AUGUST 1925 - AUGUST 1926

(Then) Colonel James B. Allison assumed command of Fort Monmouth and of the Signal Corps School in August 1925.

Allison was born in York, South Carolina, in September 1873. He graduated from the South Carolina Military Academy in 1895. He became a Second Lieutenant, 7th Infantry, Regular Army, on 4 November 1898. Allison's early assignments included a tour of duty in Alaska at the turn of the century and two tours of duty in the Philippines, on one of which he commanded a provisional battalion of Philippine Scouts. He graduated from the Army Signal School at Fort Leavenworth in 1908, and stayed there until 1909 as a company commander. For several months in 1917 he commanded the Signal Corps Training Camp at Monterey, California.

As a Colonel he became the sixth Commanding Officer of Fort Monmouth and commandant of the Signal Corps School from August 1925 to August 1926. He eventually received a promotion to Major General and was appointed Chief Signal Officer of the Army, serving in that capacity from January 1935 to September 1937.

MG Allison retired in September 1937 and died in March 1957.

Fort Monmouth dedicated Building 209 in memory of Major General Allison in September 1961. Allison Hall was completed in March 1928 as the first permanent hospital building.

5

Lieutenant Colonel John E. Hemphill
DECEMBER 1920 - AUGUST 1925

Hemphill was born in Canada in June 1867, and entered the United States Army as a Private in 1890. Within eight years he rose to the rank of First Sergeant. He received a temporary commission as a First Lieutenant in September 1899, serving until June 1901. He was again commissioned in the National Army as a Second Lieutenant of the Cavalry in July 1901. He served with the Signal Corps for four years during his early commissioned service, and finally transferred to the Signal Corps in August 1920. Lieutenant Colonel Hemphill served concurrently as post commander and as commandant of the Signal Corps School while here.

After stepping down from command in September 1925, he retired in June 1931 and died in August 1948. Hemphill Road between Hope Road and Guam Lane in the Charles Wood area was named in his honor, as were the parade grounds in the 1200 area.

The Hemphill Parade Ground, located just inside the West Gate (Johnston Gate), was dedicated on 8 June 1954 by General Orders Number 70.

LTC Hemphill proved instrumental in having Camp Alfred Vail designated as a permanent post named Fort Monmouth.

4

Colonel George W. Helms
JUNE 1918 - DECEMBER 1920

Helms was born in Virginia and graduated from the U.S. Military Academy in 1897. As an infantry officer, he served in the Philippines. He was detailed to the Signal Corps in June 1917 and commanded Camp Vail from June 1918 to December 1920. Helms was the first Commanding Officer of the post to be appointed commandant of the newly established Signal Corps School (1 October 1919 to 15 December 1920). He retired in November 1939 and died in November 1946. Fort Monmouth dedicated a semicircular residential street at the west end of Pine Brook Road in Eatontown Gardens as Helms Drive in March 1954.

Helms was the first Commanding Officer of the post to be appointed commandant of the newly established Signal Corps School.

3

Major Arthur S. Cowan

SEPTEMBER 1917 - JUNE 1918

Colonel Arthur S. Cowan first commanded Camp Alfred Vail from September 1917 - June 1918. He was the third commander of the post at that time. Cowan, a graduate of the Class of 1899, U.S. Military Academy, later returned to Fort Monmouth to become commandant of the Army Signal School in June 1929. He then became the eighth Commanding Officer of Fort Monmouth that September while concurrently serving as commandant of the School.

Colonel Cowan's tenure as the eighth Commanding Officer of Fort Monmouth spanned the worst years of the Great Depression. In spite of this, or because of this, the second phase of permanent construction began. Field officers quarters; company officers quarters; a theater (now Kaplan Hall, Communications-Electronics Museum); an additional wing to the hospital (now Allison Hall); a Headquarters Building (Russel Hall); the Signal Corps Laboratory (now Squier Hall); a fire station; an incinerator, etc., were all completed for the Signal Corps by the Construction Division of the Quartermaster Corps and the Works Projects Administration (WPA) by the end of May 1936.

Colonel Cowan completed his service at Fort Monmouth in April 1937 and was transferred to Governor's Island, New York. He retired from the Army in April 1939, but was recalled for World War II. He again retired in 1942. Colonel Cowan died in June 1957. The area east of Russel Hall, containing the flagpole, became Cowan Park in June 1961.

COL Cowan served as the 3rd and 8th commander of Fort Monmouth.

2

Major George E. Mitchell

JULY 1917 - SEPTEMBER 1917

(Then) Major George E. Mitchell became the Commanding Officer of the Signal Corps Camp at Little Silver, New Jersey, in July 1917.

He was born in Michigan in 1875 and gradated from the U.S. Military Academy in 1897.

Mitchell served in the Spanish-American War and graduated from the Army Signal School in 1910.

He served as commander, here, from July-September 1917.

Mitchell was appointed a temporary Colonel while en route to France in 1917. There, he served at General Headquarters, American Expeditionary Force, Chaumont.

Mitchell retired in October 1923 and died in May 1935.

Fort Monmouth named Mitchell Drive, a semi-circular residential street on the southeast side of Pinebrook Road, in Mitchell's honor in 1954.

1

Lieutenant Colonel Carl F. Hartmann

JUNE 1917 - JULY 1917

(Then) Lieutenant Colonel Carl F. Hartmann became the first Commanding Officer of Fort Monmouth, then known as the Signal Corps Camp at Little Silver, New Jersey, on 17 June 1917.

Hartmann was born in New York in 1868. Before entering military service he practiced law, having graduated from New York University Law School in 1893. He was commissioned as a Captain, Signal Corps, U.S. Volunteers in 1898. He received a promotion to Major in the Regular Army in 1912, and was sent to Fort Omaha, Nebraska the following year. He attended the Army War College, in Washington, D. C., in 1915.

His next assignment was as Signal Officer of the Eastern Department, with headquarters on Governors Island, New York. Promoted to Lieutenant Colonel, Regular Army on 12 April 1917, he was ordered to establish a Signal Corps Camp at Little Silver, New Jersey. He did so on 17 June 1917. He departed on 12 July 1917 to establish a similar Signal Corps Camp at Fort Leavenworth.

Hartmann subsequently served in France, where he set up a Signal Corps School at Langres, Haute Marne. He became a Colonel, Regular Army on 20 July 1920, with rank from 1 July 1920. He retired 27 September 1920 and died 8 July 1961.

Fort Monmouth's East Gate became the Hartmann Gate in June 1962.

AFTERWORD

The storied legacy of Fort Monmouth will live on in the hearts and minds of the dedicated military, civilian and contractor experts who helped to create this legacy and, perhaps as well, in those who have benefited from the technological advances and critical systems developed, acquired and sustained from here. As the command's communications and electronics mission transfers to Maryland in the years to come, the fort's loyal personnel will work tirelessly to ensure that support to the Warfighter goes on uninterrupted. The long history of advances in communications and electronics systems will be continued at Aberdeen Proving Ground by what MG Via refers to as the command's most important resource - its people. Although the relocation will entail significant challenges, fort personnel know what is at stake, and will rise to the challenge, just as they have with every conflict they have supported since WWI.

The loyalty and attachment many feel toward this unique community are evidenced by the significant number of senior personnel who have spent their entire careers here. Many originally planned to stay for a few years and move on, but in the words of one Camp Vail veteran "this place sort of gets into your blood" and their attachment to the mission and the people has only grown year after year.

This special community of scientists, engineers, program managers, logisticians and support staff has given the Army the world's best, most reliable systems for extracting, digesting, and communicating battlefield information. The capabilities these systems provide have given the American soldier and America's allies a decisive edge over their enemies and have contributed to saving countless lives from WWI to the GWOT.

Whether advancing technology, creating and modifying software, procuring hardware and services, accelerating fieldings, or rushing spare parts to the theater, the Fort Monmouth team has always gone above and beyond the call of duty. In the words of MG Via, "The command will continue to be at the forefront in providing our Nation's Warfighters with the very best command and control, communications, computers, intelligence, surveillance, reconnaissance and information systems in the world."

Appendix A

THE CHANGING FACE OF FORT MONMOUTH - AN ORGANIZATIONAL TIMELINE

1917- 1962: Signal Corps

Fort Monmouth began as a Signal Corps training camp in 1917, and continued to serve as the home of the Signal Corps until 1962. In 1949, the Signal Corps Center was established here and consolidated many existing Signal functions to include: the Signal Corps Engineering Laboratories, the Signal Corps Board, Signal School, Signal Corps Publications Agency, Signal Corps Intelligence Unit, Pigeon Breeding and Training Center, the Army portion of the Electro Standards Agency, and the Signal Corps troop units.

1962-1978: U.S. Army Electronics Command (ECOM)

In 1962, the Department of the Army abolished the Technical Services (including the Signal Corps) and established in their place the Continental Army Command (CONARC), the Combat Developments Command (CDC) and the Army Materiel Command (AMC). Though it retained its distinction as the "Home of the Signal Corps" for another decade, the materiel development and procurement functions of the Signal Corps at Fort Monmouth became the foundation for a new major subordinate command of AMC, the U. S. Army Electronics Command (ECOM). The management of Signal Corps personnel at Fort Monmouth was assigned to the Office of Personnel Operations; signal training was transferred to CONARC, and signal doctrine and combat development to the CDC.

1978-1981: CERCOM, CORADCOM

ECOM splintered into four new organizations on the recommendation of the Army Materiel Acquisition Review Committee (AMARC): the Communications-Electronics Materiel Readiness Command (CERCOM), the Communications Research and Development Command (CORADCOM), the Electronics Research and Development Command (ERADCOM), and the Avionics Research and Development Activity (AVRADA), a component of the new Aviation Research and Development Command. Only CERCOM and CORADCOM were to be headquartered at Fort Monmouth.

1981-present: U.S. Army Communications-Electronics Command (CECOM)

Reassessment of the organizational changes at Fort Monmouth concluded that, while the emphasis on research and development had increased, there was also much duplication of effort. Thus, on 1 March 1981, the Development and Readiness Command (DARCOM, formerly AMC) combined CERCOM and CORADCOM to form the new Communications-Electronics Command (CECOM), effective 1 May 1981. CECOM gained elements of ERADCOM in 1982, and AVRADA returned to CECOM in 1991.

1987-1988: Program Executive Offices (PEO) Stand-up

The Goldwater-Nichols Department of Defense Reorganiza-tion Act of 1986 is viewed as the most sweeping reform of the U.S. Military establishment since the DoD was established in 1949. Public Law 99-433, commonly called the Goldwater Nichols Reorganization Act, and National Security Decision Directive 219 formed the basis and direction of the reorganization. One of the provisions of these documents was to streamline the DoD acquisition process and to make it operate more like a commercial operation. The PEO concept grew from this guidance, restructuring the Army's organization for acquiring material. Three of the newly created PEO were at Fort Monmouth: PEO Communications Systems (PEO COMM), PEO Command and Control Systems (PEO CCS), and PEO Intelligence and Electronic Warfare Systems (PEO IEW&S).

PEO CCS merged with PEO COMM on 1 July 1995 to form the PEO for Command, Control and Communications Systems (C3S; later C3T).

**Today, PEO C3T and PEO IEW&S are part of the CECOM LCMC.

2002: Installation Management Agency (IMA)

The IMA was created on 22 August 2002. The Department of the Army mandated the transfer to IMA of all base operations support (BASOPS) functions. As a result of this decision, all CECOM personnel and spaces performing BASOPS functions, including the Fort Monmouth Garrison, transferred to IMA on 5 October 2003. IMA is now known as the Installation Management Command (IMCOM).

2002: Research, Development, and Engineering Command (RDECOM)

AMC Commander General Paul J. Kern directed the establishment of RDECOM, which stood up, provisionally, on 1 October 2002. The mission of this new command was to field technologies that sustained America's Army as the premier land force in the world. For Fort Monmouth, this resulted in the transfer of the CECOM RDEC to RDECOM in 2003. The CECOM RDEC became the RDECOM CERDEC. The RDECOM became official 1 March 2004 when the Army approved the RDECOM concept plan.

2004-present: CECOM Life Cycle Management Command (LCMC)

A 2 August 2004 memorandum of agreement between the Assistant Secretary of the Army for Acquisition, Logistics and Technology and the Commander, U.S. Army Materiel Command established a life cycle management initiative. It aligned AMC systems oriented major subordinate commands with the PEO with whom they already worked. At Fort Monmouth, this initiative formally aligned PEO IEW&S, PEO C3T, and CECOM. RDECOM remained a strategic partner in the LCMC concept. Operation order 05-01 spelled out the purpose, mission and execution of the U.S. Army Communications-Electronics Life Cycle Management Command (later renamed CECOM Life Cycle Management Command).

Appendix B

FORT MONMOUTH TENANT ACTIVITIES

CECOM LIFE CYCLE MANAGEMENT COMMAND

The mission of the CECOM Life Cycle Management Command (formerly called the C-E LCMC) is to develop, acquire, field, support and sustain superior command, control, communications, computers, intelligence, surveillance, reconnaissance and information systems for the joint Warfighter.

PROGRAM EXECUTIVE OFFICE FOR INTELLIGENCE, ELECTRONIC WARFARE AND SENSORS

The mission of PEO IEW&S is to develop, acquire, field, and provide for life cycle support of intelligence, electronic warfare, and target acquisition capabilities: integrated in the layers of the network, operationally relevant to understanding the battlefield, and enabling persistent surveillance. These capabilities are essential to set the conditions for the Joint Warfighter to control time, space, and the environment, while greatly enhancing survivability and lethality. We will accomplish this with continuous Warfighter focus to provide capability in the right place, right time, and at the best value for our Nation.

PROGRAM EXECUTIVE OFFICE FOR COMMAND, CONTROL AND COMMUNICATIONS TACTICAL

The mission of PEO Command, Control and Communications Tactical is to rapidly develop, field, and support leading edge, survivable, secure and interoperable tactical, theater and strategic command and control and communications systems through an iterative, spiral development process that results in the right systems, at the right time and at the best value to the Warfighter.

PROGRAM EXECUTIVE OFFICE FOR ENTERPRISE INFORMATION SYSTEMS

The Program Executive Office for Enterprise Information Systems has two Project Manager offices located at Fort Monmouth: the Project Manager, Network Service Center (PM NSC) and the Project Manager, Defense Communications and Army Transmission Systems (PM DCATS), as well as the Logistics Modernization Program Office and PD Common Hardware Enterprise Software Solutions. They are responsible for developing, acquiring and deploying tactical and non-tactical Information Technology systems and communications, with the goal of assuring victory through information dominance. PEO EIS provides DoD and the Army with network-centric knowledge-based business and combat service support systems and technology solutions. They provide the infrastructure and information management systems that support every Soldier, every day. PEO EIS assists with the accession and training of Soldiers, tracks the Army's personnel and medical information, provides and maintains Warfighters' equipment, and plans the movement of their supplies and assets.

COMMUNICATIONS-ELECTRONICS, RESEARCH, DEVELOPMENT AND ENGINEERING CENTER

The mission of the CERDEC is to develop and integrate Command, Control, Communications, Computers, Intelligence, Surveillance, and Reconnaissance (C4ISR) technologies that enable information dominance and decisive lethality for the networked Warfighter.

U.S. ARMY GARRISON FORT MONMOUTH

The mission of the U.S. Army Garrison Fort Monmouth is to provide base operations support, facilities, services, and well-being for the Fort Monmouth community.

PATTERSON ARMY HEALTH CLINIC

The mission of Patterson Army Health Clinic is to provide and coordinate high quality care for all of its beneficiaries in the highest tradition of military medicine, while promoting optimal health and maintaining readiness. Patterson Clinic is now home to Monmouth County's first Veterans Affairs Health Clinic.

UNITED STATES MILITARY ACADEMY PREPARATORY SCHOOL

USMAPS, established in 1945, moved to Fort Monmouth from Fort Belvoir, Virginia on 1 August 1975. The school prepares and trains selected enlisted members of the Army to qualify for admission to the United States Military Academy, and provides training that will assist them after they arrive at West Point. The school is open to enlisted members serving on active duty in the Army; to enlisted members of the Army Reserve and National Guard; and to civilians who are authorized by the Department of the Army to enlist in the Army Reserve for the purpose of attending the preparatory school. About 320 Soldiers enter USMAPS each year to compete for 170 appointments to West Point.

FEDERAL BUREAU OF INVESTIGATION

The agency's Northeast Regional Computer Support Center, Fort Monmouth, serves the FBI's largest field office--New York City--plus field offices in Albany, Boston, Newark, New Haven, Philadelphia and Richmond. The activity began with 25 personnel, mainly computer operators, and now currently has approximately 115 employees.

754TH EXPLOSIVE ORDNANCE DETACHMENT

The 754th Explosive Ordnance Disposal Detachment's mission is to train police, fire and public officials in explosive ordnance disposal and bomb threat search techniques, as well as to reduce the hazard of domestic or foreign conventional nuclear, chemical, biological and improvised explosive ordnance that personnel or outside activities may encounter.

MILITARY INTELLIGENCE DETACHMENT, ALPHA CO. 308TH M.I.B.N, 902D M.I GROUP

The 308th Military Intelligence Battalion conducts counterintelligence (CI) operations throughout CONUS to detect, identify, neutralize, and defeat foreign intelligence services (FIS) and international (IT) threats to U.S Army and selected Department of Defense forces, technologies, information and infrastructure. On order, the 308th reinforces designated unit(s) with CI and support personnel.

END NOTES

Unless otherwise stated, all resources can be found in the CE-COM LCMC Historical Archives, Fort Monmouth NJ.

[1] Rebecca Klang, *A Brief History of Fort Monmouth Radio Laboratories* (1942).

[2] Bingham, Richard. "*Fort Monmouth, New Jersey: A Concise History*" Communications-Electronics Command, (December 2002); George H. Moss Jr. and Karen L. Schnitzspahn, *Victorian Summers at the Grand Hotels of Long Branch, New Jersey* (Sea Bright: Ploughshare Press, 2000), 21.

[3] *This is Fort Monmouth* (Fort Monmouth: 1950); Moss and Schnitzspahn, *Victorian Summers*, 28; Pike and Vogel, *Eatontown*, 112.

[4] Personal letter from Major General Charles H. Corlett to Colonel Sidney S. Davis (3 December 1955).

[5] Ibid.

[6] Bingham, Richard. "Fort Monmouth, New Jersey: A Concise History." Communications-Electronics Command, (December 2002); Pike and Vogel, *Eatontown*, 80; "Passing of Monmouth Park.; Once Famous Race Course of Jersey Cut Up Into Building Lots. Special to The New York Times," *New York Times* (10 April 1910), 6; *An Archeological Overview and Management Plan for Fort Monmouth (Main Post), Camp Charles Wood and the Evans Area*, (1984); Lawrence Galton and Harold J. Wheelock, "A History of Fort Monmouth New Jersey 1917-1946," (Signal Corps Publications Agency, 1946), 12; Stenographic record of interview with COL Carl F. Hartmann, Signal Corps Retired, (26 October 1955) in the Office of the Chief Signal Officer; *An Archeological Overview and Management Plan for Fort Monmouth (Main Post), Camp Charles Wood and the Evans Area*, (1984); Historical Properties Report, Fort Monmouth, New Jersey and Sub installations Charles Wood Area and Evans Area, (July 1983).

[7] Authority of the Army Purchase Act, 25 February 1920.

[8] Stenographic record of interview with COL Carl F. Hartmann, Signal Corps Retired,(26 October 1955) in the Office of the Chief Signal Officer; Untitled manuscript, Communications Electronics Command Archives, Fort Monmouth; *History of Fort Monmouth*, 5.

[9] Ibid.

[10] "Shore Veteran Recalls Early Days of Post," *Monmouth Message* (18 May 1967), 14.

[11] Galton and Wheelock.

[12] Order 122, Office of the Chief Signal Officer, (21 August 1917). Born at Morristown NJ, in 1807, Alfred E. Vail graduated from the University of the City of New York in 1836 and early became associated with Samuel F. B. Morse. Vail's mechanical knowledge greatly expedited the first experiments in telegraphy. Some historians claim he devised the Morse alphabet of dots, dashes, and spaces. His automatic roller and grooved lever embossed on paper the characters that were transmitted. Vail was the superintendent of construction of the original telegraphy line between Washington and Baltimore. Inventor of the finger key, he received the first message successfully transmitted in 1844. In view of the great contributions made by Vail to wire communications, it was proper that his name be commemorated in a Signal Corps training camp.

[13] Stenographic record of interview with COL Carl F. Hartmann, Signal Corps Retired, (26 October 1955) in the Office of the Chief Signal Officer.

[14] Signal ROTC courses in prominent universities throughout the United States were also training radio operators and telegraphers. See *Historical Sketch of the Signal Corps*, Signal School Pamphlet No. 32 (Fort Monmouth, 1929).

[15] Galton and Wheelock, 20.

[16] Galton and Wheelock, 22-23.

[17] S.O. 139, War Department, 14 June 1918.

[18] "Pigeon that saves lost Battalion arrives home; is to receive D.S.C.," *New York Herald* (17 April 1919); Pigeon Archival Collection, CECOM LCMC Historical Office, Fort Monmouth NJ.

[19] Galton and Wheelock.

[20] "Fort Monmouth Locale Steeped in American Military Tradition," *Monmouth Message*, (18 May 1967), 12; Helen C. Phillips, *United States Army Signal School 1919-1967*, (Fort Monmouth: USA Signal Center and School, 1967).

[22] *Signal Corps Bulletin*, August 1926, 15.

[23] Phillips, 22.

[24] OC SIG O Letter to CO, FM, 12 August 1929.

[25] History Report of SC Engineer Labs, July 1930-December 1943.

[26] Galton and Wheelock, 30-31.

[27] Ibid, 31.

[28] Robyn Bennett, "Radio operators 'ham it up' to get messages through," *Monmouth Message* (29 July 2005), 7.

[29] PL 177, 69th Congress (Appropriations for Construction at Military Posts and for other Purposes).

[30] Ibid.

[31] Melissa Ziobro, "Fort Monmouth Summer Camp Trained Citizen Soldiers," *Monmouth Message* (14 September 2007).

[32] Galton and Wheelock; *History of Fort Monmouth, 1917-1953*; Bingham, Richard, "Famous Fort Monmouth Firsts," (1999).

[33] Bingham, Richard, "Famous Fort Monmouth Firsts," (1999).

[34] Phillips, 54.

[35] COL William R. Blair entered the U.S. Army in 1917. He had many tours of duty at Fort Monmouth and became Director of the Signal Corps Laboratories in 1930. Known as the Father of American Radar, Blair finally received a patent for the pulse echo technique in 1957. COL Blair retired in 1938 and died in 1962 at the age of 87.

[36] This was the first major development in the miniaturization of radio equipment.

[37] Personal papers of MG Roger B. Colton, CECOM LCMC Historical Office Archives, Fort Monmouth NJ.

[38] H. M. Davis, *History of the Signal Corps Development of U.S. Army Radar Equipment*. The field manual was WD, FM 11-25, GPO Wash: 1942.

[39] Harry H. Woodring, War Department letter to Major General Allison (2 June 1937).

[40] Harold Zahl. "Electrons Away or Tales of a Government Scientist," (New York: Vantage Press, 1968), 45-46.

[41] "History of the Radar School" written for the Army Signal Association Journal (1946).

[42] Oral history interview with Mr. Harold Tate, Edited by Robert Johnson Jr., Framingham State College (14 October 1993).

[43] Fred Carl, "SCR-270 Radar Spots Japanese Planes 50 minutes before attack," Available: Internet <http://www.infoage.org/pearl.html> [December 2005].

[44] Harold Zahl, "History of Radar," Draft Manuscript (13 April 1954), 1-47.

[45] "Col. William R. Blair, SIGC 1917-1938."

[46] "Father of Radar-It's Official," *Electronic Week* (26 August 1957)

[47] "Suit over Radar file," *AP*, date unknown.

[48] "Court gets Blair-Bendix radar suit," *Asbury Park Press* (10 April 1962).

[49] "Radar Patent granted to retired Army Officer," *Department of Defense Office of Public Information News Release No. 839-57* (20 August 1957).

[50] David Van Keuren, "The military context of early American radar," *Tracking the History of Radar* (New Jersey: IEEE Rutgers Center for History of Electrical Engineering and Deutsches Museum, 1994), 140-152

[51] Recollections of Patricia W. Blair (Daughter in law of Colonel William R. Blair) (18 September 2002).

[52] Harold A. Zahl, "One Century of Research," *U.S. Army Signal Research and Development Laboratory* (15 December 1960).

[53] "Federal Business Assn. cites Dr. Harold Zahl," *Monmouth Message* (11 June 1964).

[54] Roger B. Colton, "Lieutenant Colonel Harold Zahl, Signal Corps," *Memorandum to Commanding Officer Signal Corps Engineering Laboratories Bradley Beach, NJ* (17 October 1945).

[55] William I. Orr, "The secret tube that changed the war," *Popular Electronics* (March 1964), 57-59, 103-105.

[56] Gunter David, "Personalities in the News," *Newark Sunday News* (3 February 1963).

[57] "Impossible just a few decades away, Monmouth Researcher says," *Asbury Park Press* (31 March 1963).

[58] Harold A. Zahl, "Department of the Army nominee for National Civil Service League Career Service Award" Zahl biography file.

[59] John Marchetti Oral History interview, Conducted by Fred Carl (1999).

[60] "Army radio transmitter Pioneer closes 34 years of Government service," *Army Research and Development Newsmagazine (April 1963)*, 12.

[61] "John Joseph Slattery," *Who's Who in SCEL R&D* (1953).

[62] Roger B. Colton Distinguished Service Medal Citation.

[63] Vollum Oral History Interview, Oregon Historical Society (1980).

[64] Harold A. Zahl, *Electrons Away or Tales of a Government Scientist*, (New York: Vantage Press, 1968).

[65] Ibid.

[66] John Marchetti oral history interview, Conducted by Fred Carl (1999).

[67] Fred Carl, "From Camp Evans directly into battle: radar helped saved soldiers' lives in WWII," Available: Internet <www.Infoage .org> [11 November 2005].

[68] James A. Broderick, "Pioneer returns to Camp Evans Engineer helped develop radar," *Asbury Park Press* (11 January 1999).

[69] Jack Hansen biography file. Available in the CECOM LCMC Historical Archive, Fort Monmouth, NJ.

[70] Louis Brown, "Significant effects of radar on the second world war," *Tracking the History of Radar* (New Jersey: IEEE Rutgers Center for History of Electrical Engineering and Deutsches Museum, 1994), 121-133.

[71]Fred Carl, "Experts working in Marconi Hotel attic helped save WWII pilots," Available: Internet <www.infoage.org> [11 November 2005].

[72]John H. Bryant, "generations of Radar," *Tracking the History of Radar* (New Jersey: IEEE Rutgers Center for History of Electrical Engineering and Deutsches Museum, 1994), 1-14.

[73]Davis, *Abstract*.

[74]PL 806, 70th Cong 25 February 1929 and the Army Appropriation Act, PL 278, 71st Cong. (28 May 1930).

[75]PL 535, 71st Cong. (3 July 1930) and PL 718, 71 Cong (23 February 1931); TAGO Ltr, (29 September 1932); GO 221, (21 December 1953). HQ SC Center & FM.

[76]PL 302, 72nd Cong (Title III of the Emergency Relief and Construction Act of 1932, approved 21 July 1934; PL 67, 73rd Cong 16 Jun 1933 authorized all of the remaining permanent construction to 1936.

[77]National Defense Budget Estimates for FY 1984, Ofc Of Asst Sec Def (Comptroller) March 1983.

[78]"History of the U.S. Army Materiel Command Band," Available: Online , http://www.amc.army.mil/amc/band/CommandStaffHistory.html;

[79]Henry Kearney, "Farewell for band set next Thursday," *Monmouth Message* (5 August 1994).

[80]"Fort Monmouth Timeline," CECOM Life Cycle Management Command Archives, Fort Monmouth, New Jersey, Available: Online <http://www.monmouth.army.mil/historian/>

[81]"Federal Business Association Cites Dr. Harold A. Zahl," *Monmouth Message*, (11 June 1964).

[82]At the time, Shrewsbury Township encompassed all of what is today the Borough of Tinton Falls. Incorporation of the latter (first, as New Shrewsbury) left only the Vail Homes area to constitute Shrewsbury Township.

[83]BG D. Olmstead served as the tenth Commanding Officer of Fort Monmouth until July 1941. He was subsequently promoted to Major General and became the Chief Signal Officer of the Army from October 1941 to June 1943.

[84]GO 11, HQ Fort Monmouth.

[85]Camp Charles Wood is bound on the north by Tinton Avenue, on the east by Maxwell Place, on the south by Pine Brook Road, and on the west by Pearl Harbor Road.

[86]GO 28, (3 July 1942), HQ Fort Monmouth, and War Dept GO 58, (29 October 1942). Camp Charles Wood was named in honor of LTC Charles W. Wood, SC, and redesignated the Charles Wood Area in 1958. LTC Charles Wood was assistant executive officer of Fort Monmouth. He died suddenly on 1 Jun 1942 while on temporary duty in Washington. Wood had retired from the Army in 1937 because of illness. He was recalled to service in Oct 1940, and served as post signal property officer at Fort Monmouth and later as assistant executive officer.

[87]The Allaire Training Area (on Monmouth County Highway 524) was approximately midway between Farmingdale and Manasquan.

[88]Helen Phillips, *History of the Signal School 1919-1967*.

[89]Galton and Wheelock, 84.

[90]Melissa Ziobro, "Fort Monmouth Drafted Local Sites, Including Famed Asbury Park Convention Hall," *Monmouth Message* (8 June 2007).

[91]Fifty-two of the 490 were denied admission and reclassified.

[92]WD SO 274, 9 October 1942.

[93]Melissa Ziobro, "Gender Integration of the Army advanced at Fort Monmouth," *Monmouth Message* (9 March 2007); Jamie Mazza, "Women served America and served here," *Monmouth Message* (19 August 2005); "WAC detachment fully absorbed by Army of Sept 1," *Signal Corps Message* (3 September 1943); "Vanguard of WAAC arrives to plan invasion of Fort," *Signal Corps Message* (9 April 1943); "Post WACs, on job at Quebec Parley, Back in 15th, 803d," *Signal Corps Message* (17 September 1943); Signal Corps Information Letter (19 June 1943); Signal Corps Information Letter (21 Aug 1943);

[94]SCL 394, 30 September 1941; Field Lab No. 1, was dedicated as Camp Coles, 1 October 1942, in honor of COL Ray Howard Coles, Assistant to and Executive Officer for the C Sig O, AEF, World War I. By WD, GO 24, 6 Apr 1945, Camp Coles was redesignated Coles Signal Lab. It was rededicated Coles Area 18 December 1956, when the USA Signal Equipment Support Agency occupied the site.

[95]William R. Stevenson, *Miniaturization and Micro Miniaturization of Army Communications-Electronics 1946-1964* (Fort Monmouth: Headquarters and U.S. Army Electronics Command, 1966). Historical monograph 1, project Number AMC 21M; Max Marshall LTC, *The Story of the U.S. Army Signal Corps*, (New York: Franklin Watts Inc., 1965).

[96]Ibid.

[97]Personal Paper of MG Roger B. Colton; Dulany Terrett, *U.S. Army in World War II. The Technical Services. The Signal Corps: The Emergency* (Office of the Chief of Military History, Washington D.C., 1956).

[98]Ibid.

[99]Radio Free Europe was founded in 1949 by the National Committee for a Free Europe. It is a communications organization funded by the U.S. Congress.

[100]Pigeon collection, CECOM LCMC Historical Archive, Fort Monmouth NJ.

[101]Special to The *New York Times* from Fort Monmouth, the Division of Press Intelligence, (8 August 1941) CECOM LCMC Historical Archives, Fort Monmouth, NJ.

[102]Wendy Rejan, "Fort's foray into fighting falcons uncovered," *Monmouth Message* (25 May 2007).

[103]"Like Italian 'Signees' to visit your home? Here's how to apply," *Signal Corps Message* (17 July 1944); "Signees Pro-Ally, Not Prisoners, says Col Haskell," Signal Corps Message (4 August 1944).

[104]Kenneth J. Clifford, *Commanding Officers of Fort Monmouth,* (Fort Monmouth: Command Historian's Office, 1984).

[105]Ed Reiter, "Scientist recalls Fort's '46 Moon Project," *Asbury Park Press* (20 July 1969), 4.

[106]David G. Buchanan and John P. Johnson, *Historic Properties Report* (Draft), Building Technology and National Park Service, U.S. Department of Interior, (July 1983).

[107]"McAfee, Walter Samuel," *Who's Who in SCEL R&D* 53.

[108]Wilhemina Mitchell, "Black leaders: Many Contributed to communications achievements," *Monmouth Message* (26 February 1982); "Walter McAfee, helped boost U.S. into space," *Asbury Park Press* (21 February 1995); "Pioneer communications expert John Dewitt Jr. dies," *Reuters* (26 January 1999).

[109]Stevenson, William R., *Miniaturization and Micro-miniaturization of Army Communications-Electronics 1946-1964*, (Fort Monmouth, NJ, 1966); History of Fort Monmouth, 1917-1953; Bingham, Richard, "Famous Fort Monmouth Firsts" (1999).

[110]"Federal Business Association Cites Dr. Harold A. Zahl," *Monmouth Message*, 11 June 1964.

[111]David, Gunther. "Personalities in the News," *Newark Sunday News*, (3 February 1963).

[112]Stevenson; Marshall; Historical Report of the Signal Engineering Laboratories, for the Period July 1930 to December 1943; Richard Bingham, "Famous Fort Monmouth Firsts" (1999).

[113]Marshall; Stevenson; Historical Report of the Signal Engineering Laboratories, for the Period July 1930 to December 1943; *History of Fort Monmouth,* 1917-1953; Bingham, Richard, "Famous Fort Monmouth Firsts" (1999).

[114]Marshall; Richard Bingham, "Famous Fort Monmouth Firsts" (1999).

[115]Ibid.

[116]Stevenson; Marshall; Richard Bingham, "Famous Fort Monmouth Firsts" (1999).

[117]History Report of SCEL, July 1930 to December 1943, 6.

[118]Wendy Rejan, "Beloved, beleaguered cartoon character born here," *Monmouth Message* (15 June 2007).

[119]*Signal Corps Message*, (18 June 1943), 7.

[120]William Hewlett, Electrical Engineer, an oral history conducted in 1984 by Michael McMahon, *IEEE History Center*, Rutgers University, New Brunswick, NJ, USA.), Available: Online < http://www.ieee.org/portal/cms_docs_iportals/iportals/aboutus/history_center/oral_history/pdfs/Hewlett046.pdf>

[121]Melissa Ziobro, "Whitey Ford and the famous Fort Folk," *Monmouth Message* (4 June 2004).

[122]Thomas M. Pryor, "Army film unit's needs and other items," *New York Times* (27 July 1941).

[123]*Signal Corps Message*, (18 June 1943), 7.

[124]Melissa Ziobro, "Whitey Ford and the famous Fort Folk," *Monmouth Message* (4 June 2004).

[125]Jean Shepherd biography file, CECOM LCMC Historical Office, Fort Monmouth NJ.

[126]Phillips, 204.

[127]GO 35, DA, 3 August 1949; GO 67, HQ Fort Monmouth, NJ (22 August 1949), the last order of that Headquarters.

[128]Phillips, 219; Ibid., 220.

[129]"AN/MPQ-10A Radar set," *U.S. Army Fact Sheet Radar No. 3 Office of the Chief of Information* (May 1971).

[130]Robert Johnson, "No Short Climb: Race Workers & America's Defense Technology," Documentary (2006).

[131]Ibid., 241.

[132]Buildings 1204, 1205, 1212, now occupied by USMAPS since August 1975; Myer Hall named in honor of the founder of the Signal Corps, Chief Signal Officer, 1860-1863, 1866-1880; Myer Hall dedicated 11 September 1953; The auditorium and Myer Hall, occupied by USA Chaplain School June 1980 and USA Chaplain Board, September 1979.

[133]Patterson Army Hospital officially opened 17 March 1958 and was dedicated 17 April 1958 in honor of MG Robert Urie Patterson, Surgeon General of the Army 1931-1935.

[134]Buchanan and Johnson, 72, 81, 84.

[135]A proximity fuse is used to detonate a missile in way that provides for the greatest amount of damage to the target. For a history of the development of the proximity fuse, see William T. Moye, "Developing the Proximity Fuze, and Its Legacy," Available: Online <www.amc.army.mil/amc/ho/studies/fuze.html>

[136]Testimony of Edward J. Fister, Executive Sessions of the Senate Permanent Subcommittee on Investigations of the Committee on Government Operations, 83rd Congress 1953, Vol 3, (U.S Government Printing Office, Washington, 2003), 2186; Testimony of James Evers, Vol 3, 2253.

[137]Peter Kihss, "Monmouth security woes antedate McCarthy visits," *New York Times* (11 January 1954).

[138]Testimony of Allen Lovenstein, Executive Sessions of the Senate Permanente Subcommittee on Investigations of the Committee on Government Operations, 83rd Congress 1953, Vol 3, (U.S Government Printing Office, Washington, 2003), 2169; Testimony of Fred B. Daniels, Vol 3, 2226; Testimony of Craig Crenshaw, Vol 3, 2386.

[139]Donald A. Ritchie, Senate Historical office, Volume I, Executive Sessions of the Senate Permanente Subcommittee on Investigations of the Committee on Government Operations, 83rd Congress 1953, (U.S Government Printing Office, Washington, 2003), xix.

[140]William J. Jones, oral history, edited by Robert Johnson (2 December 1993).

[141]Testimony of William P. Goldberg, Executive Sessions of the Senate Permanente Subcommittee on Investigations of the Committee on Government Operations, 83rd Congress 1953, Vol 3, (U.S Government Printing Office, Washington, 2003), 2193.

[142]Robert Johnson, "No Short Climb: Race Workers & America's Defense Technology," Documentary (2006).

[143]Harold Tate, Oral history Interview, edited by Robert Johnson (14 October 1993), Available: Online <http://www.infoage.org/tate.html>

[144]For further information on the role of Joel Barr and Alfred Sarant, see *Engineering Communism: How Two Americans Spied for Stalin and Founded the Soviet Silicon Valley*, by Steven T. Usdin (Yale University Press, 2005).

[145]Testimony of Major General Kirk B. Lawton (14 and 16 Oct 1953), Executive Sessions of the Senate Permanente Subcommittee on Investigations of the Committee on Government Operations, 83rd Congress 1953, Vol 3, (U.S Government Printing Office, Washington, 2003), 2473-2486.

[146]Testimony of Alan Sterling Gross, Executive Sessions of the Senate Permanente Subcommittee on Investigations of the Committee on Government Operations, 83rd Congress 1953, Vol 3, (U.S Government Printing Office, Washington, 2003), 2219.

[147]David M. Oshinsky, "Fort Monmouth and McCarthy. The Victims Remember," *New Jersey History*, Vol 100, Nos 1-2, (Spring/Summer 1982); Rebecca Raines, "The Cold War Comes to Fort Monmouth, Senator Joseph R. McCarthy and the Search for Spies in the Signal Corps," *Army History*, no. 44 (Spring 1998), 8-16; J. Scott Orr and Robert Cohen, "McCarthy's Assault on Fort Monmouth. Transcripts reflect his 53' attacks," *Star Ledger* (6 May 2003); Shirley Horner, "McCarthy's purge at Fort is recalled," *New York Times* (3 July 1983); Walter McAfee oral history conducted by Robert Johnson (6 February 1994), Available: Online <http://www.infoage.org/mcafee.html>

[148]Marshall; Stevenson; Amory H. Waite Jr., *Radio Ice Depth Measurements, Papers and History Relating Thereto,* (unpublished original manuscript and collection of documents). CECOM LCMC Historical Office.

[149]DA GO 47, 24 June 1954.

[150]GO 3, HQ USASRDL, 5 February 1959.

[151]GO30, HQ USASRDL, 31 December 1958.

[152]Robert Johnson, "No Short Climb: Race Workers & America's Defense Technology," Documentary (2006).

[153]Melissa Ziobro, "Fort Monmouth touted as Black Brain Center of the U.S," *Monmouth Message* (16 February 2007); Thomas E. Daniels, "Contributions of black Americans to Electronic research development, production distribution, and training at Fort Monmouth 1940-1982," Available: Online <http://www.infoage.org/Daniels-1988-cover.html>; Robert Johnson, "No Short Climb: Race Workers & America's Defense Technology," Documentary (2006).

[154]Melissa Ziobro, "Our Deal Area First to Record Sputnik Signals," *Monmouth Message* (19 October 2007).

[155]Marshall; Leonard D. Berringer, "Atomic Time standards," *Instruments and Control Systems 39* (June 1966), 99.

[156]Release #486, Headquarters Signal Corps Center and Fort Monmouth (22 October 1951).

[157]Richard Bingham, *Fort Monmouth, New Jersey: A Concise History*, (2002).

[158]Richard Bingham, "Radio and the Digitized Battlefield," (2000).

[159]Biography files (William R. Blair), Communications-Electronics Life Cycle Management Command Archives, Fort Monmouth, New Jersey.

[160]Marshall; Stevenson; Waite, Amory H., Jr., "Radio Ice Depth Measurements, Papers and History Relating Thereto…" unpublished original manuscript and collection of documents, CECOM History Office; Bingham, Richard, "Famous Fort Monmouth Firsts" (1999).

[161]Marshall; *Historical Sketch of the United States Army Signal Corps 1860-1967*, Signal School History Office, Fort Monmouth, NJ, 1967; Richard Bingham, "Famous Fort Monmouth Firsts" (1999).

[162]Marshall; *Historical Sketch of the United States Army Signal Corps 1860-1967*; Richard Bingham, "Famous Fort Monmouth Firsts" (1999).

[163]Ibid.

[164]Melissa Ziobro, "Whitey Ford and the Famous Fort Folk," *Monmouth Message* (4 June 2004).

[165]Martin Blumenson, Reorganization of the Army 1962, (OCMH Monograph 37M, July 1965).

[166]OC SIG O to CG Fort Monmouth 9 July 1962. ECOM actually activated by OC Sig O, 23 May 1961. ECOM officially organized 1 August 1962. HQ AMC G05, 26 July 1962.

[167]HQ ECOM GO46, 26 June 1964.

[168]HQ ECOM GO54, 26 June 1964.

[169]HQ ECOM GO3, 4 January 1965; HQ ECOM GO12, 25 Feb 1965.

[170]ECOM. GO28, 27 May 1965; ECOM GO39, 27 May 1965; ECOM GO40, 27 May 1965.

[171]HQ ECOM GO29, 27 May 1965.

[172]ECOM sent an improved version of the AN/PRC-25, the AN/PRC-77, to Vietnam in 1968 by way of the 6th Armored Cavalry Regiment and the 82d Airborne Division. The AN/PRC-77 had an "X" mode for use with a new security device, the TSEC/KY-38.

[173]MO Pers Auth of Strength Rpt, Form 108-1, Force Dev; For a detailed history of ECOM's support to the Vietnam conflict see "U.S. Army Electronic Command Logistic support of the Army in Southeast Asia 1965-1970," Prepared by the Historian (23 Oct 1973) *Historical Monograph Project No. AMC 61M*.

[174]Only eighty-nine civilians elected to accompany the school to Fort Gordon; the remaining 700 retired or were reassigned to other agencies at Fort Monmouth.

[175]"Army Develops Method Which Measures Rain," *Armed Forces Press Service*, (31 December 1961).

[176]"Army develops mobile weather radar to track distant storms," *News Release No 155-62 Department of Defense Public Affairs* (31 January 1962).

[177]Marshall; *Historical Sketch of the United States Army Signal Corps 1860-1967*; CECOM LCMC Historical Archives, Fort Monmouth, New Jersey.

[178]Richard Bingham, *Fort Monmouth, New Jersey: A Concise History*, (2002).

[179]Ibid; Richard Bingham, "Famous Fort Monmouth Firsts" (1999).

[180]Ibid.

[181]Melissa Ziobro, "Infamous Mob Gangster once imprisoned at Fort Monmouth," *Monmouth Message* (24 September 2004).

[182]"A History of the United States Army Communications Electronics Command," Prepared by the CECOM Historical Office, March 1990.

[183]FY-78 Historical Review of the U.S. Army Electronics Research and Development Command, 4.

[184]MG Babers came to Fort Monmouth as the twenty-ninth CO and the second CG Of CERCOM in June 1980. He thus became the first CG of CECOM, assuming command effective 1 May 1981 (Auth para 3-1a, AR 600-20).

[185]*Monmouth Message*, 26 October 1984.

[186]CECOM News Release #064-84, 13 September 1984.

[187]*Monmouth Message*, 27 July 1984.

[188]*Monmouth Message*, 12 October 1984.

[189]*Monmouth Message*, 1 October 1984.

[190]Melissa Ziobro "Cold War Competition Heats up Innovation at Fort Monmouth: Part III," *Monmouth Message* (9 November 2007).

[191]*Monmouth Message*, 9 April 1980, 2.

[192]Richard Bingham, "Famous Fort Monmouth Firsts" (1999).

[193]Richard Bingham, *"Fort Monmouth, New Jersey: A Concise History,"* (2002).

[194]Ibid., 51.

[195]Richard Bingham, *CECOM and the War for Kuwait August 1990-March 1991*. Army Historical Program Monograph AMC 167 (May 1994); Richard Bingham, "Tradition of the Army Civilian Workforce," U.S. Army Communications-Electronics Command, Available: Online <http://www.monmouth.army.mil/historian/updates/14.htm>

[196]CECOM Annual Command History FY 90/91.

[197]Deputy Chief of Staff for Plans and Operations (DCSOPS) Business Integration Division (BID), CECOM history research conducted by BID staff (2003).

[198]Richard Bingham, "The Salomon Years 11 February 1994 - 31 March 1996." *CECOM History Program: Special Study 96-1* (11 April 1996), Available: Online <http://www.monmouth.army.mil/historian/updates/11.htm>

[199]Ibid.

[200]Ibid.

[201]Deputy Chief of Staff for Plans and Operations (DCSOPS) Business Integration Division (BID), CECOM history research conducted by BID staff (2003).

[202]Richard Bingham, "Tradition of the Army Civilian Work force," U.S. Army Communications-Electronics Command, Available: Online <http://www.monmouth.army.mil/historian/updates/14.htm>

[203]Deputy Chief of Staff for Plans and Operations (DCSOPS) Business Integration Division (BID), CECOM history research conducted by BID staff (2003).

[204]Ibid.

[205]Richard Bingham, "The Salomon Years 11 February 1994 - 31 March 1996." *CECOM History Program: Special Study 96-1* (11 April 1996), Available: Online <http://www.monmouth.army.mil/historian/updates/11.htm>

[206]Ibid.

[207]For further information see "The LRC Story: Going Beyond Expectations."

[208]Deputy Chief of Staff for Plans and Operations (DCSOPS) Business Integration Division (BID), CECOM history research conducted by BID staff (2003).

[209]Ibid.

[210]Ibid.

[211]Richard Bingham, "The Salomon Years 11 February 1994 - 31 March 1996." *CECOM History Program: Special Study 96-1* (11 April 1996), Available: Online www.monmouth.army.mil/historian/updates/11.htm

[212]Ibid.

[213]Deputy Chief of Staff for Plans and Operations (DCSOPS) Business Integration Division (BID), CECOM history research conducted by BID staff (2003).

[214]Richard Bingham, "The Salomon Years 11 February 1994 - 31 March 1996." *CECOM History Program: Special Study 96-1* (11 April 1996), Available: Online <http://www.monmouth.army.mil/historian/updates/11.htm>

[215]Ibid.

[216]*CERDEC 96*, 14; IEWD FY 95 Historical Report, 36.

[217]*CERDEC 96*, 15.

[218]LRC CCS/AV FY06 submission to the Annual Command History.

[219]Richard Bingham, "Tradition of the Army Civilian Workforce," U.S. Army Communications-Electronics Command, Available: Online <http://www.monmouth.army.mil/historian/updates/14.htm>

[220]Deputy Chief of Staff for Plans and Operations (DCSOPS) Business Integration Division (BID), CECOM history research conducted by BID staff (2003).

[221]Richard Bingham, "Tradition of the Army Civilian Workforce," U.S. Army Communications Electronics Command, Available: Online <http://www.monmouth.army.mil/historian/updates/14.htm>

[222]Richard Bingham, "The Salomon Years 11 February 1994 - 31 March 1996." *CECOM History Program: Special Study 96-1* (11 April 1996), Available: Online <http://www.monmouth.army.mil/historian/updates/11.htm>

[223]"U.S. Army Communications-Electronics Command Support of Operation Uphold Democracy," *CECOM Historical Office* (September 1994), Available: Online <http://www.monmouth.army.mil/historian/updates/16.htm>

[224]"U.S. Army Communications Electronics Command Support of Operation Vigilant Warrior," CECOM Historical Office (October 1994), Available: Online <http://www.monmouth.army.mil/historian/updates/15.htm>

[225]Deputy Chief of Staff for Plans and Operations (DCSOPS) Business Integration Division (BID), CECOM history research conducted by BID staff (2003).

[226]Richard Bingham, "Tradition of the Army Civilian Workforce," U.S. Army Communications-Electronics Command, Available: Online <http://www.monmouth.army.mil/historian/updates/14.htm>

[227]*CECOM News Report*, VII, i (Winter 1994), 7.

[228]Austin Chadwick, "Night Vision Shows the Way in Life-Fire Demo," *Monmouth Message*, (13 May 1994). The NVESD team included David Moody, Darryl Phillips, Wayne Smoot, David Tompkinson, and Taeho Jo under the leadership of Tom Bowman and with the support of contract employees from QuesTech and Fibertek; U.S. Army Communications-Electronics Command Annual Command History Fiscal Year 1994, 206.

[229]CECOM 94: *Research, Development, and Engineering Center*, n.d., 10; U.S. Army Communications-Electronics Command Annual Command History Fiscal Year 1994, 230.

[230]Sheila Greenwood, "First Milstar satellite launched," *Monmouth Message* (18 March 1994); Cleo Zizos, "General, Soldier first in Army to bounce voices over Milstar satellite," *Monmouth Message* (13 May 1994).

[231]Ibid.

[232]Information Systems Engineering Command, *OEF/OIF history report*, (1 June 2003).

[233]Debbie Sheehan, "Fort Monmouth is center of excellence," *Monmouth Message* (21 November 2003).

[234]*Monmouth Message*, October 2001-August 2003.

[235]CECOM OEF/OIF history and lessons learned reports (2003).

[236]DCSOPS BID, CECOM history research conducted by BID staff (2003).

[237]CECOM Annual Command History FY03.

[238]Ibid.

[239]Ibid.

[240]Ibid.

[241]Michael Mazzucchi, "LCMC Announcement from Major General Mazzucchi," 2 February 2005; Operation Order 05-01 (Form the Communications-Electronics Life Cycle Management Command – CE-LCMC), 2005.

[242]Fragmentary Order 07 to Operation Order 05-01, CECOM LCMC Fort Monmouth NJ (September 2007).

[243]Statistics updated by Rosemary Dellera, CECOM LCMC G3, *E Mail Correspondence* (11 September 2007).

[244]"Team C4ISR Overview," *Team C4ISR Knowledge Center*, (19 October 2006).

[245]*CECOM and the Global War on Terror.*

[246]"Team C4ISR Overview," *Team C4ISR Knowledge Center*, (19 October 2006).

[247]Heather McCooey, "LARS, Reset," *PPT briefing*, (2 November 2006).

[248]LRC, "Logistics Managers Meet at Ft. Monmouth," *Team C4ISR Knowledge Center* (11 May 2006).

[249]TYAD FY05 Historical submission to the Annual Command History.

[250]CECOM OEF/OIF history and lessons learned reports (2003).

[251]Harrison Donnelly, "Lifecycle Manager: Bringing Together All Facets of the C4ISR Lifecycle," *Military Information Technology Online Archives* (7 March 2005), Available: Online <http://www.military-information-technology.com/article.cfm?DocID=827>

[252]William White, CECOM LCMC LAR, LRC Readiness Directorate, LA Division, *E Mail Correspondence* (10 September 2007).

[253]"Logistics and Readiness Center Logistics and Engineering Operations Division Reset Program Mission," *Team C4ISR Knowledge Center*, (4 June 2004).

[254]"Team C4ISR Overview," *Team C4ISR Knowledge Center*, (19 October 2006).

[255]Ibid.

[256]Timothy Rider, "Army structures current, future networks into single program," *Fort Monmouth News Release #07-14* (27 June 2007).

[257]Rejan, Wendy. *CECOM and the Global War on Terror* (2003).

[258]CERDEC Nomination for the 2002 DA Research and Development Organization of the Year and RDO Awards.

[259]CERDEC, "Top Ten! AMC lauds center for one of the best inventions," *Monmouth Message* (12 December 2003).

[260]Rejan, Wendy. *CECOM and the Global War on Terror* (2003).

[261]"A brief history of AMC, 1962-2000," Available: Online <http://www.amc.army.mil/amc/ho/brief_history.htm>

[262]Herbert A. Leventhal, "Project Management in the Army Materiel Command 1962-1987," U.S. Army Materiel Command (1992), ix-xi.

[263]PEO COMM Annual Historical Review FY89.

[264]CECOM Historical Office, *Annual Historical Review of the Program Executive Offices, FY88* (Fort Monmouth: CECOM, 1988), 1.

[265]C-E LCMC Op Ord 05-01

[266]PEO COMM Annual Historical Review FY89.

[267]PEO CCS Annual Historical Review FY90.

[268]"Army Digitization Master Plan," GloabSecurity.org (2005), Available: Online <http://www.globalsecurity.org/military/library/report/1995/admp95-adoch1.htm>

[269]MG Steven W. Boutelle and Emerson Keslar, "New website combines technological wizardry and meaningful content," *Technological Innovations* (November-December 2000), Available: Online <http://www.dau.mil/pubs/pm/pmpdf00/boutn-d.pdf>

[270]Julius Simchick, 60 CECOM RCS CSHIS-G(R3) Annual Historical Review of the Program Executive Offices (1 Oct 1987- 30 Sep 1988).

[271]PEO IEW FY91 submission to the Annual Command History; PM J-STARS FY91 submission to the Annual command History.

[272]PM Signals Warfare FY91 submission to the Annual Command History.

[273]CECOM History Office, "PEO IEW Annual Historical Review" (1 Oct 1989-30 Sept 1990)

[274]Wendy Rejan, *C-E LCMC Annual Command History FY05* (September 2006).

[275]Program Executive Office Intelligence, Electronic Warfare & Sensors Bi-Annual Report (1999 + 2000).

LANDMARKS AND PLACE NAMES END NOTES

Unless otherwise stated, all resources can be found in the CE-COM LCMC Historical Archives, Fort Monmouth NJ.

[1] Arthur S. Cowan biography file; program from dedication ceremony; Memorialization and Tradition Committee, Fort Monmouth Box 1: 1945-1949, folder "Landmarks and Place Names 1945-1962."

[2] William H. Dean biography file. See also Memorialization and Tradition Committee, Fort Monmouth Box 1: 1945-1949, folder "Landmarks and Place Names 1945-1962."

[3] Henry Dunwoody biography file; see also Memorialization and Tradition Committee, Fort Monmouth Box 1: 1945-1949, folder "Landmarks and Place Names 1945-1962."

[4] See Memorialization and Tradition Committee, Fort Monmouth Box 1: 1945-1949, folder "Landmarks and Place Names 1945-1962."

[5] A.W. Greely biography file; Memorialization and Tradition Committee, Fort Monmouth Box 4: 1954-1960, folder "Spanish American War reunion."

[6] Sources conflict: he assumed command either December 15th or 16th. John E. Hemphill biography file. See also Memorialization and Tradition Committee, Fort Monmouth Box 1: 1945-1949, folder "Landmarks and Place Names 1945-1962."

[7] ECOM Information Office, "Release #31170643 (Clean up of Husky Brook Pond)," 3 November 1970.

[8] H.R. Jagger biography file. The Historical Office has very little information on this monument. It is mentioned in the oldest available "Fort Monmouth Landmarks and Place Names" book, dated 1954, as well as 1935 and 1936 maps of the post.

[9] See Memorialization and Tradition Committee, Fort Monmouth Box 1: 1945-1949, folder "Landmarks and Place Names 1945-1962;" and Memorialization and Tradition Committee, Fort Monmouth Box 5: 1957-1962, folder "Misc. correspondence 1960."

[10] Albert James Myer biography file.

[11] The dedication occurred in either early May 1950 or November 1951. Note that the 1961 Landmarks and Place Names booklet dates the dedication to November 26, 1951, in disagreement with the Memorialization and Tradition Committee records. See Memorialization and Tradition Committee, Fort Monmouth Box 1: 1945-1949, folder "Landmarks and Place Names 1945-1962;" and Memorialization and Tradition Committee, Fort Monmouth Box 1 (sic): 1949-1951, folder "Signal Corps Tradition Fund 1950."

[12] Alvin C. Voris biography file. See also Memorialization and Tradition Committee, Fort Monmouth Box 1: 1945-1949, folder "Landmarks and Place Names 1945-1962."

[13] Edwin Daniel Augenstine biography file.

[14] See Memorialization and Tradition Committee, Fort Monmouth Box 1: 1945-1949, folder "Landmarks and Place Names 1945-1962."

[15] Timothy Rider, "Dedication of Battle of the Bulge Memorial set Sunday," CECOM Media Release 01-14, May 1, 2001.

[16] Verified by the program.

[17] Verified by the program.

[18] Verified by the program.

[19] Phone conversation between Wendy Rejan and Kathy Dorry of the Holocaust Committee on Thursday 26 April 2007.

[20] Debbie Sheehan, "Survivors remember Holocaust in different ways," *Monmouth Message*, 13 May 2005.

[21] These are the property of U.S. Military Academy and any inquiries should be addressed to the West Point Museum.

[22] See Memorialization and Tradition Committee, Fort Monmouth Box 1988-1998, folder "1994."

[23] Debbie Sheehan, "Tree dedicated in former leader's honor," *Monmouth Message*, 14 September 2005.

[24] Date verified by the program.

[25] See Memorialization and Tradition Committee, Fort Monmouth Box 1998-2007, folder "2003."

[26] Email from Mindy Rosewitz to George Fitzmaier, 15 March 2007.

[27] Email Correspondence from Tom Hoffman to Mindy Rosewitz, 21 March 2007.

[28] Henry Dunwoody biography file.

[29] This monument is the property of U.S. Military Academy and any inquiries should be addressed to the West Point Museum.

[30] Renita Foster, "Project a 'natural' for fort engineer," *Monmouth Message*, 15 June 2001.

[31] Date verified by the program.

[32] Verified by the program. General orders number 60, dated June 18, 1962, mandated the designation (three days after the dedication ceremony). See Memorialization and Tradition Committee, Fort Monmouth Box 1: 1945-1949, folder "Landmarks and Place Names 1945-1962."

[33] No program or General Orders can be found from the original dedication, however, the 1961 Landmarks and Place Names book includes the Johnston Gate, while the 1958 version does not. A 22 August 1961 Memorialization and Tradition Committee document states that the West Gate will be named for COL Gordon Johnston. 1986 re-dedication verified by the program.

[34] Ibid.

[35] See Memorialization and Tradition Committee, Fort Monmouth Box 6: 1954-1958, folder "Ten Year Program of Memorialization at Fort Monmouth 1958."

[36] "National Register Nomination Case Studies," see http://aec.army.mil/usaec/cultural/nhc_04.doc http://aec.army.mil/usaec/cultural/nhc_03.doc.

[37] See Memorialization and Tradition Committee, Fort Monmouth Box 6: 1954-1958, folder "Ten Year Program of Memorialization at Fort Monmouth 1958."

[38] Ibid.

[39] Ibid.

[40] See Memorialization and Tradition Committee, Fort Monmouth Box 1: 1945-1949, folder "Landmarks and Place Names 1945-1962."

[41] Verified by the program. See also Memorialization and Tradition Committee, Fort Monmouth Box 1: 1945-1949, folder "Landmarks and Place Names 1945-1962;" and See Memorialization and Tradition Committee, Fort Monmouth Box 6: 1954-1958, folder "Ten Year Program of Memorialization at Fort Monmouth 1958;" "National Register Nomination Case Studies," see http://aec.army.mil/usaec/cultural/nhc_04.doc http://aec.army.mil/usaec/cultural/nhc_03.doc.

[42] "National Register Nomination Case Studies," see http://aec.army.mil/usaec/cultural/nhc_04.doc http://aec.army.mil/usaec/cultural/nhc_03.doc.

[43] "National Register Nomination Case Studies," see http://aec.army.mil/usaec/cultural/nhc_04.doc http://aec.army.mil/usaec/cultural/nhc_03.doc.

[44] "National Register Nomination Case Studies," see http://aec.army.mil/usaec/cultural/nhc_04.doc http://aec.army.mil/usaec/cultural/nhc_03.doc.

[45] Verified by the program.

[46] William Blair biography file.

[47] "National Register Nomination Case Studies," see http://aec.army.mil/usaec/cultural/nhc_04.doc http://aec.army.mil/usaec/cultural/nhc_03.doc.

[48] See Memorialization and Tradition Committee, Fort Monmouth Box 1: 1945-1949, folder "Landmarks and Place Names 1945-1962;" See Memorialization and Tradition Committee, Fort Monmouth Box 6: 1954-1958, folder "Ten Year Program of Memorialization at Fort Monmouth 1958."

[49] "National Register Nomination Case Studies," see http://aec.army.mil/usaec/cultural/nhc_04.doc http://aec.army.mil/usaec/cultural/nhc_03.doc.

[50] See Memorialization and Tradition Committee, Fort Monmouth Box 1: 1945-1949, folder "Landmarks and Place Names 1945-1962."

[51] "National Register Nomination Case Studies," see http://aec.army.mil/usaec/cultural/nhc_04.doc http://aec.army.mil/usaec/cultural/nhc_03.doc.

[52] His dates, date of designation and GO number verified by the GO itself. See also Memorialization and Tradition Committee, Fort Monmouth Box 1: 1945-1949, folder "Landmarks and Place Names 1945-1962;" "National Register Nomination Case Studies," see http://aec.army.mil/usaec/cultural/nhc_04.doc http://aec.army.mil/usaec/cultural/nhc_03.doc.

[53] "National Register Nomination Case Studies," see http://aec.army.mil/usaec/cultural/nhc_04.doc http://aec.army.mil/usaec/cultural/nhc_03.doc.

[54] See Memorialization and Tradition Committee, Fort Monmouth Box 6: 1954-1958, folder "Ten Year Program of Memorialization at Fort Monmouth 1958."

[55] "National Register Nomination Case Studies," see http://aec.army.mil/usaec/cultural/nhc_04.doc http://aec.army.mil/usaec/cultural/nhc_03.doc.

[56] See Memorialization and Tradition Committee, Fort Monmouth Box 1: 1945-1949, folder "Landmarks and Place Names 1945-1962."

[57] "National Register Nomination Case Studies," see http://aec.army.mil/usaec/cultural/nhc_04.doc http://aec.army.mil/usaec/cultural/nhc_03.doc

[58] See Memorialization and Tradition Committee, Fort Monmouth Box 1: 1945-1949, folder "Landmarks and Place Names 1945-1962."

[59] "National Register Nomination Case Studies," see http://aec.army.mil/usaec/cultural/nhc_04.doc http://aec.army.mil/usaec/cultural/nhc_03.doc.

[60] Verified by the program.

[61] Albert Smarr biography file; also verified by program.

[62] See Memorialization and Tradition Committee, Fort Monmouth Box 1: 1945-1949, folder "Landmarks and Place Names 1945-1962."

[63] See Memorialization and Tradition Committee, Fort Monmouth Box 5: 1957-1962, folder "Men to be memorialized, officers, 1959, A-G.."

[64] Frank Moorman biography file; F.W. Moorman biography file.

[65] Verified by the program.

[66] See Memorialization and Tradition Committee, Fort Monmouth Box 6: 1954-1958, folder "Ten Year Program of Memorialization at Fort Monmouth 1958."

[67] See Memorialization and Tradition Committee, Fort Monmouth Box 1: 1945-1949, folder "Landmarks and Place Names 1945-1962."

[68] Verified by the program.

[69] See Memorialization and Tradition Committee, Fort Monmouth Box 1: 1945-1949, folder "Landmarks and Place Names 1945-1962."

[70] Verified by the program.

[71] See Memorialization and Tradition Committee, Fort Monmouth Box 6: 1954-1958, folder "Ten Year Program of Memorialization at Fort Monmouth 1958."

[72] Ibid.

[73] Ibid.

[74] See Memorialization and Tradition Committee, Fort Monmouth Box 1: 1945-1949, folder "Landmarks and Place Names 1945-1962."

[75] "VA Clinic," *Fort Monmouth Public Webpage,* http://www.monmouth.army.mil/C4ISR/vaclinic.shtml.

[76] See Memorialization and Tradition Committee, Fort Monmouth Box 1: 1945-1949, folder "Landmarks and Place Names 1945-1962."

[77] Aldred Pruden biography file.

[78] Verified by the program.

[79] Verified by the program; see also Memorialization and Tradition Committee, Fort Monmouth Box 1: 1945-1949, folder "Landmarks and Place Names 1945-1962."

[80] Verified by the program,

[81] See Memorialization and Tradition Committee, Fort Monmouth Box 1: 1945-1949, folder "Landmarks and Place Names 1945-1962."

[82] Ibid.

[83] Ibid.

[84] Ibid.

[85] Ibid.

[86] Ibid.

[87] Ibid

[88] Ibid.

[89] Ibid.

[90] Ibid.

[91] Ibid.

[92] Ibid.

[93] Ibid.

[94] Ibid.

[95] Ibid.

[96] Ibid.

[97] Ibid.

[98] Ibid.

[99] Ibid.

[100] Ibid.

[101] Ibid.

[102] Ibid.

[103] Ibid.

[104] Richard Bingham, *Fort Monmouth, New Jersey: A Concise History*. Fort Monmouth, New Jersey: U.S. Army Communications-Electronics Command, 2002).

[105] Helen Phillips, *United States Army Signal School 1919 – 1967*. Fort Monmouth, New Jersey: U.S. Army Signal Center and School, 1967.

[106] See Buildings III, folder "Make It Happen Center."

[107] See Memorialization and Tradition Committee, Fort Monmouth Box 1: 1945-1949, folder "Landmarks and Place Names 1945-1962."

[108] Ibid.

[109] Ibid.

[110] Ibid.

[111] Ibid.

[112] Richard Bingham, *Fort Monmouth, New Jersey: A Concise History*. Fort Monmouth, New Jersey: U.S. Army Communications-Electronics Command, 2002).

[113] See Memorialization and Tradition Committee, Fort Monmouth Box 1: 1945-1949, folder "Landmarks and Place Names 1945-1962;" William Strong, "Gibbs Hall," Fort Monmouth, New Jersey: U.S. Army Communications-Electronics Materiel Readiness Command, 1981; George Sabin Gibbs biography file.

[114] Richard Bingham, *Fort Monmouth, New Jersey: A Concise History*. Fort Monmouth, New Jersey: U.S. Army Communications-Electronics Command, 2002).

[115] Richard Bingham, "Fort Monmouth: Sketches for a Windshield Tour," Fort Monmouth, NJ: Communications – Electronics Command, 1997 and 2002.

[116] Albert Myer biography file.

[117] Ibid.

[118] Ibid.

[119] Verified by the program.

[120] See Memorialization and Tradition Committee, Fort Monmouth Box 1: 1945-1949, folder "Landmarks and Place Names 1945-1962."

[121] See Memorialization and Tradition Committee, Fort Monmouth Box 1: 1945-1949, folder "Landmarks and Place Names 1945-1962."

[122] Richard Bingham, *Fort Monmouth, New Jersey: A Concise History*. Fort Monmouth, NJ: Communications-Electronics Command, 2002.

[123] "Deal Test Area Facilities to Move to Evans Area" Monmouth Message 5 October 1972.

[124] See Memorialization and Tradition Committee, Fort Monmouth Box 1: 1945-1949, folder "Landmarks and Place Names 1945-1962."

GLOSSARY

AARS	Army Amateur Radio Service
ABCS	Army Battle Command Systems
ACALA	Armament and Chemical Acquisition and Logitics Agency
ACDE	Army Common Operating Environment
ACS	Aerial Common Sensor
ACTD	Advanced Concept Technology Demonstration
ADCS	Air Defense and Control Systems
ADO	Army Digitization Office
ADP	Automatic Data Processing Laboratory
AEF	American Expeditionary Force
AFATDS	Advanced Field Artillery Tactical Data System
ALC	Anticipatory Logistics Cell
ALT	Acquisition, Logistics and Technology
ALT	Administrative Lead Time
AMARC	Army Materiel Acquisition Review Committee
AMC	Army Materiel Command
AMCOM	Aviation and Missile Command
AOR	Area of Responsibility
APG	Aberdeen Proving Ground
APS	Automated Procurement System
ARL	Airborne Reconnaissance Low
ARL	Army Research Laboratory
ARPA	Advanced Research Project Agency
ASAALT	Assistant Secretary of the Army for Acquisition, Logistics and Technology
ASARDA	Assistant Secretary of the Army for Research, Development and Acquisition
ASAS	All Source Analysis System
ASE	Aircraft Survivability Equipment
AT&T	American Telephone and Telegraph Company
ATACS	Army Tactical Communications Systems
ATCOM	Aviation and Troop Command
ATRJ	Advanced Threat Radar Jammer
BCS3	Battle Command Sustainment and Support System
BCT	Brigade Combat Team
BDI	Balkan Digitization Initiative
BITS	Battlefield Information Transfer System
BRAC	Base Realignment and Closure
C2SID	Command, Control and Systems Integration Directorate
C3S	PEO Command, Control and Communications Systems
C3T	PEO Command, Control and Communications-Tactical
C4ISR	Command, Control, Communications, Computers, Intelligence, Surveillance, and Reconnaissance
CAC	Control Analysis Centers
CAISI	Combat Service Support Automated Information System Interface
CALS	Computer Aided Acquisition and Logistics Support
CCS	Command and Control Systems
CDC	Combat Development Command

CEAC	CECOM Executive Advisory Committee
CECOM	Communications-Electronics Command
CECOM LCMC	Communications-Electronics Life Cycle Management Command
CENCOMS	Center for Communications Systems
CENSEI	Center for Systems Engineering and Integration
CENTACS	Center for Tactical Computer Systems
CENTCOM	United States Central Command
CERCOM	Communications-Electronics Materiel Readiness Command
CERDEC	Communications-Electronics Research, Development and Engineering Center
CG	Commanding General
CGS	Common Ground Station
CGSC	Command and General Staff College
CHS	Common Hardware/Software
CI	Counterintelligence
CIMMD	Close-in Manportable Mine Detector
CINC	Commander in Chief
CIPO	CINC Interoperability Program Office
CIPO	Combatant Commander Interoperability Program Office
CMTC	Citizens Military Training Camp
CMWS	Common Missile Warning System
CNVEO	Center for Night Vision and Electro Optics
COB	CECOM Office Building
COMINT	Communications Intelligence
COMM	Communications Systems
COMSEC	Communications Security
CONARC	U.S. Continental Army Command
CONUS	Continental United States
CORADCOM	Communications Research and Development Command
CREW	Counter RCIED Electronic Warfare
CSA	Communications Systems Agency
CSC	Computer Sciences Corporation
CSLA	Communications Security Logistics Agency
CSSCS	Combat Service Support Control System
CSW	Center for Signal Warfare
CTSF	Central Technical Support Facility
DA	Department of the Army
DARCOM	United States Army Materiel Development and Readiness Command
DBS	Direct Broadcast Satellite
DCATS	Defense Communications and Army Transmission Systems
DCGS	Distributed Common Ground System
DEA	Drug Enforcement Agency
DIL	Digital Integrated Laboratory
DISA	Defense Information Systems Agency
DISC4	Director for Information Systems, Command, Control, Communications and Computers
DLA	Defense Logistics Agency
DoD	Department of Defense
DOIM	Directorate of Information Management
DORS	Defense Outplacement Referral System
DSC	Distinguished Service Cross
DSCR	Defense Supply Center, Richmond
DSCS	Defense Satellite Communications Systems

	Operations Center
DTSR	Digital Temporary Storage Recorders
DVE	Driver's Vision Enhancer
EBBS	Electronic Bulletin Board System
ECM	Electronic Counter Measure
ECOM	Electronics Command
EDWAA	Economic Dislocated Workers Adjustment Assistance
EIS	PEO Enterprise Information Systems
ELINT	Electronics Intelligence
EMPRS	En-route Mission Planning and Rehearsal System
EMRA	Electronics Materiel Readiness Activity
EOC	Emergency Operations Center
EPLRS	Enhanced Position Location Reporting System
ERADCOM	Electronics Research and Development Command
ESC	Electronics Systems Center, Air Force
ESCS	Eastern Signal Corps School
ESE	Enterprise Systems Engineering
ESSC	Electronic Sustainment Support Center
ETDL	Electronics Technology and Devices Laboratory
EW	Electronic Warfare
EXFOR	Experimental Force
EXMP	Experimental Master Plan
FATDS	Field Artillery Tactical Data Systems
FBCB2	Force XXI Battle Command Brigade and Below
FBI	Federal Bureau of Investigation
FCS	Future Combat System
FIS	Foreign Intelligence Services
FLIR	Forward Looking Infrared Radar
FMS	Foreign Military Sales
FRA	Forward Repair Activity
FRIAR	First Response Intelligence Analyst Resource
FSR	Field Service Representative
GHQ	General Headquarters
GPS	Global Positioning System
GSA	General Services Administration
GWOT	Global War on Terror
HCLOS	High Capacity Line-of-Sight
HISA	Headquarters Installation Support Activity
HLS	Homeland Security
HMMWV	High Mobility Multi-purpose Wheeled Vehicle
IBOP	Interagency Interactive Business Opportunities Page
ICMD	Improved Countermeasure Dispenser
ID	Infantry Division
IDIQ	Indefinite Delivery/Indefinite Quantity
IED	Improvised Explosive Device
IEW	Intelligence and Electronic Warfare
IEW&S	Intelligence, Electronic Warfare and Sensors
IEWD	Intelligence and Electronic Warfare Directorate
IFN	Items for Negotiation
IGY	International Geophysical Year
ILS	Integrated Logistics Support

IMA	Installation Management Agency
IMCOM	Installation Management Command
IMETS	ntegrated Meteorological System
IMMC	Intelligence Materiel Management Center
INS	Immigration and Naturalization Service
INSCOM	Intelligence and Security Command
IPT	Integrated Product Teams
IR&D	Independent Research and Development
ISC	Information Systems Command
ISEC	Information Systems Engineering Command
ISMA	Information Systems Management Agency
ISR	Intelligence Surveillance and Reconnaissance
ISSC	Information Systems Software Center
IT	Information Technology
ITT	International Telephone and Telegraph
JACTD	Joint Advanced Concept Technology Demonstration
JCFAWE	Joint Contingency Force Advanced Warfighting Experiment
JIEO	Joint Information Engineering Organization
JNN	Joint Network Node
JSTARS	Joint Surveillance Target Attack Radar System
JTDS	Joint Tactical Data System
KBLPS	Knowledge-based Logistics Planning Shell
LACV	Lighter Air Cushioned Vehicle
LAD	Logistics Archive Desk
LAN	Local Area Networks
LAR	Logistics Assistance Representative
LCMC	Life Cycle Management Commands
LCMR-A	Lightweight Counter Mortar Radar-Army
LCU	Landing Craft Utility
LLDR	Lightweight Laser Designator Rangefinder
LMP	Logistics Modernization Program
LRAS	Long Range Advanced Scout Surveillance System
LRC	Logistics and Readiness Center
LSE	Logistics Support Element
LSS	Lean Six Sigma
LSSC	Logistics Systems Support Center
LVRS	Lightweight Video Reconnaissance System
MACV	Military Assistance Command Vietnam
MARS	Military Affiliate Radio System
MASS	Modular Ammunition Solar Shades
MCS	Maneuver Control System
MEP	Mobile Electric Power
MI	Military Intelligence
MSAT-Air	Multi-Sensor Aided Targeting-Air
MSC	Major Subordinate Command
MSCS	Multi-Service Communications Systems
MSE	Mobile Subscriber Equipment
MTMC	Military Traffic Management Command
NASA	National Aeronautics and Space Administration
NATO	North Atlantic Treaty Organization
NAVCON	Navigation Control System
NDI	Non-developmental Item

NES	Network Encryption System
NETCOM	Network Enterprise Technology Command
NICP	National Inventory Control Point
NSC	Network Service Center
NTC	National Training Center
NVEO	Night Vision and Electro-Optics
OC-SigO	Office of the Chief Signal Officer
OCS	Officer Candidate School
OEF	Operation Enduring Freedom
OIF	Operation Iraqi Freedom
OPO	Office of Personnel Operations
OPTADS	Operations Tactical Data Systems
OSD	Office of the Secretary of Defense
PACT	Procurement Administrative Lead Time
PEO	Program Executive Office/Officer
PLGR	Precision Lightweight GPS Receivers
PLPS/TIDS	Position Location Reporting System/Tactical Information Distribution System
PRAG	Performance Risk Assessment Analysis Group
PSDS2	Persistent Surveillance and Dissemination System of Systems
PTDS	Persistent Threat Detection System
PWD	Proximity Warning Device
QIP	President's Quality Improvement Prototype Award
QRC	Quick Reaction Capability
R&D	Research and Development
RCIO	Regional Chief Information Officer
RDEC	Research, Development and Engineering Center
RDECOM	Research, Development and Engineering Command
RD&J	Radar Deception and Jamming
REMBASS	Remotely Monitored Battlefield Sensor System
RFPI	Rapid Force Projection Initiative
ROTC	Reserve Officer Training Corps
RSTA	Reconnaissance, Surveillance and Target Acquisition
RUS	Robotics and Unmanned Sensors
S&TC	Space and Terrestrial Communications Directorate
SAS	Survivable Adaptive Systems
SATCOM	Satellite Communications
SBCCOM	Soldier and Biological Chemical Command
SBCT	Stryker Brigade Combat Team
SCAMP	Single Channel Anti-Jam Manpack Terminal
SCI	Sensitive Compartmented Information
SCOTT	Single Channel Objective Tactical Terminal
SCR	Senior Command Representative
SDI	Strategic Defense Initiative
SES	Senior Executive Service
SIGINT	Signals Intelligence
SINCGARS	Single Channel Ground and Airborne Radio System
SMART-T	Secure Mobile Anti-Jam Reliable Tactical Terminal

SOMA	Signal Organization and Missile Alignment
SOTAS	Standoff Target Acquisition System
SOUTHCOM	Southern Command
SPAWAR	Space and Naval Warfare Systems Command
SPIL	Software Prototyping and Integration Laboratory
SPT	Standardization Program Team
SSF	Single Stock Fund
STAR	System Test Bed for Avionics Research
STARS	Software Technology for Adaptable, Reliable Systems
STO	Science and Technology Objectives
SW	Signals Warfare
SWA	Southwest Asia
T2S2	Translation/Transcription Support System
TACOM	Tank-Automotive Command
TASA	Television –Audio Support Activity
TCMO	Theater COMSEC Management Office
TESAR	Tactical Endurance Synthetic Aperture Radar
TIGER	Tactical Intelligence Gathering and Relay
TIM	Transformation of Installation Management
TMDE	Test, Measurement and Diagnostic Equipment
TOE	Tables of Organization and Equipment
TQM	Total Quality Management
TRADOC	United States Training and Doctrine Command
TRCS	Tactical Radio Communications Systems
TRI-TAC	Joint (Tri-Service) Tactical Communications
TYAD	Tobyhanna Army Depot
UNICOM/STARCOM	Uniform Communications/Strategic Army Communications Systems
USACOM	United States Atlantic Command
USACHCS	United States Army Chaplain Center and School
USAEUR	United States Army Europe
USARV	United States Army Vietnam
USASCRDL	United States Army Signal Corps Research and Development Laboratory
UAV	Unmanned Aerial Vehicle
USMAPS	United States Army Military Academy Preparatory School
VERA/VSIP	Voluntary Early Retirement Program/Voluntary Separation Incentive Program
VHFS	Vint Hill Farms Station
WIN-T	PM Warfighter Information Network-Tactical

INDEX

A

Abbey, Tech 5th Grade Claude W., 129
Aberdeen Proving Ground (APG), MD, 18, 31, 54, 95,101-102
Abramowitz, LTC Reuben, 121
Abrams Tank, 72, 107
Abrams, GEN Creighton, 42
Acquisition and procurement reforms, 82-88, 102
Acquisition Center, CECOM, 83-88, 90, 104
Acquisition, Logistics and Technology (ALT), 109
Adams, John G.
Adelphi, MD, 49, 63
Adjutant General of the Army, 2, 6, 17
Administrative Lead Time (ALT), 87
Advanced Concept Technology Demonstration (ACTD), 75,77-78, 85
Advanced Field Artillery Tactical Data System (AFATDS), 110
Advanced Research Project Agency (ARPA), 40
Advanced Technology Demonstrations (ATD), 61-63, 74-77, 78
Advanced Threat Radar Jammer (ATRJ) 85
Advanced Warfighting Experiment, 70-75, 97. See also Task Force XXI.
Aerial units (World War I)
    122d Aero Squadron, 4
    504th Aero Squadron, 4
Afghanistan, 98-99, 104, 106, 107, 112, 116
African-Americans at Fort Monmouth, 37-39
Airborne Reconnaissance Low (ARL), 114-15
Airborne units
    82d Airborne, 59, 77, 99
    101st Airborne, 59
Aircraft Survivability Equipment (ASE), 115
Albert J. Myer Center, 32. See also Myer Center.
Alexander, LT E.P. (later BG), 128
All Source Analysis System (ASAS), 55, 77, 80, 105, 108, 111
Allaire Combat Training Area, 20
Allen, BG James, 128
Allensworth, LTC Allen, 125
Allison, COL James B., 9, 151
Al Qaeda, 98
Alsace-Lorraine, France, 5
American Expeditionary Force (AEF), 4
American Flying Services Foundation, 27
American Telephone and Telegraph Company (AT&T), 1, 3
Anthony, Michael P., 104
Anti-aircraft Artillery and Guided Missile firing systems, 31
Anti-Semitism, 34-35
Anticipatory Logistics Cell (ALC), 103-104
Apache Longbow attack helicopter, 72, 75, 115
Area of Responsibility (AOR), 101
Armament and Chemical Acquisition and Logistics Agency (ACALA), 79
Armed Services Electric Standards Agency, 31
Armistice, World War I, 6
Armored Cavalry regiments, 77, 106
    1st Cavalry Division, 57-58
Armstrong, MAJ Edwin H., 12, 123
Army Acquisition Corps, 69, 71
Army Acquisition Executive, 109-10
Army Amateur Radio Service (AARS), 9
Army 389th band, 18
Army Battle Command Systems (ABCS) 77, 108, 111, 112
Army Common Operating Environment (ACDE), 77
Army Digitization Office (ADO), 71-72, 110
Army Knowledge Management, 111
Army maneuvers (1930s), 11
Army Materiel Acquisition Review Committee (AMARC), 49-50, 156
Army Materiel Command (AMC), 18, 156
    Origin of, 41
    Roles of, 44-45, 49, 55
    Changes in, 63, 67, 68-69, 71, 73
    Organization and modernization, 79, 81, 83
    Activities of, 88-95
    Reorganization of, 100, 103
    And Project Managers, 109-10
Army Pigeon Service Agency, 26

Army Research Laboratory (ARL), 68
Army Tactical Command and Control System, 77
Army Tactical Communications Systems (ATACS), 51
Army TMDE Activity, 61
Army Transformation, 100
Army War College, 7, 76
Artillery Locating Radar, 106. See also Firefinder.
Asbury Park, NJ, 21
Assistant Secretary of the Army for Acquisition, Logistics and Technology (ASAALT), 52
Assistant Secretary of the Army for Research, Development and Acquisition (ASARDA), 109
Astro-Electronics Division, 37
Atmosphere Sciences Laboratory, 49
Automated Procurement System (APS), 87
Automatic Data Processing (ADP) Laboratory, 42, 45
Avenue of Memories, 117, 127
Aviation and Missile Command (AMCOM), 100
Aviation and Troop Command (ATCOM), 63, 66, 79, 93
Aviation Research and Development Command (ARDCOM), 49, 156
Aviation Systems Command, 49
Avionics, 43-45, 49, 63, 75
Avionics Laboratory, 42-45, 47, 49
Avionics Research and Development Activity (AVRADA), 156

B

Babers, MG Donald M., 51, 142
Baghdad, 106, 112
Bain, Edward, 87
Baker, SGT George, 30
Baledogle, Somalia, 93
Balkan Digitization Initiative (BDI), 91
Balkans, 88-91
Baltimore, MD, 33
Barbados, 78
Barker, Cadet Ernest S., 128
Barker Circle, Fort Monmouth, 9, 128
Barr, Joel, 35
Barton, LTC David B., 130
Base Realignment and Closure (BRAC), 63-66, 68, 70, 80, 101-102, 103, 135-36
Batteries, lithium, 47, 99, 101, 104
    Importance of, 58-59, 60, 101
Battle Command Brigade and Below System (BCB2), 104, 105, 108
Battle Command Sustainment and Support System (BCS3), 111
Battle Laboratories, 70, 72, 73-74, 77, 82
Battlefield Information Transfer System (BITS), 71
Battles: of the Atlantic, 16; of Britain, 16; of Monmouth Courthouse, 1, 7
Bayonne, NJ, 66
Bell Laboratories, 13
Bell Telephone, 24. See also American Telephone & Telegraph Company.
Belmar, 29. See also Wall Township.
Berkson, Bradley, 101
Bijur, CAPT Arthur H., 124
Black, BG Garland C., 124
Blair, MAJ (later COL), William R., 12-14, 121
Blue Force Tracking, 103
Bolton, Claude M., Jr., 101
Bosnia, 77-78, 88-89, 90-91
Boston, MA, 98
Boylan, Robert J., 10
Bradley Beach, NJ, 24
Bradley Fighting Vehicle, 107, 111
Brigade Combat Team (BCT), 106, 112, 114-15
British radar, 16
Brockel, Kenneth, 87
Brohm, MG Gerard P., 67, 73, 78-79, 82, 140
Brookdale Community College, 69, 136
Bureau of Standards, 4, 8
Burma, 26
Burns, SGT Kenneth P., 131
Bush, Pres. George W., 92, 98
Business and Information Systems, 83

C

Cameron, Thomas, 83

Taliban, 98
Tank-Automotive Command (TACOM) 63, 79, 95
Tarbell, Dr. Alan, 59
Targets and technology, 74-75
Task Force XXI (Advanced Warfighting Experiment), 70-76, 104, 108, 111.
 See also PM Task Force XXI.
Tate, Harold, 35, 37-38
Team C4IEWS, 68, 72
Team C4ISR, 68, 97, 100
Team Fort Monmouth, 90
Technical Control and Analysis Center (TCAC), 114
Telecommunications Center, 93
Television –Audio Support Activity (TASA), 57
Test, Measurement and Diagnostic Equipment (TMDE), 51, 61, 64
Theater COMSEC Management Office (TCMO), 57
Thermal Weapon Sight (TWS), 115
Thomas, MG Billy M., 141
Thomas, SGT Kevin V., 94
Thompson, Dr., 3
Tindall, MAJ Richard G., 130
Tinton Falls, NJ, 63, 65, 73, 80, 135
TIROS weather satellites, 129, 135
Tobyhanna Army Depot, PA (TYAD), 64-65, 69-70, 80, 89, 91, 103-104
Todd, SFC Robert J., 130
Total Quality Management (TQM), 78, 83
Townes, William D., 31, 34-35, 38
Trakowski, COL Albert C., 21
Transformation of Installation Management (TIM), 100
Transition team, BRAC, 65-66
Translation/Transcription Support System (T2S2), 92
Transportation Battalion (403d), 58
Transportation equipment, 105-106
Tri-band Super High Frequency Tactical Satellite Terminal (TRIBAND), 84-86
Trojan SPIRIT II, 76
Truman, President Harry S, 37-38
TRW, Inc., 71, 73, 89, 91
Tuzla, Hungary, 78, 91

U

UNICOR, 88
Uniform Communications/Strategic Army Communications Systems
 (UNICOM/STARCOM), 42
United States Air Force, 40, 59, 70, 93, 99
United States Army, 1, 4, 7, 9, 12, 15, 19, 42, 44
 In post-war era, 60, 92-93, 97, 114
United States Army Air Force, 4
United States Army Chaplain Center and School (USACHCS), 53, 63, 66
United States Army Communications Electronics Museum, 5, 17, 25
United States Army Communications-Electronics Command (CECOM), 46, 156, 158. See also Communications-Electronics Command
United States Army Corps of Engineers, 13, 39, 64
United States Army Electronics Command (ECOM)
United States Army Europe (USAEUR), 77-78, 89
United States Army Garrison Fort Monmouth, 55, 156, 158
United States Army Information Systems Command (ISC), 52-53
United States Army Information Systems Management Activity
 (ISMA), 53-54
United States Army Installation Management Agency (IMA)
United States Army Intelligence and Security Command, 54
United States Army Materiel Development and Readiness Command
 (DARCOM), 50, 55. See also Army Materiel Command.
United States Army Military Academy Preparatory School (USMAPS),
 102, 158
United States Army Missile Command, 61
United States Army Ordnance Corps, 13
United States Army Signal Corps Research and Development Laboratory
 (USASCRDL), 37
United States Army Vietnam (USARV), 45
United States Atlantic Command (USACOM), 78, 94
United States Central Command (CENTCOM), 59, 101, 105
United States Continental Army Command (CONARC), 26, 106, 156
United States Customs Service, 92
United States Marine Corps, 93, 107, 114
United States Navy, 12, 40, 60, 70, 93
United State Special Operations Command, 82

United States Training and Doctrine Command (TRADOC), 46, 66, 70, 93, 108
University of Pennsylvania, 69
Unmanned Aerial Vehicle (UAV), 75, 85

V

Vail, Alfred E., 3, 125
Vail Hall, 53
Vail, Theodore N., 3
Valachi, Joseph Michael, 46
Van Deusen, COL (later BG) George L., 19, 52, 123, 149
Van Keuren, Melvin, 2
Van Kirk 1LT John Sewart, 118
Via, MG Dennis L., 101, 138
Vietnam War, 42-47, 99
Vint Hall Farms Station (VHFS),
 VA, 49, 52, 55, 113
 Closure of, 63-66
Vollum, Howard, 15
Voluntary Early Retirement Program/
Voluntary Separation Incentive Program (VERA/VSIP), 65, 67
Voris, COL Alvin C., 118, 150

W

Wade, 1LT LaVerne L., 131
Wall Township, NJ, 13, 29, 136
Wallace, GEN William S., 112
Wallington, COL Martin G., 128
War Department, 2, 24
Warfighter Information Network-Tactical (WIN-T), 108, 112
Warlord, 77
War on Drugs, 91-92
War on Terror, 98. See also Global War on Terror.
Warrior Focus and Prairie Warrior, 77
Washington, GEN George, 1
Watson Laboratories, 27. See also Eatontown Signal Laboratories.
Watters, Chaplain/MAJ Charles J., 53, 125
Watters Hall (later Mallette Hall), 53, 65-66, 125
Wentworth, Edwin W., 59
West Point, NY, 102
Western Electric, 1, 13-14
Westmoreland, GEN William, 42
White Sands Missile Range, NM, 49
Whitehurst, CPL Carl L., 2-3
Whitesell, Tech4 Joseph L., 127
Whittlesey, MAJ, 5
Wholesale Logistics Modernization Program, 81. See also Logistics
 Modernization Program.
Wideband Communications System, 42
Wiese, H.C., 23
Williams, COL Grant, 12
Wilson, CPL, 129
Wittner, Lawrence S., 56
Women's Army Auxiliary Corps (WAAC). 22-23
Women's Army Corps (WAC), 22-23
Works Projects Administration (WPA), 17
World Trade Center, 97-98
World War I, 1, 3-8
World War II, 6, 12, 15-16
 U.S. mobilization, 19-21, 23
 Use of pigeons in, 25-26
Wright, Constance, 38
Wright, E. Frederic, 119
Wright Field, Dayton, OH, 8, 15
Wright-Patterson Air Force Base, 113

Y

Y2K compliance, 95
Yuma Proving Ground, AZ, 47-48, 113

Z

Zahl, Dr. Harold, 13-15, 39-40
Zenith Radio Corporation, 16